Springer Undergraduate Mathematics Series

The Springer Undergraduate Mathematics Series (SUMS) is a series designed for undergraduates in mathematics and the sciences worldwide. From core foundational material to final year topics, SUMS books take a fresh and modern approach. Textual explanations are supported by a wealth of examples, problems and fully-worked solutions, with particular attention paid to universal areas of difficulty. These practical and concise texts are designed for a one- or two-semester course but the self-study approach makes them ideal for independent use.

More information about this series at http://www.springer.com/series/3423

Robert Magnus

Fundamental Mathematical Analysis

 Springer

Robert Magnus
Faculty of Physical Sciences
University of Iceland
Reykjavik, Iceland

ISSN 1615-2085   ISSN 2197-4144   (electronic)
Springer Undergraduate Mathematics Series
ISBN 978-3-030-46320-5   ISBN 978-3-030-46321-2   (eBook)
https://doi.org/10.1007/978-3-030-46321-2

Mathematics Subject Classification (2020): 26, 40

This Springer imprint is published by the registered company Springer Nature Switzerland AG
The registered company address is: Gewerbestrasse 11, 6330 Cham, Switzerland

To Jórunn Erla

Preface

This text has developed from courses that I have taught on analysis to university students of mathematics in their first semester. However, it has grown to be much more than a course for first year students. Although the 11 chapters (beginning at Chap. 2) include the material usually to be found in beginning courses of analysis, I have also had further objectives that are not usually communicated to beginning students.

In my view mathematics underwent a transformation in the late nineteenth century and early twentieth, indeed a revolution, that could not have been foreseen. Previously there had not been general agreement about standards of proof except perhaps in classical Euclidean geometry. The logical basis of arguments used to prove results about calculus and infinite series, indeed about most mathematics since the Renaissance, was not understood, and there was some anxiety as to whether the edifice of mathematical knowledge might come crashing down. Some mathematicians issued warnings, but the speed and momentum of new discoveries was fortunately unstoppable, and the results seemed correct; they certainly passed all empirical tests of correctness. The idea that analysis was transformed, from a subject with shaky foundations, to become a flagship of correct mathematical argument, and that this change occurred over a rather short period, is something that I regard as important for understanding its nature. Suddenly there was general agreement over what constituted a correct proof, provided only that the details could be taken in by a human reader. Historically, analysis was a great success story.

The objective implied by the last paragraph, to communicate to the reader the success of analysis in overcoming previously held doubts, is not attained by pedantic rigour or a painstaking level of formality. Nor does it call for any genuine attempt to recount the history of analysis or follow a historical development of the subject. Nor does it call for any novel approach to any topic. It does, though, colour the way the topics are presented. It calls for clarity and meaning in the proofs, honesty about what has been achieved and the ever present awareness that the reader is an intelligent adult who genuinely is trying to understand what analysis is about and why it is important.

Set theory has played a basic role in the evolution of analysis. A text of this kind has to present a certain amount of set theory at the outset. But a proper axiomatic treatment of set theory would alienate many readers. The alternative to this has usually been the awfully named naive set theory, involving a careless approach to difficult ideas. But this too can alienate readers (though probably not the same group of readers). Some middle approach is needed that is honest about set theory but not pedantically detailed. The reader should be made aware that there is a need for clear principles for building sets, including many sets that mathematicians take for granted, even if these principles are not all explained in detail. How can analysis, as it is usually understood, exist unless it is accepted that an infinite set exists? This a major stumbling block for those not trained in mathematics and a "look it's obvious" approach will not win any converts. And why should it? In an accurate treatment an infinite set is introduced by a set-building axiom. When set theory is naive, non-mathematicians can appear foolish, and mathematicians can appear doctrinaire.

Every text of this kind has its red lines, introduced by the hackneyed phrase 'beyond the scope of this book'. The word 'fundamental' of the title is supposed to be taken seriously and construed as meaning a certain portion of analysis. What bits of analysis are fundamental? They must include the following items: an accurate description of the real numbers, limits, infinite series, continuity, derivatives, integrals and the elementary transcendental functions. These are standard contents of a first university course in analysis. Very broadly, the boundaries of fundamental analysis lie where the key to further progress requires certain far-reaching theories that are introduced to students after a first course, typically, complex analysis, metric spaces, multivariate calculus or the Lebesgue integral.

In this text there is no discussion of countability versus uncountability for sets. There are no open or closed sets (apart from intervals), and therefore no topology or metrics; and certainly no Heine–Borel theorem, though we go dangerously close to requiring it. This means that we stop short of a nice, necessary and sufficient condition for integrability. The integral is Riemann-Darboux; though I freely confess my view that the Lebesgue integral is the greatest advance in analysis of the twentieth century. There is no general treatment of any class of differential equations; though some very special and important equations appear at crucial places in the narrative. There is no complex analysis (though there is a chapter introducing complex numbers) and almost no functions of several variables; certainly no final chapter, so beloved by authors of analysis texts, entitled 'Extension to several variables'. Surely several variables deserve a book of their own.

Missing is any construction of the real numbers or the complex numbers. My view is clear: neither construction is needed for analysis. They serve only two purposes: logically, to prove that the axioms of analysis are consistent; and pedagogically, to answer students who obstinately want to know what the square root of two and the square root of minus one are in reality, and who are not necessarily convinced by the answers. Moreover, giving prominence to constructions tends to suggest that there is only one right way to understand real numbers or complex numbers.

After reading two paragraphs devoted to what is not included, the reader may well wonder what the author considers fundamental analysis to be. To see what is included the reader is referred to the rather thorough list of contents.

Analysis is like the trunk of a great tree that gives rise to branches, some small, some large and some still growing. I have included a number of sections marked with the symbol ($\diamondsuit$) and referred to as nuggets (as in nugget of wisdom, though I'm actively searching for a different name). These take up fascinating topics that can be explored using fundamental analysis, but can be omitted without losing the main thread. They go in some cases far beyond what a beginning student would ordinarily encounter. They serve to enrich the narrative and often point to a whole subject area that springs out of the main trunk of the tree. They are not needed in the main text of sections not so marked; however, they may be needed for some of the exercises. Some push the boundaries of the main text and encroach on areas where one really starts to need complex analysis or multivariate calculus to make significant progress. Mostly they can be omitted in a first course of analysis. These sections, and exercises elsewhere that may need material from them, are marked with the nugget symbol ($\diamondsuit$). Most conclude with a short subsection called 'Pointers to further study' listing topics or whole subject areas that the reader can look up if they wish to pursue the topic of the nugget further.

Advice for Instructors

This text began life as lectures for a first course of analysis, which was taught a number of times to mathematics students in their first year at the University of Iceland, and consisted of 23 lectures of 80 min each. Material from all 12 chapters was covered in the lectures, in the same order of presentation, omitting the content of sections marked with the nugget symbol ($\diamondsuit$). Some of the topics that ended up in the nuggets were assigned to students as study projects on which they were required to give a presentation.

Thus, in spite of a considerable expansion, and because the additional and more demanding material is clearly marked, the text can be used as the basis for a course. The instructor would only have to agree with the author on a number of key pedagogical issues, that can give rise to heated disputes and to which the answer is a matter of personal preference. For example, over whether or not to construct the real numbers; or over whether to present sequences and series before functions of a real variable; or whether uniform convergence should be covered in a first course; or how to define the circular functions. The first issue is discussed in the nugget (Sect. 3.10) 'Philosophical implications of decimals' and elsewhere in this preface; the elementary transcendental functions are rigorously defined and studied at the earliest point in the text at which it is possible in a practical and meaningful manner.

The text contains many exercises mostly collected together into exercise sections. However, some isolated exercises interrupt the text with the purpose of inviting the reader to engage immediately and constructively with the material. It will be noticed that many of the exercises are challenging and some of them present results of independent interest. It is expected that the instructor can provide additional routine exercises for the purpose of practising the basic rules.

Scattered throughout the text are some pictures. The philosophy behind them is that they may help the reader to visualise an idea or a proof, but are never a necessary part of the discourse. They were hand-drawn using the free graphics software IPE and are intended to resemble nice impromptu sketches that a teacher might make in class.

Reykjavik, Iceland   Robert Magnus
November 2019   Professor Emeritus

Contents

Chapter 1
Introduction

1.1 What Is Mathematical Analysis?

Mathematical analysis, or simply analysis, is the study of limits, series, functions of a real variable, calculus (differential calculus and integral calculus) and related topics, on a logical foundation and using methods that are considered acceptable by modern standards. The aims of analysis were attained by two main achievements: an exact formulation of the properties of the real numbers, and a correct definition of the notion of limit. Analysis is both very reliable, in that its conclusions are considered correct with considerable confidence, and very useful; modern science would be unthinkable without calculus, for example.

John von Neumann wrote: "The calculus was the first achievement of modern mathematics and it is difficult to overestimate its importance. I think it defines more unequivocally than anything else the inception of modern mathematics; and the system of mathematical analysis, which is its logical development, still constitutes the greatest technical advance in exact thinking."

1.2 Milestones in the History of Analysis

The dates in the following list are approximate.

c. 300 BC Euclid publishes a proof that $\sqrt{2}$ is not a rational number (Elements, theorem 117, book 10).

c. 300 BC Euclid publishes Eudoxus' theory of irrational numbers (Elements, book 5).

c. 250 BC Archimedes solves several problems, such as that of calculating the area of a parabolic segment, using methods that foreshadow integration.

1660–1690 Newton and Leibniz invent calculus (differential and integral calculus). They base it on the idea of infinitesimals. These are quantities that are smaller than any positive real number, yet are still positive and non-zero.

© The Editor(s) (if applicable) and The Author(s), under exclusive
license to Springer Nature Switzerland AG 2020
R. Magnus, *Fundamental Mathematical Analysis*, Springer Undergraduate
Mathematics Series, https://doi.org/10.1007/978-3-030-46321-2_1

1660–present The sensational solution of the Kepler Problem is only the first of an inexhaustible supply of problems in applied mathematics, science and technology that are solved using calculus.

1734 George Berkeley criticises the foundations of calculus. He asks what infinitesimals really are, and writes: "May we not call them the ghosts of departed quantities?"

1600–1800 Infinite series are used although a clear definition of convergence is lacking. Many exciting results are obtained by using infinite series as if they were finite sums (most noteworthy is the work of Euler). It is known even so that uncritical use of series can lead to contradictions, such as that $1 = 2$.

1817 Bolzano gives a definition of limit not based on infinitesimals. It attracts little attention.

1820 Fourier introduces Fourier series which make it possible to represent a discontinuous function as an infinite series of trigonometric functions. The problem is that there is still no clear concept of function, no clear concept of continuity or the convergence of a series.

1828 Abel expresses the view that in all mathematics there is not a single infinite series whose convergence has been established by rigorous methods, and comments that the most important parts of mathematics lack a secure foundation. He gives the first proof establishing the sum of the binomial series that is recognisably correct.

1820–1850 Cauchy takes on the task of placing calculus on a secure foundation. He uses some form of the modern ε, δ arguments but still frequently relies on infinitesimals. He invents complex analysis and proves Taylor's theorem. In 1821 he publishes *Cours d'Analyse*, the forerunner of all modern books on mathematical analysis. He famously makes the mistake of claiming that the limit of a sequence of continuous functions is continuous.

1872 Dedekind defines Dedekind sections. He carries out an exact study of the nature of the real numbers, the first since Eudoxus. The notion of irrational number is clarified. In this way a secure foundation is obtained for analysis.

1860–1880 Weierstrass completes the work of Cauchy. The concept of uniform convergence is clarified. He creates famous counterexamples that show the dangers that lurk in uncritical thinking. One such is a continuous function that is nowhere differentiable.

1860–1890 Dedekind and Cantor create set theory. It proves to be the correct language in which to express the conclusions of analysis.

Chapter 2
Real Numbers

As professor in the Polytechnic School in Zurich I found myself for the first time obliged to lecture upon the elements of the differential calculus and felt, more keenly than ever before, the lack of a really scientific foundation for arithmetic.

R. Dedekind. *Essays on the theory of numbers*

2.1 Natural Numbers and Set Theory

Natural numbers are used to count the members of finite sets. They are 0, 1, 2, 3 and so on. They constitute a set denoted by $\mathbb{N}$. We consider 0 a natural number; it is needed to count the members of the empty set. The set of positive natural numbers (that is 1, 2, 3, etc.) is denoted by $\mathbb{N}_+$.

We cannot undertake an exact treatment of analysis without set theory. We assume the reader understands the following formulas concerning sets:

(a) $x \in A$ This says that x is an element of the set A. Its negation, the statement that x is not an element of A, is written $x \notin A$.

(b) $A \subset B$ This says that the set A is a subset of the set B, that is, every element of A is an element of B. It does not preclude the possibility that $A = B$.

(c) $A \cup B$ The union of the sets A and B, that is, the set of elements x such that $x \in A$ or $x \in B$.

(d) $A \cap B$ The intersection of the sets A and B, that is, the set of all elements x such that $x \in A$ and $x \in B$.

(e) $\emptyset$ The empty set, mentioned above as the set with 0 elements.

If $A \subset B$ we also say that B includes A, but never that B contains A. The latter would always mean that the set A is an element of the set B. In this text we shall

R. Magnus, *Fundamental Mathematical Analysis*, Springer Undergraduate Mathematics Series, https://doi.org/10.1007/978-3-030-46321-2_2

mention sets of sets only rarely. Principally, the sets we need have as their elements numbers of some kind, for example real numbers or natural numbers.

It is worth reminding the reader that the logical disjunction of two propositions p and q, written symbolically $p \vee q$ and read "p or q", is true when either proposition is true, including when both are. Thus $A \cup B$ includes the set $A \cap B$.

Union and intersection are simple ways to build new sets out of old. But we will need much more powerful ways which we will not try to justify. For example we need the natural numbers to form a set. Only if this is readily believable can we proceed to analysis. It is a common experience of mathematicians that attempts to explain analysis to someone without mathematical training get stranded on this point. The interlocutor does not accept the existence of an infinite set, although they may readily accept the fact that the natural numbers are infinitely many, apprehending that there is an unlimited supply of them. They may find it hard to admit that there is any sense in treating the natural numbers as a totality, as a completed whole. This is not foolish; that the natural numbers form a set cannot be proved; it requires an axiom of set theory (axiom of infinity). Without this fact there is no analysis. We will see ample confirmation of this, in the great emphasis placed on sequences for example.

We can also form the intersection and union of infinitely many sets (the latter requires an axiom in proper accounts of set theory). We will look at these constructions, and other operations on sets, if and when they are needed, but they are readily acceptable as common sense once the notion of an infinite set is accepted.

In simple cases we can list the elements of a set, enclosing them in curly brackets. For example
$$A = \{0, 1, 4, 16\}$$

builds a set whose elements are 0, 1, 4 and 16. The order of the elements does not count, nor do repetitions. Thus the sets
$$\{1, 2\}, \quad \{2, 1\}, \quad \{2, 1, 2\}$$

are all equal to each other. The reason is that two sets A and B are defined to be equal when they have the same members, that is, when $x \in A$ if and only if $x \in B$.

We might try to build the set of all even numbers by writing
$$B = \{0, 2, 4, 6, 8, \ldots\}$$

and such formulations are readily understandable if used sensibly. However, the set of all even numbers is more correctly formed using a prescription called specification, that builds the set of all elements of a given set that have a specified property. Using the property "n is divisible by 2", we can form the set B of even numbers by
$$B = \{n \in \mathbb{N} : n \text{ is divisible by } 2\}.$$

The general form of this construction is

$$B = \{x \in E : P(x)\}.$$

Here E is a given set and $P(x)$ is a property (called by logicians a predicate) which may or may not be true for different assignments of elements of E to the variable x. So we are singling out the set of all elements x of E for which $P(x)$ is true.

Using this we can build lots of subsets of the set $\mathbb{N}$ of natural numbers; for example, the set of even numbers, odd numbers, prime numbers, square numbers and so on. But we never get anywhere near all the possible subsets of $\mathbb{N}$ by this means because we cannot form enough predicates. We may want to make a statement about all subsets of $\mathbb{N}$. These form the elements of a set so vast that it defeats our imagination to encompass it (though that is nothing compared to the really big sets of set theory). Although we will not need it in this text, the set of all subsets of $\mathbb{N}$ is an important set for analysis, so it had better exist (again it requires a special axiom of set theory, the power-set axiom).

We now state an important property of the natural numbers. We shall consider it an axiom. It refers to completely arbitrary subsets of $\mathbb{N}$.

Principle of induction. If A is a set of natural numbers (that is, $A \subset \mathbb{N}$) such that $0 \in A$, and such that $x + 1 \in A$ whenever $x \in A$, then $A = \mathbb{N}$.

Axiomatic set theory usually has an axiom that makes infinite sets possible (the axiom of infinity mentioned earlier). It posits the existence of a set that contains 0, and for all natural numbers x, it contains $x + 1$ if it contains x. From this one can construct the set $\mathbb{N}$ by defining it as the intersection of all sets that have the property stated in the previous sentence. If one accepts this definition of $\mathbb{N}$, then the principle of induction follows as a theorem about $\mathbb{N}$. However, it is also quite natural to take the existence of $\mathbb{N}$ as a set for granted, and the principle of induction as an axiom that singles out the essential nature of the natural numbers: that you can reach any natural number from 0 by successively adding 1; and anything that you reach in this way is a natural number.

Another version of the induction principle, one that has more practical value, uses predicates that can be applied to the natural numbers, instead of subsets. An example of a predicate could be "n is divisible by 5".

Principle of induction with predicates. Let $P(n)$ be a predicate applicable to the natural numbers (that is, it is true or false for each substitution of a natural number for n). Assume that $P(0)$ is true, and that for every n it is the case that $P(n)$ implies $P(n + 1)$. Then $P(n)$ is true for all natural numbers n.

The principles are equivalent. Suppose that we assume the principle of induction. Let $P(n)$ be a predicate with the properties that $P(0)$ is true and that $P(n)$ implies $P(n + 1)$. We form the set $A = \{n \in \mathbb{N} : P(n)\}$ and see that $0 \in A$, and, for every x, if $x \in A$ then $x + 1 \in A$. We deduce by the principle of induction that $A = \mathbb{N}$, that is, that $P(n)$ is true for all n, thus establishing the principle of induction with predicates.

Conversely, let us assume the principle of induction with predicates. Let $A \subset \mathbb{N}$. We apply the principle of induction with predicates to the predicate "$n \in A$" to obtain the principle of induction.

Instead of assuming that $P(0)$ is true we could assume that there is a natural number k such that $P(k)$ is true. Then the conclusion would be that $P(n)$ is true for all natural numbers $n \geq k$.

Our first proposition says something immensely important about all possible subsets of $\mathbb{N}$, in their unimaginable variety.

Proposition 2.1 *Every non-empty subset of* $\mathbb{N}$ *contains a lowest element.*

More informally: in every collection of natural numbers there is a smallest number. This result is often used without being explicitly mentioned; and the present work is probably no exception.

Proof Given the non-empty subset A of $\mathbb{N}$, let B be the set

$$B = \{x \in \mathbb{N} : x \leq y \text{ for all } y \in A\}.$$

The specification says that B is the set of all natural numbers x, with the property that $x \leq y$ for all y in A. Elements of B are called lower bounds for A.

Now $0 \in B$. On the other hand B is not all of $\mathbb{N}$, since there exists some $z \in A$ (as A is not empty) and then $z + 1 \notin B$. Now we turn the induction principle on its head and conclude that there exists u in B such that $u + 1$ is not in B (for otherwise B would be all of $\mathbb{N}$). So u is a lower bound for A but $u + 1$ is not.

Now u must be in A. For $u \leq y$ for all $y \in A$, so that if u is not in A we would have $u < y$ for all y in A. Since we are only considering integers we would have $u + 1 \leq y$ for all y in A and $u + 1$ would be a lower bound—which it is not. We conclude that $u \in A$ and hence u is the lowest element of A. $\square$

2.1.1 Exercises

1. Show that the following predicates are true for all natural numbers:

    (a) Either n is divisible by 2 or $n + 1$ is divisible by 2.
    (b) $2^n > n$.

2. (a) Prove that $2^n > n^2$ whenever n is a natural number greater than or equal to 5.
    (b) Prove that $2^n > n^3$ whenever n is a natural number greater than or equal to 10.
3. Let a be a natural number. Prove the following by induction:

    (a) For all natural numbers n, the number $(a + 1)^n - 1$ is divisible by a.
    (b) For all even natural numbers n, the number $(a - 1)^n - 1$ is divisible by a.

    Note. This follows most naturally from the binomial theorem, which will be the topic of several exercises in the coming pages.
4. Prove that the following version of the induction principle, which appears to have a weaker premise than the usual one, follows from the usual induction principle:

Let $P(n)$ be a predicate for the natural numbers. Assume that $P(0)$ is true, and that whenever $P(k)$ is jointly true for all of $k = 0, 1, 2, ..., n$ then $P(n + 1)$ is true. Then $P(n)$ is true for all n.

Hint. Consider the predicate $Q(n)$ that says that $P(k)$ is jointly true for all natural numbers k such that $0 \le k \le n$.

Note. Use of this rule is called proof by complete induction.

5. The *locus classicus* of proof by complete induction. Show that every natural number greater than or equal to 2 is divisible by some prime number.

6. The Fermat numbers are defined by the formula $a_n = 2^{2^n} + 1$, for $n = 0, 1, 2,....$

    (a) Show that if $n \ne m$ then the numbers a_n and a_m have no common prime divisor. In short, they are coprime. Note that 1 is not considered a prime.
    (b) Deduce that the set of primes is infinite.

Note. Euclid published a proof that the set of primes is infinite that the reader is probably more familiar with.

2.2 Axioms for the Real Numbers

We shall describe the set $\mathbb{R}$ of real numbers by axioms, listing the properties that it should have from which we are confident that it is possible to derive the whole of analysis. We shall not attempt to build such a set, although this can be done using set theory. Intuitively, the real numbers model a line in Euclidean geometry, or more precisely, a coordinate line, like the x-axis of coordinate geometry standing alone. This picture is the reason why we often refer to a real number as a point. The basic intuition is that the line is marked off by a selection of real numbers for the purpose of measurement, and most importantly, any degree of accuracy can be attained by increasing the density of markings.

We shall not make any essential use of this picture. Instead we set out properties of the real numbers in the form of axioms. These fall into three distinct groups. We will introduce them in stages interspersed with deductions of familiar rules, together with reasons why further axioms are needed.

2.2.1 Arithmetic Axioms

The first group of axioms concerns arithmetic.

Axioms A. $\mathbb{R}$ is a *field*.

These axioms specify the algebraic operations we can perform with real numbers, together with their properties. There are two binary operations, $x + y$ (addition) and

$x \cdot y$ (multiplication), and two *distinct* constants $\bar{0}$ and $\bar{1}$, that satisfy the six axioms in the following list:

(A1)   (Commutative laws) For all x and y we have

$$x + y = y + x \quad \text{and} \quad x \cdot y = y \cdot x.$$

(A2)   (Associative laws) For all x, y and z we have

$$(x + y) + z = x + (y + z) \quad \text{and} \quad (x \cdot y) \cdot z = x \cdot (y \cdot z).$$

(A3)   (Neutral elements) For all x we have

$$x + \bar{0} = x \quad \text{and} \quad x \cdot \bar{1} = x.$$

(A4)   (Additive inverses) For all x there exists y, such that

$$x + y = \bar{0}.$$

(A5)   (Multiplicative inverses) For all x not equal to $\bar{0}$ there exists y, such that

$$x \cdot y = \bar{1}.$$

(A6)   (Distributive law) For all x, y and z we have

$$x \cdot (y + z) = (x \cdot y) + (x \cdot z).$$

Later we shall identify $\bar{0}$ and $\bar{1}$ with the natural numbers 0 and 1. For the moment it seems possible that they could have properties quite unlike 0 and 1. This is the reason for placing bars over them.

From axioms A1–A5 we can derive some common algebraic rules:

 (i) (Cancellation in sums)   If $x + y = x + z$ then $y = z$.
(ii) (Cancellation in products)   If $x \cdot y = x \cdot z$ and $x \neq \bar{0}$ then $y = z$.

It follows from the cancellation rules that, given an element x, an element y that satisfies $x + y = \bar{0}$ is uniquely determined by x. It therefore makes sense to denote it by $-x$. Similarly an element y that satisfies $x \cdot y = \bar{1}$, given that $x \neq \bar{0}$, is uniquely determined by x. We denote it by x^{-1}.

Exercise Prove the cancellation rules from the axioms.

Further familiar rules can be derived with the help of axiom A6 also:

 (iii) (Multiplication by $\bar{0}$)   For all x we have $x \cdot \bar{0} = \bar{0}$.
 (iv) (Multiplication by $-\bar{1}$)   For all x we have $(-\bar{1}) \cdot x = -x$.

Exercise Prove the stated rules from the axioms.

Now we define subtraction by

$$x - y := x + (-y),$$

and division by

$$\frac{x}{y} := x \cdot y^{-1},$$

given of course that $y \neq \bar{0}$. As is usual the colon in the equation signifies that the right-hand side is the definition of the left-hand side, although we are not overly insistent over its use.

The sum and product of n real numbers,

$$x_1 + x_2 + \cdots + x_n, \qquad x_1 \cdot x_2 \cdot \ldots \cdot x_n,$$

are well defined, without the need for brackets, because of the associative laws. Strictly speaking this should be proved. The proof, using induction and a good helping of patience, is surprisingly long. We will simply accept these facts.

Let n be a positive natural number and x an element of $\mathbb{R}$. We define

$$x^n := \overset{n \text{ factors}}{x \cdot x \cdot \ldots \cdot x}$$

$$nx := \overset{n \text{ summands}}{x + \cdots + x}.$$

Up to now there is no way to prove that $2\bar{1}$ (that is, $\bar{1} + \bar{1}$) is not equal to $\bar{0}$. A set of elements that satisfy axioms A1–A6 is called a field. There are many examples of fields that are nothing like the real numbers. For example there exists a field with only 7 elements. In such a field we must have $7\bar{1} = \bar{0}$.

Real numbers must therefore possess some other defining properties in addition to axioms A1–A6.

2.2.2 Axioms of Ordering

The second group of axioms concerns the ordering properties of the real numbers.

Axioms B. $\mathbb{R}$ is an *ordered field*.

There exists an order relation that can apply to pairs of elements of $\mathbb{R}$, written (when applicable) $x < y$. This relation satisfies the following axioms:

(B1) (Trichotomy) For each x and y exactly one of the following three possibilities must hold:
$$x = y, \qquad x < y, \qquad y < x.$$

(B2) (Transitivity) For all x, y, z, if $x < y$ and $y < z$ then $x < z$.

(B3) For all x, y, z, if $x < y$ then $x + z < y + z$.

(B4) For all x, y, z, if $\bar{0} < z$ and $x < y$ then $x \cdot z < y \cdot z$.

Note that axioms B3 and B4 relate ordering and algebraic properties.

We define further relations:

$$x > y \text{ means } y < x,$$
$$x \leq y \text{ means } (x < y \text{ or } x = y),$$
$$x \geq y \text{ means } (x > y \text{ or } x = y).$$

We define the concepts of positive and negative. If $x > \bar{0}$ we say that x is positive. If $x < \bar{0}$ we say that x is negative.

The following familiar rules are consequences of the axioms:

(i) x is positive if and only if $-x$ is negative.

(ii) $x > y$ if and only if $x - y$ is positive.

(iii) $\bar{1}$ is positive.

(iv)   If x is positive then x^{-1} is positive.

(v)   For all x not equal to $\bar{0}$ the number x^2 is positive.

(vi)   If $x < y$ and $z < \bar{0}$ then $x \cdot z > y \cdot z$.

We shall give the proof of rule (iii), leaving the others as exercises.

Proof of Rule (iii) We note that by trichotomy either $\bar{0} < \bar{1}$ or $\bar{1} < \bar{0}$ (since equality is ruled out by the assumption of distinctness in axioms A). Assume if possible that $\bar{1} < \bar{0}$. Then $-\bar{1}$ is positive (by rule (i), which we suppose was already proved) and we have $(-\bar{1}) \cdot \bar{1} < (-\bar{1}) \cdot \bar{0}$ by axiom B4, that is $-\bar{1} < \bar{0}$, leading to $\bar{0} < \bar{1}$. This contradicts the assumption $\bar{1} < \bar{0}$ and proves rule (iii).   □

Exercise Prove the remaining rules.

An ordered field includes a copy of $\mathbb{N}$. Because $\bar{1} > \bar{0}$ we have, for each x, that $x < x + \bar{1}$. This leads to a strictly increasing sequence $\bar{0}$, $\bar{1}$, $2\bar{1}$, $3\bar{1}$, ..., that is,

$$\bar{0} < \bar{1} < 2\bar{1} < 3\bar{1} < \cdots < n\bar{1} < (n+1)\bar{1} < \cdots$$

We can never reach $\bar{0}$, because if we did, we could conclude that $\bar{0} < \bar{0}$, which is impossible. By the same argument any two terms in the sequence are distinct.

The set $\{n\bar{1} : n \in \mathbb{N}_+\} \cup \{\bar{0}\}$, that is to say all elements $n\bar{1}$ where $n = 1, 2, ...$, together with $\bar{0}$, is therefore a copy of the natural numbers included in $\mathbb{R}$. So we can identify $n\bar{1}$ with the natural number n. From now on we write 0 instead of $\bar{0}$, 1 instead of $\bar{1}$, 2 instead of $2\bar{1}$ and so on. We will not distinguish between the natural number n and the real number n. We will also usually omit the dot in the product $x \cdot y$, writing it instead as xy.

This is beginning to look like the familiar coordinate line. Next we must fill the gaps between the integers.

2.2.3 Integers and Rationals

Elements of the set $\mathbb{N} \cup -\mathbb{N}$, that is, of

$$\{0, 1, 2, 3, ...\} \cup \{-1, -2, -3, ...\},$$

are called integers. The set of integers is denoted by $\mathbb{Z}$.

Elements of the set

$$\mathbb{Q} := \left\{ \frac{m}{n} : m \in \mathbb{Z},\ n \in \mathbb{N}_+ \right\}$$

are called rational numbers. The (first) colon signifies as usual that the right-hand member is the definition of the left-hand member. The notation mixes specification and the listing of a set's elements within curly brackets. A more correct, but less transparent formulation is

$$\{x \in \mathbb{R} : \text{there exist integers } m, \text{ and } n > 0,\ \text{such that } x = m/n\}.$$

A remarkable fact emerges: $\mathbb{Q}$ *is an ordered field.* The reader should check that all the axioms A and B are satisfied by $\mathbb{Q}$. It is virtually enough to show that the sum, product, difference and quotient of two rational numbers are rational. An even more remarkable conclusion follows:

There is no way to show that $\mathbb{R}$ contains elements other than those in $\mathbb{Q}$, by means of the axioms A and B alone.

2.2.4 $\mathbb{Q}$ is Insufficient for Analysis

It was an early discovery that Euclidean geometry is inconsistent with the assumption that all segments have rational length. It was found that the diameter of a unit square is not rational. So $\mathbb{Q}$ is not enough for practical tasks like drawing up plans for a new kitchen. The following proposition is in Euclid's Elements.

Proposition 2.2 *The real number 2 has no square root in $\mathbb{Q}$. In other words, there does not exist in $\mathbb{Q}$ any number x that satisfies $x^2 = 2$.*

Proof We are going to use an elementary fact of number theory, that any two natural numbers have a highest common divisor. This implies that a positive element of $\mathbb{Q}$

may be written as m/n, where m and n are positive integers with highest common divisor 1.

Suppose, if possible, that 2 has a rational square root. We express it in the form m/n, where m and n are positive integers with highest common divisor 1. Then $m^2/n^2 = 2$, so that $m^2 = 2n^2$. Then 2 divides m^2, and therefore also m, as is easy to see. Hence we can write $m = 2k$ where k is a natural number. We find that $4k^2 = 2n^2$, so that $2k^2 = n^2$. By the same argument 2 divides n also. This is a contradiction because m and n have highest common divisor 1. $\square$

This conclusion is a special case of a theorem of Gauss on polynomial equations, usually called Gauss's lemma. See the exercises in this section.

A real number that is not rational is called an irrational number. By using only the axioms for an ordered field it is impossible to show that any irrational number exists, for the simple reason that $\mathbb{Q}$ satisfies all the axioms for an ordered field. As we saw, any number that is a square root of 2 must be irrational.

It would be inconvenient if no square root of 2 existed. We would not be able to assign a length, as a real number, to the diagonal of a unit square. So $\mathbb{R}$ is not merely an ordered field. We need a new axiom to ensure that irrational numbers, such as the square root of 2, exist. To express this axiom we need to look at subsets of $\mathbb{R}$.

2.2.5 Dedekind Sections

The set of all subsets of $\mathbb{R}$ is dizzyingly large, but it seems to enter analysis in an essential way. The existence of this set is guaranteed by an axiom of set theory, the power set axiom. For fundamental analysis we can mostly get by without needing the set of all subsets of $\mathbb{R}$, since many important subsets can be defined by specification. To keep things simpler we first look at certain subsets of $\mathbb{R}$ with a simple structure.

Definition A Dedekind section of $\mathbb{R}$ is a partition of $\mathbb{R}$ into two disjoint sets, D_l and D_r (left set and right set), such that $\mathbb{R} = D_l \cup D_r$, neither D_l nor D_r is empty, and for all $x \in D_l$ and $y \in D_r$ we have $x < y$.

Recall that a set can be empty. We are explicitly excluding that D_l or D_r can be the empty set $\emptyset$. The definition requires that D_l and D_r have no common elements. This is expressed by the formula $D_l \cap D_r = \emptyset$, or in words: D_l and D_r are disjoint. The syntax is conventional but confusing, as this is not a property of the sets individually but a property of the pair of sets.

In order to give an example of a Dedekind section it is convenient to use a new binary operation on sets, the set difference $A \setminus B$. This is the set of all elements of A that are not elements of B. In the case that $A = \mathbb{R}$ and $B \subset \mathbb{R}$, it is usual to call $A \setminus B$ the *complement* of the set B.

Example of a Dedekind Section Let D_l be the set of all positive real numbers x that satisfy $x^2 < 2$, together with all negative real numbers and 0. Let $D_r = \mathbb{R} \setminus D_l$.

Exercise Check that this defines a Dedekind section.

2.2.6 Axiom of Completeness

The final axiom for the real numbers is expressed in terms of Dedekind sections.

Axioms C. $\mathbb{R}$ is a complete ordered field.

There is one additional axiom that turns an ordered field into a complete ordered field:

(C1) For each Dedekind section $\mathbb{R} = D_l \cup D_r$ there exists a real number t, such that $x \leq t$ for all $x \in D_l$ and $t \leq y$ for all $y \in D_r$.

We know that t is either in D_l or in D_r, but not in both. That means that either t is the highest element of D_l or it is the lowest element of D_r.

Exercise Show that t is uniquely defined by the Dedekind section.

Often the definition of Dedekind section includes a stipulation that D_l has no highest element. This is to ensure that there is a one-to-one correspondence between Dedekind sections and real numbers.

2.2.7 Square Root of 2

The trouble began with $\sqrt{2}$. We can settle the issue straight off.

Consider the Dedekind section that we defined before:

$$D_l = \{x \in \mathbb{R} : x \leq 0\} \cup \{x \in \mathbb{R} : x > 0, \ x^2 < 2\}$$
$$D_r = \{x \in \mathbb{R} : x > 0, \ x^2 \geq 2\}.$$

According to axiom C1 there exists t, that is either the highest element of D_l or the lowest element of D_r. It is clear that $t > 0$, this being obvious if t is the lowest element of D_r, and if t is the highest element of D_l then $t \geq 1$ since $1 \in D_l$.

We intend to show that $t^2 = 2$ by excluding the possibilities $t^2 < 2$ and $t^2 > 2$. Then axiom B1 gives $t^2 = 2$. We show in detail that $t^2 < 2$ is impossible. The reader is asked to give the details for excluding $t^2 > 2$ (Exercise 1 below).

Suppose that $t^2 < 2$. Then t is in D_l, and is therefore its highest element. We shall produce a contradiction by exhibiting a real number z such that $z > t$ but $z^2 < 2$. Such a number z would be an element of D_l higher than t. The tricky thing is that we have to do this without assuming that the number $\sqrt{2}$ exists.

The least cunning approach is to let $z = t + \varepsilon$ where $\varepsilon > 0$. We want if possible to choose ε so that $(t + \varepsilon)^2 < 2$, and this is equivalent to

$$t^2 + 2\varepsilon t + \varepsilon^2 < 2,$$

which in turn is equivalent to

$$\varepsilon < \frac{2 - t^2}{2t + \varepsilon}.$$

Let us agree to look for ε in the interval $0 < \varepsilon < 1$. For such ε we have

$$\frac{2 - t^2}{2t + \varepsilon} > \frac{2 - t^2}{2t + 1}.$$

Therefore it suffices if ε also satisfies

$$\varepsilon < \frac{2 - t^2}{2t + 1}.$$

The conclusion is that the number

$$\varepsilon = \frac{1}{2} \min\left(1, \frac{2 - t^2}{2t + 1}\right)$$

will work, where $\min(a, b)$ denotes the lower of the two numbers a and b. The number ε is positive and quite explicitly defined.

2.2.8 Exercises

1. Exclude the possibility that $t^2 > 2$ in the above argument for the existence of $\sqrt{2}$ using a similar approach. This concludes the proof that $\sqrt{2}$ exists.
2. Suppose that $t > 0$ and let $s = (2t + 2)/(t + 2)$. Show that if $t^2 < 2$, then $t < s$ and $s^2 < 2$; moreover if $t^2 > 2$, then $t > s$ and $s^2 > 2$. This gives another, and purely algebraic way, to prove that the number t of the previous section is $\sqrt{2}$.
3. We take the idea of the previous exercise a step further. Let $t_1 = 1$ and for $n = 1, 2, 3, \ldots$ we set

$$t_{n+1} = \frac{2t_n + 2}{t_n + 2}.$$

Calculate t_n using a calculator up to t_7 (or further if you have the patience). The results suggest that t_n approaches $\sqrt{2}$ with increasing n and could be used to approximate $\sqrt{2}$ by rational numbers. The germ of the limit concept is apparent here.

4. Prove that if n is a positive integer, and x and y are real numbers satisfying $0 < x < y$, then $x^n < y^n$. This shows that the function x^n is strictly increasing on the domain of all positive real numbers.
5. Show that in any ordered field the inequality $x + x^{-1} \geq 2$ holds for all positive x.
6. Let F be a field, that is, F is a set with operations, and constants $\bar{0}$ and $\bar{1}$, that satisfy the axioms A1–A6. Let P be a subset of F that has the following properties:

   (a) $\bar{0}$ is not in P.
   (b) If x and y are in P then $x + y$ is in P.
   (c) If x and y are in P then $x \cdot y$ is in P.
   (d) If $x \neq \bar{0}$ then either x is in P or $-x$ is in P.

   Define a relation in F, written $x < y$, as follows: $x < y$ shall mean that $y - x \in P$. Show that this is an order relation that makes F into an ordered field for which P is the set of all positive elements.

2.2.9 The Functions Max, Min, and Absolute Value

The functions min (used in the last section) and max are examples of functions of *two real variables*. They are defined by

$$\min(a, b) := \begin{cases} a, & \text{if } a < b \\ b, & \text{if } b \leq a, \end{cases}$$

$$\max(a, b) := \begin{cases} a, & \text{if } a > b \\ b, & \text{if } b \geq a. \end{cases}$$

The absolute value of x, denoted by $|x|$, is defined by

$$|x| := \begin{cases} x, & \text{if } x \geq 0 \\ -x, & \text{if } x < 0. \end{cases}$$

Some useful rules for manipulating absolute value and the functions max and min are

(i) $|ab| = |a||b|$
(ii) $|a + b| \leq |a| + |b|$ (triangle inequality for real numbers)
(iii) $|a - b| \geq \big||a| - |b|\big|$
(iv) $\max(a, b) = \frac{1}{2}(a + b + |a - b|)$
(v) $\min(a, b) = \frac{1}{2}(a + b - |a - b|)$.

Often the correct approach when dealing with absolute values, or max and min, is to consider cases. For example, if $a > 0$ then $|a| = a$; if $a < 0$ then $|a| = -a$; if $a > b$ then $\max(a, b) = a$; if $a < b$ then $\max(a, b) = b$ and so on.

Rule (ii) can be proved by cases but a more elegant approach is available. It begins by noting that $a \leq |a|$.

Exercise Prove rules (i–v).

The functions $\max(a, b)$ and $\min(a, b)$ can be extended by induction to any finite non-zero number of real variables. Thus for $n = 2, 3, 4, 5, \ldots$ and so on, we set

$$\max(x_1, x_2, \ldots, x_{n+1}) = \max\left(\max(x_1, x_2, \ldots, x_n), x_{n+1}\right)$$
$$\min(x_1, x_2, \ldots, x_{n+1}) = \min\left(\min(x_1, x_2, \ldots, x_n), x_{n+1}\right).$$

Defining a sequence of objects by induction means that the nth object is defined in terms of the preceding ones and we shall see more examples of this in the next chapter.

The sum of n numbers and the product of n numbers are also, strictly speaking, defined by induction, though we drew no attention to this when introducing them. Thus correctly, the definitions for all natural numbers n of the sum $x_1 + \cdots + x_n$ and the product $x_1 \ldots x_n$ should be given by a scheme which could be written as

$$x_1 + \cdots + x_{n+1} = (x_1 + \cdots + x_n) + x_{n+1}, \quad (n \geq 1)$$
$$x_1 \ldots x_{n+1} = (x_1 \ldots x_n) x_{n+1}, \quad (n \geq 1)$$

or in a more formal notation that avoids the dots:

$$\sum_{k=1}^{n+1} x_k = \sum_{k=1}^{n} x_k + x_{n+1}, \quad (n \geq 1)$$
$$\prod_{k=1}^{n+1} x_k = \left(\prod_{k=1}^{n} x_k\right) x_{n+1}, \quad (n \geq 1).$$

Given a sequence of numbers x_1, x_2, x_3 and so on, the expression $\sum_{k=1}^{n} x_k$ introduced here denotes the sum of all the numbers x_k as k runs from 1 up to, and including, n. Similarly, $\prod_{k=1}^{n}$ denotes the product of the numbers x_k as k runs from 1 up to, and including n.

The notation for sums and products, doubtless familiar to the reader, is capable of some flexibility. For example the expression $\sum_{k=m}^{n} x_k$ denotes the sum of all numbers x_k as k runs from m up to n, inclusive. This would normally be used in cases where $m \leq n$. If this is not known beforehand we can still use the expression but interpret it to be 0 if $m > n$. In short, the sum of an empty set of numbers is 0. The corresponding convention for product is that the product of an empty set of numbers is 1.

Given the n numbers $x_1, \ldots, x_n$ it should be obvious that $\max(x_1, \ldots, x_n)$ picks out their maximum, whilst $\min(x_1, \ldots, x_n)$ picks out their minimum. In fact the inductive definition even gives us a nice algorithm for computing these values. This makes it obvious that these quantities do not depend on the ordering of the numbers. We can even define $\max(x) = x = \min(x)$, thus defining these quantities for one variable.

The rather obvious fact that, of a finite set of real numbers, there is one that is highest and one that is lowest is often very important; and it is equally important that of an infinite set of numbers there may be no highest, or no lowest.

2.2.10 Mathematical Analysis

Mathematical analysis is the branch of mathematics that is based on the axiom of completeness of the real numbers, axiom C1. Most common functions, including those used in school mathematics, can only be adequately defined using mathematical analysis. A thorough treatment of functions and their role in analysis really starts in Chap. 4, but we give here a derivation of the nth root function, where n is a natural number, and more generally, the function x^a, where a is a rational exponent. The basic role of the completeness axiom will be apparent.

We saw that by means of axiom C1 we could prove that the equation $x^2 = 2$ has a positive root; and that without it, or more precisely, on the basis of axioms A and B alone, we could not. It is easy to see that there is only one positive root. Indeed we can observe that the function x^2 is strictly increasing for positive values of x. So if it takes the value 2 it can do so only once for a positive x.

We can use the same logic to show that the equation $x^2 = y$ has a unique positive root for each positive real number y. We can therefore define the function $\sqrt{y}$, the square root of y, as follows: for each positive real number y the number $\sqrt{y}$ is the unique positive root of $x^2 = y$.

In the same way we can define the nth root of y (where n is a natural number greater than 1 and $y > 0$) as the unique positive real number x that satisfies $x^n = y$. One can imitate the method used to show that $\sqrt{2}$ exists, defining the left-hand set of a Dedekind section as the set of all real numbers x such that $x > 0$ and $x^n < y$, together with all non-positive numbers. The binomial theorem is useful to finish the argument (see Exercises 11, 12 and 13).

Later in this text, in Chap. 4, an argument of a more general nature will be presented, based on continuity and deploying a powerful result called the intermediate value theorem. This is preferable to the rather clumsy approach using Dedekind sections. However it is done, the outcome is a function $\sqrt[n]{y}$ that produces the positive n^{th} root of the positive number y.

It now turns out that $\sqrt[k]{y^\ell}$ depends only on the ratio ℓ/k (see the exercises). We denote it by $y^{\ell/k}$, thus defining a fractional power of y. The scheme is extended to negative powers by setting $y^{-a} = (y^a)^{-1}$. And to the zeroth power $y^0 = 1$. In this way we can define y^a for rational a and positive y. Although certain powers of negative numbers make sense, for example $(-2)^{1/3}$, it is safest to assume that $y > 0$.

The laws of exponents (such as $y^{a+b} = y^a y^b$) are satisfied, and one can piece together a cumbersome proof of them (Exercise 19). It may be thought better to wait until arbitrary real powers have been defined (Chap. 7) and give a nice, smooth proof of these.

2.2.11 Exercises (cont'd)[1]

7. Prove the distributive rules for max and min:

$$\max(x, \min(y, z)) = \min(\max(x, y), \max(x, z)),$$

$$\min(x, \max(y, z)) = \max(\min(x, y), \min(x, z)).$$

8. Let a and b be rational numbers such that $\sqrt{a} + \sqrt{b}$ is rational. Show that $\sqrt{a}$ and $\sqrt{b}$ are both rational.

9. The Fibonacci numbers are the sequence of integers 1, 1, 2, 3, 5, 8, 13, ... and so on, in which each integer from 2 onwards is the sum of the two that precede it. Sequences will be studied thoroughly in Chap. 3. The sequence of Fibonacci numbers satisfies the recurrence relations

$$a_{n+2} = a_{n+1} + a_n, \quad n = 1, 2, 3, ... \tag{2.1}$$

There are many different sequences that satisfy these relations. The Fibonacci numbers are distinguished by the initial conditions $a_1 = 1$, $a_2 = 1$. The purpose of this exercise is to develop a formula for the nth Fibonacci number. You will see that it requires the use of irrational numbers. You can use the following steps:

(a) Show that there exist exactly two distinct real numbers, r_1 and r_2, such that if $\lambda = r_1$ or $\lambda = r_2$ then the sequence $a_n = \lambda^n$, ($n = 1, 2, 3, 4, ...$) satisfies the recurrence relations (2.1). Find r_1 and r_2 and show that they are irrational.

(b) Show that further solutions of the recurrence relations can be obtained by setting $a_n = Ar_1^n + Br_2^n$ where A and B are real constants.

(c) The Fibonacci numbers form the uniquely defined sequence of natural numbers that satisfy the recurrence relations together with the initial values $a_1 = a_2 = 1$. Find A and B such that the solution $a_n = Ar_1^n + Br_2^n$ satisfies $a_1 = a_2 = 1$, thus producing a formula for the nth Fibonacci number.

10. The binomial coefficients $\binom{n}{k}$ where n and k are natural numbers and $0 \leq k \leq n$ are defined as

$$\binom{n}{k} = \frac{n!}{k!(n-k)!}.$$

---

[1]This is the second group of exercises in Sect. 2.2. For this reason the numbering is continued from the previous set.

The exclamation mark used here is the factorial symbol, doubtlessly familiar to the reader. For completeness we recall that given a positive integer n we denote by $n!$ the product of all positive integers less than or equal to n. We define $0!$ to be 1.

Prove the addition formula:

$$\binom{n+1}{k} = \binom{n}{k} + \binom{n}{k-1}, \quad 1 \le k \le n.$$

Note. The formula to be proved is the basis of Pascal's triangle, a nice way to compute the binomial coefficients, and an obvious indication that they are all integers.

11. Prove the binomial rule: for all natural numbers n and real a and b we have

$$(a+b)^n = \sum_{k=0}^{n} \binom{n}{k} a^{n-k} b^k.$$

Hint. Induction is one possibility, making use of the previous exercise.

12. Suppose that $y > 0$, $x > 0$ and $x^n < y$, where n is a positive integer. Without using the existence of the nth root of y, show that there exists z, such that $z > x$ and $z^n < y$.

    Hint. You can let $z = x + \varepsilon$ and imitate the argument that established the existence of $\sqrt{2}$. The binomial rule can be useful. The result is essentially a proof that $\sqrt[n]{y}$ exists. See the next exercise.

13. Let $y > 0$. Write out an argument for the existence of $\sqrt[n]{y}$, where n is a positive integer, based on the Dedekind section of $\mathbb{R}$ whose left set is the set

$$D_l = \{x \in \mathbb{R} : x \le 0\} \cup \{x \in \mathbb{R} : x > 0, \ x^n < y\}.$$

14. Prove that if the square root of a natural number n is rational then the square root is a natural number.

    Hint. You will need a theorem of number theory: if a is a natural number greater than 1 and a prime p divides a^2 then p divides a.

15. Prove Gauss's lemma. If a polynomial with leading coefficient 1

$$x^n + c_{n-1}x^{n-1} + c_{n-2}x^{n-2} + \cdots + c_0$$

has integer coefficients, then every rational root is an integer.

    Hint. If a is an integer and a prime p divides a^n then p divides a.

    Note. This generalises the result of the previous exercise. It also leads to a simple algorithm for finding all the rational roots of such a polynomial: if $c_0 \neq 0$ test all integral divisors of c_0 for roothood.

16. Prove the formulas:

    (a) $\sum_{k=1}^{n} k = \frac{1}{2}n(n+1)$

    (b) $\sum_{k=1}^{n} k^2 = \frac{1}{6}n(n+1)(2n+1)$

(c) $\sum_{k=1}^{n} k^3 = \frac{1}{4}n^2(n+1)^2$
(d) $\sum_{k=1}^{n} (2k-1) = n^2$.

17. Prove the Cauchy–Schwarz inequality. Let a_k and b_k be real numbers for $k = 1, 2, \ldots, n$. Then

$$\sum_{k=1}^{n} a_k b_k \leq \left(\sum_{k=1}^{n} a_k^2 \right)^{\frac{1}{2}} \left(\sum_{k=1}^{n} b_k \right)^{\frac{1}{2}}.$$

Equality holds if and only if the two n-vectors $(a_1, \ldots, a_n)$ and $(b_1, \ldots, b_n)$ are linearly dependent; putting it differently, equality holds if and only if either $b_k = 0$ for all k or else there exists t such that $a_k = tb_k$ for all k.
Hint. Let $P(t) = \sum_{k=1}^{n}(a_k + tb_k)^2$ and note that $P(t) \geq 0$ for all t, whilst, unless all the b_k are 0, $P(t)$ is a second-degree polynomial. Recall some school algebra about second-degree polynomials that never take negative values.

18. Show that $\sqrt[k]{y^\ell}$ depends only on the ratio ℓ/k. Here we assume that y is a positive real number, and that k and l are positive integers.

19. Let $x > 0$, $y > 0$ and let a and b be rational numbers. Prove the three laws of exponents

$$y^{a+b} = y^a y^b, \quad (y^a)^b = y^{ab}, \quad x^a y^a = (xy)^a.$$

Hint. Planning is everything. First prove the laws in the case that a, b and c are natural numbers. That's just a matter of counting factors and using the associative and commutative laws. Next do the case when the powers are the reciprocals of natural numbers, after that the case of positive rational powers, finally rational powers; or else wait until arbitrary powers, possibly irrational, have been defined in Chap. 7.

20. Prove the following useful properties of rational powers:

   (a) If $a > 0$ and $0 < x < 1$ then $0 < x^a < 1$.
   (b) If $a > 0$ and $x > 1$ then $x^a > 1$.
   (c) If $a > 1$ and $0 < x < 1$ then $x^a < x$.
   (d) If $a > 1$ and $x > 1$ then $x^a > x$.
   (e) If $0 < a < 1$ and $0 < x < 1$ then $x^a > x$.
   (f) If $0 < a < 1$ and $x > 1$ then $x^a < x$.

2.3   Decimal Fractions

Every positive real number has a representation typified by

$$x = 1001.3835104779\ldots$$

where the digits to the right of the point, taken from the numbers 0, 1, 2, 3, 4, 5, 6, 7, 8, 9, continue indefinitely. Conversely, every expression of this kind defines a real number. Negative numbers are included by writing a minus sign at the front.

These facts are consequences of axiom C1. The concept of limit is needed to make the nature of the decimal representation precise. The reader is certainly familiar with decimal fractions[2] and they do provide a valuable basis for one's intuition about real numbers. The following discussion, which logically should come later, uses the notion of an infinite series, to be fully explained in a subsequent chapter. The purpose for giving it here is to forge an immediate link between the real numbers and objects of the reader's experience. The subject of decimals will be taken up again after infinite series have been properly introduced.

A *repeating decimal* (also called a recurring decimal) is one of the form (to give an explanatory example by way of definition)

$$x = 1001.3835104779477947794779...$$

the recurrence being that the string 4779 is supposed to repeat indefinitely. We specify this by the notation

$$x = 1001.3835104\overline{779}.$$

A *terminating decimal* is one of the form

$$x = 1001.383510\overline{0}$$

and is written more simply

$$x = 1001.38351.$$

2.3.1 Practical and Theoretical Meaning of Decimals

Practically, decimals are invaluable. They make it possible to calculate using real numbers. Theoretically, a decimal is a sequence of rational approximations to a real number, that are of the form $a/10^m$, where a and m are integers. As an example consider

$$\sqrt{2} = 1.4142135623...$$

where more digits can be found by a simple algorithm which may be repeated indefinitely. It is an example of an *infinite series*

$$\sqrt{2} = 1 + \frac{4}{10} + \frac{1}{10^2} + \frac{4}{10^3} + \frac{2}{10^4} + \frac{1}{10^5} + \frac{3}{10^6} + \frac{5}{10^7} + \frac{6}{10^8} + \frac{2}{10^9} + \cdots$$

---

[2]Often the term "decimal fraction" is used to mean a rational number whose denominator is a power of 10. We use the term to mean a real number between 0 and 1 in its decimal representation.

Truncating to a finite number of terms gives a rational approximation to $\sqrt{2}$. A commonly used approximation to $\sqrt{2}$ is $1.414 = 1414/1000$, obtained by truncating the series after the fourth term.

The rational numbers stand out in this scheme. They are the repeating decimals. Examples are

$$\frac{1}{16} = 0.0625 = 0.0625\overline{0}$$

$$\frac{3}{52} = 0.05\overline{769230}.$$

2.3.2   Algorithm for Decimals

Let x be a positive real number. The following algorithm produces the decimal representation of x.

First we find the highest natural number less than or equal to x. The existence of this natural number is a consequence of axiom C1 as we shall see in a later section. Write this natural number in the decimal system (we assume that the procedure for this is known) and place a decimal point after it, for example

$$1001.$$

Subtract this from x. There remains a number x_1 in the interval $0 \leq x_1 < 1$. We call this the first remainder. Now we have $0 \leq 10x_1 < 10$. Let d_1 be the integer part of $10x_1$, that is, the highest integer that is less than or equal to $10x_1$. It is one of the numbers $0, 1, 2, 3, 4, 5, 6, 7, 8, 9$. Call it the first decimal digit and write it after the decimal point, thus

$$1001.3$$

Subtract it from $10x_1$. This leaves the second remainder $x_2 = 10x_1 - d_1$, also in the interval $0 \leq x_2 < 1$. Next form $10x_2$, then its integer part, called the second decimal digit d_2, and the third remainder $x_3 = 10x_2 - d_2$ and so on.

In each step the remainder determines the next decimal digit and the next remainder, according to the scheme

$$d_n = [10x_n], \quad x_{n+1} = 10x_n - d_n.$$

In this notation $[y]$ denotes in general the highest integer (in this case it must be a natural number) less than or equal to y. The algorithm ends if the remainder is 0 at some step.

2.3.3 Decimal Representation of Rational Numbers

Why is the decimal representation of a rational number a repeating decimal?

If $x = m/n$ (with positive integers m and n) the remainder at each step is a rational number of the form a/n in the interval $0 \leq a/n < 1$. But there are only n of these, namely

$$0, \frac{1}{n}, \frac{2}{n}, \ldots, \frac{n-1}{n}.$$

When the $n + 1$st remainder is reached two of the remainders that have already been calculated must be equal, for example $x_k = x_l$. The decimal digits that are determined by $x_k, \ldots, x_{l-1}$ then repeat themselves.

If the decimal representation thus obtained is not terminating, the remainders are never 0. The length of the repeating string, called the period, is therefore at most $n - 1$; similarly the length of the initial string (before the repeats commence) cannot exceed $n - 1$.

2.3.4 Repeating Decimals and Geometric Series

Repeating decimals are examples of geometric series (about which more later). This is why they represent rational numbers. Look at the example

$$0.05\overline{769230}$$

This is the number x that satisfies

$$100x = 5.\overline{769230}$$

$$= 5 + 769230 \times \left(\frac{1}{10^6} + \frac{1}{10^{12}} + \frac{1}{10^{18}} + \cdots \right)$$

(if you don't know how to sum the series don't worry; you'll learn this later)

$$= 5 + 769230 \times \frac{1}{10^6} \times \left(\frac{1}{1 - \frac{1}{10^6}} \right)$$

$$= 5 + \frac{769230}{999999}$$

$$= \frac{5769225}{999999}$$

Perhaps unexpectedly this simplifies drastically to $\frac{75}{13}$. So that $x = \frac{75}{1300} = \frac{3}{52}$. This typifies the conversion of repeating decimals to ordinary fractions.

2.3.5   *Exercises*

1. Convert the following vulgar fractions to decimals.

   (a) $\frac{1}{11}$
   (b) $\frac{1}{99}$
   (c) $\frac{1}{32}$
   (d) $\frac{3}{17}$

   Note. The perspicacious reader may have noticed that the decimal algorithm is really the same as the long division algorithm as it was taught in primary schools two generations ago. This provides a convenient way to organise the calculation with pencil and paper.

2. If you know how to sum a geometric series you can practise by converting the following repeating decimals to vulgar fractions.

   (a) $0.\overline{10}$
   (b) $0.5\overline{10}$
   (c) $0.55\overline{101}$

2.4   Subsets of $\mathbb{R}$

Subsets of the real line play a fundamental role in analysis. Most of the many equivalent versions of the completeness axiom involve properties of subsets of $\mathbb{R}$. On a more elementary level, most functions treated in calculus are defined on a subset of $\mathbb{R}$, which in many cases is not all of $\mathbb{R}$.

2.4.1   *Intervals*

A subset of $\mathbb{R}$ is called an *interval* if it is of one of the following ten types (take note of the notation introduced in each line):

   (1) $]a, b[:= \{x \in \mathbb{R} : a < x < b\}$
   (2) $[a, b] := \{x \in \mathbb{R} : a \leq x \leq b\}$
   (3) $]a, b] := \{x \in \mathbb{R} : a < x \leq b\}$
   (4) $[a, b[:= \{x \in \mathbb{R} : a \leq x < b\}$
   (5) $]a, \infty[:= \{x \in \mathbb{R} : a < x\}$
   (6) $]-\infty, b[:= \{x \in \mathbb{R} : x < b\}$
   (7) $[a, \infty[:= \{x \in \mathbb{R} : a \leq x\}$
   (8) $]-\infty, b] := \{x \in \mathbb{R} : x \leq b\}$
   (9) $]-\infty, \infty[:= \mathbb{R}$
   (10) $\emptyset$ (the empty set).

The numbers a and b are called the endpoints of the interval (though not $\pm\infty$; they are not numbers). *Open intervals* are ones that contain no endpoints, although they may *have* endpoints (items 1, 5, 6, 9, 10). *Closed intervals* are ones that contain all their endpoints (items 2, 7, 8, 9, 10), and they do this trivially if they *have none*.

The notation for open intervals varies somewhat in the literature. The use of the reverse bracket is sometimes disparaged. Instead of writing "$]a, b[$" it is probably more usual to prefer "(a, b)", particularly in the English speaking world, with similar changes for the other types of interval. In fact both notations are specified in ISO standard 31-11. Moreover the reverse bracket is used by that most authoritative author Bourbaki. None of this particularly recommends it of course. However, there is a long history of using the expression "(a, b)" to denote the point in a coordinate plane with first coordinate a and second coordinate b. The same notation is customarily used for an ordered pair in set theory.

From a practical point of view, the risk of misunderstanding the reverse bracket notation is near to zero. That is why it is preferred in this text.

2.4.2 The Completeness Axiom Again

The notation of intervals enables us to express the completeness axiom in a concise form.

Axiom C1 restated

Let a Dedekind section be given with left set D_l and right set D_r. Then there exists a real number t such that, either

$$D_l = \,]-\infty, t], \quad D_r = \,]t, \infty[$$

or

$$D_l = \,]-\infty, t[, \quad D_r = [t, \infty[.$$

2.4.3 Bounded Subsets of $\mathbb{R}$

As examples of subsets of $\mathbb{R}$ we have seen the various intervals. From them we can form more subsets by intersection and union. Even so the variety of the set of all subsets of $\mathbb{R}$ is mind-dazzling, and there are many unsolved problems about them.

We now define some important properties that subsets of $\mathbb{R}$ may possess; they involve the ordering of the real numbers.

(a) A subset A of $\mathbb{R}$ is said to be *bounded above* if there exists $y \in \mathbb{R}$ such that $x \leq y$ for all $x \in A$. A number y that has this property is called an *upper bound of A*.

(b) A subset A of $\mathbb{R}$ is said to be *bounded below* if there exists $y \in \mathbb{R}$ such that $y \leq x$ for all $x \in A$. A number y that has this property is called a *lower bound of A*.
(c) A subset A of $\mathbb{R}$ is said to be *bounded* if it is both bounded above and bounded below.

2.4.4   Supremum and Infimum

We prove an important consequence of axiom C1 that applies to arbitrary subsets of $\mathbb{R}$ that are either bounded above or bounded below.

Proposition 2.3

(1) Let A be a non-empty subset of $\mathbb{R}$ that is bounded above. Then the set of all upper bounds of A is an interval of the form $[u, \infty[$.
(2) Let A be a non-empty subset of $\mathbb{R}$ that is bounded below. Then the set of all lower bounds of A is an interval of the form $]-\infty, v]$.

Proof It suffices to prove the result for case 1, from which case 2 is a simple deduction. Let the non-empty set A be bounded above. Then there exists an upper bound y. Every $z > y$ is also an upper bound. However, not every number is an upper bound. Since A is not empty there exists $x \in A$, and then $x - 1$ is not an upper bound. Let D_r be the set of all upper bounds and let D_l be the complement of D_r. It is clear that the pair D_l and D_r form a Dedekind section. By axiom C1 there exists a number u, such that either u is the lowest element of D_r or else it is the highest element of D_l.

However, D_l cannot have a highest element. There cannot be a number which is the highest of all numbers that are not upper bounds of A. For if v is not an upper bound of A there must exist $x \in A$, such that $v < x$; and then the number $\frac{1}{2}(v + x)$ lies between v and x, is equal to neither, is not an upper bound of A, but is higher than v.

We conclude that u is the lowest element of D_r. The set of upper bounds is then the interval $[u, \infty[$. □

Exercise Derive case 2 of Proposition 2.3 from case 1.

We can paraphrase Proposition 2.3 as follows. If a non-empty set of real numbers is bounded above, then among all upper bounds there is one that is lowest, often called the *least upper bound*. If a non-empty set of real numbers is bounded below, then among all lower bounds there is one that is highest, often called the *greatest lower bound*.

Definition Let A be a non-empty subset of $\mathbb{R}$ that is bounded above. The lowest upper bound of A is called the *supremum* of A. It is denoted by sup A.

Definition Let A be a non-empty subset of $\mathbb{R}$ that is bounded below. The highest lower bound of A is called the *infimum* of A. It is denoted by inf A.

The notions of supremum and infimum, which Proposition 2.3 enables us to define, are immensely important. The proposition is a consequence of axiom C1, but in most applications we use the proposition rather than the axiom. In fact this is probably the last time we use the axiom. Many mathematicians actually use Proposition 2.3 as an axiom, an alternative to axiom C1 (to which it is in any case equivalent). More precisely it is the first part (stating that *every non-empty set that has an upper bound has a least upper bound*) that is often used as an axiom, the second part being an obvious corollary of the first.

Exercise Show that the statement italicised in the last paragraph implies axiom C1.

There are dozens (and that can be taken literally) of equivalent formulations of the completeness of the real numbers. Probably the most common is the postulate that every non-empty subset of $\mathbb{R}$, that is bounded above, has a supremum. It is a matter of psychology which is preferred, but in this text axiom C1 is preferred. It is doubtful if anyone has a mental picture of an arbitrary subset of $\mathbb{R}$. Maybe a sort of diffuse, one-dimensional cloud. Moreover an axiom that says something about all bounded subsets of real numbers seems to require the set of all subsets of $\mathbb{R}$, which is a set vastly bigger than $\mathbb{R}$ itself. An axiom about all Dedekind sections is much simpler. A Dedekind section of $\mathbb{R}$ is both intuitive and highly graphic: we just cut the x-axis with a transversal line.

2.4.5 Exercises

1. Show that A is bounded if and only if there exists $k > 0$ such that $|x| \le k$ for all $x \in A$.
2. Determine which of the intervals listed 1–10 are bounded, bounded above, or bounded below. In each case find all upper bounds and all lower bounds.
3. Let $u < x$. Prove that $u < \frac{1}{2}(u + x) < x$. (This was a key step in the proof of Proposition 2.3.)
4. It is important to acquire the feeling that Proposition 2.3 is not obvious. To help with this acquisition we can consider replacing $\mathbb{R}$ by $\mathbb{Q}$. The notions of upper bound and lower bound can be defined for subsets of $\mathbb{Q}$ just as they were for subsets of $\mathbb{R}$. Consider the set $A = \{x \in \mathbb{Q} : x^2 < 2\}$. Show that A is bounded but has neither a least upper bound nor a greatest lower bound (in $\mathbb{Q}$ that is).

2.4.6 Supremum or Maximum?

Let us consider some examples of supremum and infimum.

(a) Let $A = [0, 1]$. Then $\sup A = 1$.
(b) Let $B = [0, 1[$. Then $\sup B = 1$ also.

In the first case 1 is the highest element of A, or the maximum of A. In the second case B has no highest element, but it has a supremum.

A non-empty set that is bounded above always has a supremum, but not necessarily a maximum, though if a maximum exists they are equal. The same applies to infimum and minimum. Confusion of the two caused many problems in the early days of analysis, and still does today among students of mathematics. Even the greatest mathematicians have come unstuck; Riemann's failed proof of the existence of a solution to the Dirichlet problem involved the confusion of infimum and minimum.

2.4.7   Using Supremum and Infimum to Prove Theorems

Much of analysis consists of proofs that certain desirable objects exist. These are often solutions to equations. We have already considered the square root of 2. This is the positive solution to the equation $x^2 = 2$. It is just a simple example of a polynomial equation. The treatment we gave was rather clumsy; necessarily so since useful tools were yet to be introduced.

Supremum and infimum are precisely such tools that enable us to mobilise the completeness axiom. When we want to prove for example that an equation has a solution, these tools can hand us a number that is a candidate for a solution.

Arguments using supremum or infimum often proceed in the following way. Suppose that A is a set of real numbers that is bounded above. Let $t = \sup A$. There are two important things we can say about t and in most applications both are needed. Firstly, for all $x \in A$ we have $x \leq t$ (because t is an upper bound). Secondly, suppose that $\varepsilon > 0$. Then $t - \varepsilon$ is not an upper bound (since t is the lowest one). This means that there exists $x \in A$ such that $t - \varepsilon < x$. Another way of expressing this is to say that if $y < t$ there exists $x \in A$ such that $y < x$.

The reader is invited to convince themselves that $\sup A$ is precisely the unique number t that has these two properties.

Consider again the square root of 2. Let us run through the argument again but this time we use least upper bound. We can let A be the set defined by

$$A := \{x \in \mathbb{R} : x > 0, \ x^2 < 2\}.$$

Then A is bounded above; it is easily seen for example that 2 is an upper bound. So let $t = \sup A$. Now we can show that $t^2 = 2$. If $t^2 > 2$ then we can find by the argument of Sect. 2.2 (under "Square root of 2") a number z such that $0 < z < t$ and $z^2 > 2$. It follows that z is an upper bound of A but is lower than t, in contradiction to the assumption that t was the least upper bound. Again if $t^2 < 2$ we can find by the argument of the same section a number y such that $t < y$ and $y^2 < 2$. Then $y \in A$, so that t is not even an upper bound of A, again a contradiction. The only conclusion available (by trichotomy) is that $t^2 = 2$.

Later we shall exploit continuity to carry out arguments like this one in a streamlined fashion.

2.4.8 The Archimedean Property of $\mathbb{R}$

The property considered in this section further supports our intuition that the axioms adequately convey the notion of the real coordinate line. It may be viewed as the theoretical underpinning of the measuring tape.

Proposition 2.4 *For every real number x there exists a natural number n such that $n > x$.*

Proof Assume, to the contrary, that there exists a real number x, such that every natural number n satisfies $n \leq x$. The set $\mathbb{N}$ is then bounded above in $\mathbb{R}$ and therefore has a supremum t. But then there exists a natural number m such that $m > t - 1$ (otherwise t would not be the least upper bound of $\mathbb{N}$). It follows that $m + 1 > t$, and we have the contradiction that t is not an upper bound of $\mathbb{N}$. $\qquad\qquad\square$

Here we have a result which, like Proposition 2.3, is apt to seem obvious and not needing proof. However it is impossible to prove the Archimedean property of $\mathbb{R}$ by using the axioms A and B alone. To show this one has to produce an example of an ordered field, a so-called non-Archimedean field, in which the set of natural numbers (always included in an ordered field) is bounded above. This can be done.

Although axiom C1 was used in the proof of Proposition 2.4, it is not equivalent to it. Putting it differently we cannot use the Archimedean property as an axiom replacing axiom C1 and expect to produce all the desired properties of the real numbers, although dozens of usable replacements for axiom C1 are known. In fact the field $\mathbb{Q}$ alone has the Archimedean property, but contains no irrational numbers, and we have seen that at least some irrationals are needed for analysis.

Nevertheless Proposition 2.4 supports the intuition that the real numbers, based on the axioms A, B and C, satisfactorily describe the coordinate line, in as far as the latter contains no points lying entirely to the right of all points with integer coordinates.

We can go further in producing a satisfactory model of the coordinate line as we understand it intuitively. A simple consequence of Proposition 2.4 is that if x is a positive real number, there exists a highest natural number, usually denoted by $[x]$, that is lower than or equal to x. This number is $n - 1$, where n is the lowest natural number strictly greater than x (which exists by Propositions 2.4 and 2.1). Finding it is the first step in constructing the decimal representation of a number x. Furthermore it is easy to extend the function $[x]$ to all real numbers (see the following exercises). For every real number x there exists a unique element m of $\mathbb{Z}$, such that $m \leq x < m + 1$. Every real number falls within an interval whose endpoints are consecutive integers.

2.4.9 Exercises (cont'd)

5. Find the supremum, infimum, maximum and minimum of the following sets of numbers, whenever they exist, justifying your conclusions:

(a) $[0, 1]$
(b) $[0, 1[$
(c) $\{x \in \mathbb{R} : x^2 < 2 \text{ and } x \text{ is rational}\}$
(d) $\{x \in \mathbb{R} : x^2 \leq 2 \text{ and } x \text{ is rational}\}$
(e) $\{x \in \mathbb{R} : x^2 < 4 \text{ and } x \text{ is rational}\}$
(f) $\{x \in \mathbb{R} : x^2 \leq 4 \text{ and } x \text{ is rational}\}$
(g) $\{x \in \mathbb{R} : x^2 + x + 1 < 0\}$
(h) $\{x \in \mathbb{R} : x^2 + x - 1 < 0\}.$

6. Let A and B be subsets of $\mathbb{R}$, both bounded, and suppose that $A \subset B$. Show that
    $\sup A \leq \sup B$ and $\inf A \geq \inf B$.
7. Let A be a set of real numbers with the following property: whenever a and b are
    distinct points of A and $a < x < b$ then x is in A. Prove that A is an interval.
    Hint. In the case when A is bounded and non-empty one can let $u = \inf A$ and
    $v = \sup A$. Show that A is then one of the intervals with endpoints u and v. This
    result is often useful.
8. Show that if we work only with the rational numbers (for which intervals can be
    defined by the same formulas as were used in the case of the real numbers) then
    the conclusion of the previous exercise fails in general.
    Note. The statement of the previous exercise has sometimes been proposed as a completeness
    axiom, that is, as a replacement for axiom C1.
9. Extend the function $[x]$ (as defined above for positive x) to all real arguments.
    Show that one may unambiguously define $[x]$ as the unique integer such that
    $[x] \leq x < [x] + 1$.
10. Let $a > 0$. Show that for every real x there exist a unique integer m and a unique
    number r, such that $0 \leq r < a$ and $x = ma + r$.
11. Let $A \subset \mathbb{Z}$. Show that if A is bounded above it has a highest element, and if A
    is bounded below it has a lowest element.

2.5   Approximation by Rational Numbers

We can interpret Proposition 2.4 as saying that there are no infinitely large real numbers, that is, no numbers that exceed every natural number. The counterpart to this
is that there are no infinitely small numbers either, that is, no real numbers that are
at the same time positive and are smaller than $1/n$ for every natural number n. This
is essentially a denial that infinitesimals exist in the real realm.

Proposition 2.5 *For each real number $\varepsilon > 0$ there exists a natural number n such
that $0 < 1/n < \varepsilon$.*

Proof By Proposition 2.4 there exists a natural number n such that $n > 1/\varepsilon$. But
then $0 < 1/n < \varepsilon$.   □

Proposition 2.6 *Each real number x is the supremum of the set of rational numbers
that are lower than x.*

Proof Let $A = \mathbb{Q} \cap \,]-\infty, x[$ (that is, A is the set of rational numbers lower than x). It is clear that x is an upper bound of A and we want to show that it is sup A. Suppose, on the contrary, that sup A, let us call it t, is actually lower than x, and let us derive a contradiction. Then $x - t > 0$ and we can find a natural number n by Proposition 2.5 such that $0 < 1/n < x - t$. Since $t = \sup A$, there exists a number $r \in A$, such that $t - 1/n < r \le t$. But then $r + 1/n$ is rational, $r + 1/n > t$ and $r + 1/n \le t + 1/n < x$. That is, $r + 1/n$ is an element of A that is strictly above t. This contradicts the fact that t is an upper bound for A. □

Proposition 2.7 *Between any two distinct real numbers there exist infinitely many rational numbers.*

Proof Let $x < y$. We create an increasing sequence of rationals between x and y in order to build the required infinite set. Sequences will be studied thoroughly in the next chapter.

We first show that there exists a rational number r_1 such that $x < r_1 < y$. Consider the midpoint $z = \frac{1}{2}(x + y)$, which satisfies $x < z < y$. By Proposition 2.6, z is the supremum of the set of all rational numbers less than z. Hence there exists a rational number r_1 such that $x < r_1 \le z$. We therefore have $x < r_1 < y$. Repeating this argument using r_1 instead of x we find a rational r_2 such that $r_1 < r_2 < y$, then a rational r_3 such that $r_2 < r_3 < y$, and so on. In this way we construct an increasing sequence of rationals r_n, $n = 1, 2, 3, ...$, such that $x < r_n < y$. The elements of this sequence form an infinite set. □

These propositions assure us that we may always approximate a real number by rational numbers, and make the error of the approximation as small as we like. This leads into the notions of limit and limit point, taken up in the next chapter. They reinforce our impression that the real numbers are a good model for the coordinate line, where adding more points increases accuracy of measurement. Technically we say that the rational numbers are dense in the set of real numbers.

All of this is important for practical calculations with real numbers. One way to approximate a real number is to use the decimal representation (or the representation in another number base). However, there are other ways that may be more precise. The subject of approximation by rational numbers is called diophantine approximation and is considered a branch of number theory.

2.5.1 Exercises

1. Let a and b be real numbers such that $a < b$. Show that a rational number between a and b can be obtained by the following argument, which is perhaps more intuitive than the proof of Proposition 2.7. Let n be a natural number, such that $0 < 1/n < b - a$, and let m be the lowest integer (possibly negative) such that $na < m$. Show that $a < m/n < b$.

2. Before the advent of calculus, arguments were used that involved neglecting quantities small with respect to other quantities. So for example, if a quantity h appeared in a formula, and h was small (on some scale) then h^2 might be neglected as being super small. This led to useful methods of approximation.

As an example, if x is an approximation to $\sqrt{a}$, we can write $x + h$ for the actual $\sqrt{a}$. Then $(x + h)^2 = a$ so that $x^2 + 2hx + h^2 = a$. If we suppose that the term h^2 can be neglected we might conclude that h is approximately $(a - x^2)/2x$, or rather suggestively

$$\frac{h}{x} \approx \frac{1}{2} \frac{a - x^2}{x^2}.$$

In words we have the well known rule that the proportional deficit of x is approximately half the proportional deficit of x^2.

   (a) Derive a similar rule for approximating a cube root: the proportional deficit of x is approximately one-third the proportional deficit of x^3.
   (b) Approximate the cube root of 1729. With 12 as a first approximation use mental arithmetic to get a better approximation, which should prove to have three correct digits after the decimal point and the fourth very nearly correct.

   Note. This example is the basis of an amusing anecdote in "Lucky numbers" in the book "Surely you must be joking Mr Feynman".

3. Let k_1 and k_2 be real numbers, neither of which is 0.

   (a) Show that if the set of all numbers of the form $mk_1 + nk_2$, as m and n range over the integers, contains a lowest positive number a, then k_1 and k_2 are both integer multiples of a.
   (b) Suppose that k_1/k_2 is irrational. Prove the following: for each $\varepsilon > 0$, there exist integers m and n such that $0 < mk_1 + nk_2 < \varepsilon$.

4. There exist ordered fields (non-Archimedean fields) in which the natural numbers are bounded above. Of course they do not satisfy axiom C1. To describe one we will need a little algebra. The field we shall describe, let us denote it by F, consists of all rational functions in one variable with real coefficients. These have the form of fractions

$$\frac{p_m x^m + p_{m-1} x^{m-1} + \cdots + p_0}{q_n x^n + q_{n-1} x^{n-1} + \cdots + q_0}$$

where the coefficients $p_0, \ldots, p_m, q_0, \ldots, q_n$ are real. The algebraic operations are the usual ones for fractions. The element x should not be thought of as a number; it is an *indeterminate*, and outside the system of real numbers. To describe the ordering we suppose that the fraction is written *in lowest terms*, that is, the numerator and denominator have no common factors except scalars (polynomials of degree 0). The fraction f/g is defined to be positive if the leading coefficients p_m and q_n have the same sign. Then $f < g$ is supposed to mean that $g - f$ is positive.

(a) Show that the relation $f < g$ is an ordering in the sense of axioms B.
   Hint. It can help to use Sect. 2.2 Exercise 6.
(b) The natural number ℓ is identified with the zero-degree polynomial ℓ. Show that for all natural numbers ℓ we have $\ell < x$. Hence the element x is an upper bound for the set $\mathbb{N}$.
(c) Show that for all natural numbers ℓ we have $0 < 1/x < 1/\ell$. So we can think of $1/x$ as an infinitesimal.
(d) Find a non-empty subset of the ordered field F that is bounded above but does not have a least upper bound.

Chapter 3
Sequences and Series

To those who ask what the infinitely small quantity in mathematics is, we answer that it is actually zero. Hence there are not so many mysteries hidden in this concept as there are usually believed to be.

L. Euler

3.1 Sequences

What are sequences? How do we talk about them? We have already seen examples of sequences, such as the Fibonacci numbers (studied in Sect. 2.2 Exercise 9), or the decimal digits of a real number. As always for the most important ideas, we have a choice of ways to express ourselves.

3.1.1 The Notion of Sequence

A scientist might make a table of measurements such as

1	2	3	4	5	6	7
0.707	0.577	0.5	0.447	0.408	0.378	0.354

We can easily read from this the first measurement or value, the second and so on. We can read the nth value for each natural number n from 1 to 7. These natural numbers have the role of place numbers. Each measurement has a place number.

We can also exhibit this example as a coordinate vector, without showing the place number, but allowing us to infer the place number by counting from left to right:

© The Editor(s) (if applicable) and The Author(s), under exclusive license to Springer Nature Switzerland AG 2020
R. Magnus, *Fundamental Mathematical Analysis*, Springer Undergraduate Mathematics Series, https://doi.org/10.1007/978-3-030-46321-2_3

$$(0.707, \ 0.577, \ 0.5, \ 0.447, \ 0.408, \ 0.378, \ 0.354).$$

For each natural number n from 1 to 7 we let a_n denote the number with place number n in the above table. We have here an example of a sequence with 7 terms. We express this as follows: $(a_n)_{n=1}^{7}$ is a sequence for which a_n is given by the value with place number n in the table. The use of ordinary parentheses, with subscript and superscript showing the range of the place numbers, is quite typical here. Curly brackets are sometimes used, but should be discouraged, as they usually specify sets, in which the order of elements is immaterial and repetitions irrelevant (recall the discussion in Sect. 2.1).

The value with place number n is called the nth term of the sequence. In a finite sequence there is a first term, a second term, a third term and so on up to some place number N. Different terms may have the same value, though the place numbers are different.

We can also specify a sequence using a formula. For example, let $(a_n)_{n=1}^{7}$ be a sequence for which $a_n = 1/\sqrt{n+1}$ for each natural number n from 1 to 7. We have a formula instead of a table. In this example there is no reason to stop at 7 terms. We can continue indefinitely. We then have an infinite sequence $(a_n)_{n=1}^{\infty}$.

It bears repeating that a sequence should not be confused with a set. For example, the coordinates of a point in coordinate geometry form a sequence. The points $(1, 2)$ and $(2, 1)$ in a coordinate plane are distinct sequences each with two terms. Their values constitute the same set of numbers $\{1, 2\}$. Again, the sequence $(1, 1)$ is not the same as the set $\{1, 1\}$; the latter is the same as $\{1\}$ and has only one element. This demonstrates again the different uses of parentheses and curly brackets—for the former the order matters; for the latter neither order nor repetitions make any difference.

We can give a reasonable working definition of sequence as follows. Here it clearly appears why the natural numbers must constitute a set.

Definition A sequence is an assignment of an element to each natural number of a given range, called the index set.

The index set is usually the set of all integers n such that $1 \leq n \leq N$, giving rise to a *finite sequence*, or the set $\mathbb{N}_{+}$, giving rise to an *infinite sequence*. We shall often speak of a sequence $(a_n)_{n=1}^{N}$ (in the case of a finite sequence), or a sequence $(a_n)_{n=1}^{\infty}$ (in the case of an infinite sequence), prior to explaining how the value a_n is to be assigned to the place number, or index, n.

We are being deliberately vague over the precise meaning of assignment. It could be a formula. It could be a table. However, we want to admit other cases when we do not have a practical method to calculate a_n for a given n. Therefore we avoid speaking of a "rule", which has connotations of a formula, suggesting a practical procedure.

We saw that a sequence is not the same as a set. Values appearing as terms in a sequence can be repeated. Sometimes we wish to consider the set of all values that appear as terms in a sequence $(a_n)_{n=1}^{N}$ or $(a_n)_{n=1}^{\infty}$. There is a notational trick. We write the set as

$$\{a_n : 1 \leq n \leq N\} \quad \text{or} \quad \{a_n : 1 \leq n \leq \infty\}$$

depending on the case. When considered as a set, the elements have forgotten their place numbers.

Consider some examples of sequences, in order of decreasing practicality. The first is the sequence

$$(1, \ 1.4, \ 1.41, \ 1.414, \ 1.4142, \ ...)$$

of approximations to $\sqrt{2}$ obtained by truncating its decimal representation. It is not easy to write down a formula for the nth term of this sequence, but the term can be calculated by a fairly easy algorithm.

The second is the sequence $(a_n)_{n=1}^{\infty}$ where $a_n = 1$ if n is prime and $a_n = 0$ if n is composite. This is an infinite sequence because the index set is infinite. Even so each term is either 0 or 1, so that the values form a finite set. There seems at present to be no practical way to calculate a_n since it asks us to determine whether or not n is prime, a problem that for large n is still prohibitively costly to solve on a computer.

The third example is to let $a_n = 1$ if the equation $x^n + y^n = z^n$ has a solution for which x, y, z are positive integers, and $a_n = 0$ if there is no such solution. Before 1995 it was known that $a_1 = 1$, $a_2 = 1$ and $a_n = 0$ for a known but finite number of places n. It was suspected that $a_n = 0$ for all $n \geq 3$ but no practicable way was known to calculate a_n for a given n. In 1995 Andrew Wiles published the proof of Fermat's Last Theorem, as a result of which we now know that $a_n = 0$ for all $n \geq 3$.

A common variant of sequence, that we shall make much use of, is to allow the place numbers to start at $n = 0$, with a_0 instead of a_1. We then have a sequence $(a_n)_{n=0}^{\infty}$ for example. This is useful when n denotes discrete time, beginning at the instant $n = 0$.

A further variation is to admit negative place numbers. For example $(a_n)_{n=-10}^{10}$, or a sequence infinite on both sides, $(a_n)_{-\infty}^{\infty}$.

In a later chapter we will define functions, like sequences, by using the undefined term "assignment". It turns out that the notion of assignment can be defined by set theory, and then what assignments are admissible depends on what set building axioms we wish to allow, or whether we want to allow some really weird sets. It is therefore still possible for two people to disagree over what constitutes an admissible assignment. But it does mean that sequences and functions are both examples of the same thing. Our mental pictures may still be quite different.

3.1.2 Defining a Sequence by Induction

We can define a_n in terms of a_k for $k = 1, 2, ..., n - 1$. This could have more practical value than a formula for a_n. We need to specify some initial values to get the process started.

Example The Fibonacci numbers constitute a sequence $(a_n)_{n=1}^{\infty}$ defined by letting $a_1 = 1, a_2 = 1$ and

$$a_n = a_{n-1} + a_{n-2}, \quad (n = 3, 4, 5, \ldots).$$

Using this one can generate the Fibonacci numbers more simply than by using the formula for the nth Fibonacci number obtained in Sect. 2.2 Exercise 9.

Example Let $a_1 = 1$ and

$$a_n = \sqrt{a_{n-1} + 1}, \quad (n \geq 2).$$

In these examples the sequence is said to be defined by induction, or by recursion. We have a rule for finding a_n from the terms a_k with place numbers $k < n$, and an appropriate set of initial values to start the ball rolling.

Many important operations on a finite number of objects, that we often take for granted, are defined formally by induction. We can name the sum of n numbers, the product of n numbers, the union of n sets and the intersection of n sets.

3.1.3 Infinite Series

Let $(a_n)_{n=1}^{\infty}$ be an infinite sequence of real numbers and define inductively a new sequence $(s_n)_{n=1}^{\infty}$ by

$$s_1 = a_1, \quad \text{and} \quad s_n = s_{n-1} + a_n, \quad (n \geq 2).$$

Informally we sometimes write

$$s_n = a_1 + a_2 + \cdots + a_n$$

but the formal symbol (which we have already used several times and is doubtless familiar to the reader) is

$$s_n = \sum_{k=1}^{n} a_k.$$

But what can

$$\sum_{k=1}^{\infty} a_k$$

mean? It seems to ask us to add infinitely many numbers and on the face of it means nothing. Whether or not it means anything such an expression is called an infinite series.

3.2 Limits

One of the greatest achievements of analysis was to give the expression $\sum_{k=1}^{\infty} a_k$ a precise meaning so that mathematicians could obtain valid and trustworthy conclusions. The key to this is the concept of limit. After the elucidation of the nature of the real numbers, the notion of limit is the second main foundation stone of analysis. So it is worth devoting some time and effort to understanding it. We first define the notion of limit of a sequence, and develop it at some length before applying it to infinite series.

Let $(a_n)_{n=1}^{\infty}$ be an infinite sequence of real numbers.

Definition A number t is called the limit of the sequence $(a_n)_{n=1}^{\infty}$ if the following condition is satisfied:

For every $\varepsilon > 0$, there exists a natural number N, such that $|a_n - t| < \varepsilon$ for all $n \geq N$.

The inequality $|a_n - t| < \varepsilon$ appearing in this definition is equivalent to

$$t - \varepsilon < a_n < t + \varepsilon,$$

and for practical purposes (such as carrying out a proof) this may be by far the most convenient form. It may also be written, using set theory, as

$$a_n \in \,]t - \varepsilon, t + \varepsilon[.$$

We write

$$\lim_{n \to \infty} a_n = t$$

to denote that t is the limit of the sequence $(a_n)_{n=1}^{\infty}$.

The limit t may, or may not, be equal to a_n for some n. There may even be infinitely many places n, such that $a_n = t$; or there may be none at all. This may seem an obvious point, but early thinking about limits often assumed that a limit might not be a value of the sequence of which it is a limit. The sequence was supposed to approach arbitrarily near to its limit without ever reaching it. This thinking was an early obstacle to finding the correct definition of limit.

Not all sequences have limits. A sequence that has a limit is said to be convergent. We also say that a_n *converges to t*, or that a_n *tends to t*, and sometimes write $a_n \to t$. Thus when we say that a sequence $(a_n)_{n=1}^{\infty}$ is convergent, we mean: there exists a real number t, such that $\lim_{n \to \infty} a_n = t$.

Often we are dealing with particular sequences where we have a formula. Examples are

$$a_n = \frac{1}{n}, \quad a_n = 2^{-n}, \quad \text{or} \quad a_n = \frac{n+1}{n}.$$

In such cases we do not write "$\lim_{n\to\infty} a_n$ where $a_n = 1/n$", and so on (though we might), but instead (for these examples)

$$\lim_{n\to\infty} \frac{1}{n}, \quad \lim_{n\to\infty} 2^{-n}, \quad \lim_{n\to\infty} \frac{n+1}{n}.$$

A sequence that does not have a limit, that is, a sequence that is not convergent, is often said to be divergent. Strictly speaking, when we are presented with a sequence $(a_n)_{n=1}^{\infty}$, it is not good mathematical grammar to write the formula "$\lim_{n\to\infty} a_n$" until we have ascertained that the limit exists. The expression "$\lim_{n\to\infty}$" is not a function symbol that can be stuck in front of an arbitrary sequence. This is perhaps a little awkward, but there are ways to alleviate this (see the nugget "Limit inferior and limit superior").

3.2.1 Writing the Definition of Limit in English, and in Logic

The property of a sequence embodied in the definition of limit involves quantifiers, that is, the expressions "there exists" and "for all", in fact three of them. It seems that there is no escaping this, and it is perhaps the reason why the correct definition of limit was so long emerging. It is also a significant challenge to define the property in ordinary language.

The definition of $\lim_{n\to\infty} a_n = t$ is written in natural English that contains some in-line mathematical formulas. It is not hard to produce a non-mathematical sentence, again in natural English, that has the same syntactical structure and the same logical structure:

In every class was a pupil, who obtained full marks in all their examinations.

The definition of limit can be expressed entirely in a mathematical language called first-order logic. It has the form (slightly simplified for greater ease of reading):

$$(\forall \varepsilon > 0)(\exists N \in \mathbb{N})(\forall n \in \mathbb{N})(n \geq N \Rightarrow |a_n - t| < \varepsilon).$$

A literal translation of this into rather unnatural English, keeping to the same phrase order, might be

For all $\varepsilon > 0$ there exists a natural number N, such that, for all natural numbers n, $n \geq N$ implies $|a_n - t| < \varepsilon$.

It will be noticed that in the logical sentence all quantifiers ("there exists", "for all", symbolised by "$\exists$", "$\forall$") are placed at the front, whereas in idiomatic English one of them is placed at the end. The order of the quantifiers is most important.

Let us examine the example of the pupils who obtained full marks. This may be expressed using formal quantifiers by

$$(\forall a)(\exists b)(\forall c)(F(a, b, c)). \tag{*}$$

Here we are using "a" for a variable that ranges over all classes, "b" for a variable that ranges over all pupils and "c" for a variable that ranges over all examinations. The expression "$F(a, b, c)$" stands for

b was in class a, and if b took the examination c they achieved full marks in it.

The conditional clause (introduced by "if") is needed because it is not implied that all the pupils took the same examinations. But it has a slightly odd implication. In logic the claim:

if b took the examination c they achieved full marks in it

which is a part of $F(a, b, c)$ is true if b did not in fact take the examination c. So the sentence (*) is true in the (perhaps exceptional and possibly unintended) circumstances that in every class was a pupil who took no examinations at all, because it is true that such a pupil got full marks *in all that they took*. In mathematics one has to be on the lookout for such circumstances, where a statement can be true *by default* because it directs one to test something of which there are no instances, like the examinations taken by that absentee pupil.

Whilst the phrasing of the definition of limit given in the last section was common in the early twentieth century (Hardy, Whittaker and Watson, Burkill), in the latter half of the century there began a tendency to write the quantifiers in the English sentence in the position in which they occur in the formal logical definition. It even seems that the older phrasing of the definition is disparaged. It may be thought that placing a quantifier at the end of the sentence, as is common in idiomatic English, is ambiguous, as it is arguably not clear where it should be placed in the precise first-order logical sentence that is supposed to express the same idea. A misunderstanding could arise that the intended logical sentence is

$$(\forall n \in \mathbb{N})(\forall \varepsilon > 0)(\exists N \in \mathbb{N})(n \geq N \Rightarrow |a_n - t| < \varepsilon).$$

This proposition is completely different from the condition for a limit. It is trivially true, since given n and ε we can choose N so that $n < N$, and then the proposition "$n \geq N \Rightarrow |a_n - t| < \varepsilon$" is true, because in propositional logic an assertion of the form "$p \Rightarrow q$" is true if p is false, as we saw in the case of the pupil who took no examinations, but got full marks in all that they took. It therefore says nothing about the sequence $(a_n)_{n=1}^{\infty}$.

In fact the risk of this misunderstanding of the definition is negligible, and is rendered even more so by placing a comma after "N", which ties the quantifier "for all $n \geq N$" firmly to the formula "$|a_n - t| < \varepsilon$".

Nevertheless let us look at ways that have been used to make the literal translation of the logical formula more natural whilst keeping the order of its mathematical elements.

The problem lies in the awkward juxtaposition of two formulas "n" and "$n \geq N$". This must be avoided. Here is one way:

For all $\varepsilon > 0$ there exists a natural number N, such that, for all natural numbers $n \geq N$ we have $|a_n - t| < \varepsilon$.

The phrase *"we have"* is perhaps acceptable but is rather artificial. Even so we shall sometimes use this type of formulation for the sake of variation. Here is another way:

For all $\varepsilon > 0$ there exists a natural number N such that, for all natural numbers n, if $n \geq N$ then $|a_n - t| < \varepsilon$.

The juxtaposition *"n, if n"* jars but is perhaps acceptable. Some writers adhere to this form but reduce the jarring effect by using a centred display, even going so far as to enclose it in a box (though not shown here):

For all $\varepsilon > 0$ there exists a natural number N, such that, for all natural numbers n,

$$\text{if } n \geq N \text{ then } |a_n - t| < \varepsilon.$$

This is again rather artificial. Yet another way is to drop the third quantifier entirely, whilst implying its presence:

For all $\varepsilon > 0$ there exists a natural number N, such that if $n \geq N$ then $|a_n - t| < \varepsilon$.

This is acceptable if we make the quite reasonable assumption that the reader understands from the context that *"if $n \geq N$ then $|a_n - t| < \varepsilon$"* means *"for all n, if $n \geq N$ then $|a_n - t| < \varepsilon$"*. However failure to understand it in this way is hardly less plausible than a misunderstanding of the older phrasing of the definition.

In fact nobody would dream of saying:

In every class was a pupil, such that in all their examinations they got full marks.

3.2.2 Limits are Unique

If a sequence has a limit then it has only one. This was implicit in the way we talked about limits in the last section. We must show that two distinct numbers t and s cannot both be limits of the same sequence $(a_n)_{n=1}^{\infty}$. In fact if $s \neq t$ we can choose $\varepsilon > 0$ smaller than $|s - t|/2$. Then there is no number common to the intervals $]t - \varepsilon, t + \varepsilon[$ and $]s - \varepsilon, s + \varepsilon[$. So it is impossible that there could exist N_1 and N_2 such that $a_n \in]t - \varepsilon, t + \varepsilon[$ for all $n \geq N_1$ and $a_n \in]s - \varepsilon, s + \varepsilon[$ for all $n \geq N_2$.

3.2.3 Exercises

1. Write the argument of the last paragraph using inequalities, that is, show that $|a_n - t| < \varepsilon$ and $|a_n - s| < \varepsilon$ cannot be simultaneously true if ε is chosen as we described.

2. Check the assertion that the inequality $|x - a| < \varepsilon$ is equivalent to the two inequalities $a - \varepsilon < x < a + \varepsilon$. This reformulation is often helpful.
3. Find some other examples of non-mathematical assertions that require three quantifiers.

3.2.4   Free Variables and Bound Variables

In the formula

$$\lim_{n \to \infty} \frac{n+1}{n}$$

the variable n can be replaced by any other letter without changing the sense or value. Thus

$$\lim_{n \to \infty} \frac{n+1}{n} = \lim_{\beta \to \infty} \frac{\beta+1}{\beta},$$

provided it is understood that the variables n and β range over the natural numbers. It is not just that the quantities are equal, the expressions have the *same meaning*.

In the formula $1/n$ we can put in values for n (from the realm of natural numbers, for example) and calculate in this way instances of the formula. The variable n is free. But putting "$\lim_{n \to \infty}$" before "$1/n$" ties n down. It makes no sense to ask: "What is the value of $\lim_{n \to \infty} 1/n$ for $n = 5$?" The variable n has become bound by the prefix "$\lim_{n \to \infty}$". It is for this reason that any other letter may replace "n" without changing the meaning.

Another example of bound and free variables is the expression

$$\sum_{k=1}^{n} k^p.$$

Here k is bound but n and p are free. We may replace k by any other letter, except for n and p as they are in use, and the meaning remains unchanged. But although we may ask about the value of this expression for $p = 2$ and $n = 10$, there is no sense in asking about its value for $p = 2$, $n = 10$ and $k = 5$.

The test for a free variable is to ask yourself whether the following question makes sense:

What is the value of the expression if I substitute 1 (or some other value within the allowed range) for the variable?

Here are some other commonly occurring expressions involving bound variables:

(a) $\max\limits_{1\leq k\leq n} a_k$ The maximum of the numbers $a_1, a_2, ...,a_n$; the variable k is bound, but n is free.

(b) $\min\limits_{1\leq k\leq n} a_k$ The minimum of the numbers $a_1, a_2, ...,a_n$; the variable k is bound, but n is free.

(c) $\sup\limits_{1\leq n\leq\infty} a_n$ The supremum of the set of terms of the infinite sequence $(a_n)_{n=1}^{\infty}$.

(d) $\inf\limits_{1\leq n\leq\infty} a_n$ The infimum of the set of terms of the infinite sequence $(a_n)_{n=1}^{\infty}$.

A slightly different case is that of logical statements involving quantifiers. An example is

$$(\forall x)(x < y \Rightarrow x < 1).$$

This says, in words, that for all x, if x is a real number less than y, then x is less than 1. Here x is a bound variable; any other letter may replace it. But y is a free variable, for it makes sense to ask whether the statement is true or false for given values of y. For example, is it true or false for $y = 0$, or 1, or 2?

3.2.5 Proving Things Using the Definition of Limit

In order to prove that $\lim_{n\to\infty} a_n = t$ directly from the definition of limit, we have to produce, or show that it is possible to produce, for each $\varepsilon > 0$, an integer N, such that for all $n \geq N$ we can show that $|a_n - t| < \varepsilon$ (or something equivalent to this inequality). What does this mean in practice? What is required in an acceptable argument that is supposed to justify that a sequence has the limit t?

It is extremely important to write out the argument so that the reader (equipped with enough mathematical knowledge) can convince themselves that it is correct. It is often not enough for the writer to be convinced. This applies to all mathematical writing of course.

The most important thing is to demonstrate that something is possible for every positive real number ε. To accomplish this it is best to write clearly at the beginning of the argument:

"Let $\varepsilon > 0$."

(including the full stop), and if not at the beginning (see below for a discussion of this), at least starting a new paragraph or in some way giving it prominence.

The next thing is to produce an N. Some ingenuity may be needed, or some informed guessing. In any case one must verify that just in virtue of $n \geq N$ it will follow that $|a_n - t| < \varepsilon$.

Finally it must be made apparent that you can produce such an N for any given positive ε. Here it is generally enough to make it clear that no special assumptions about ε are needed (apart from its being positive).

Actually that last remark can be modified in the light of good sense. For example it might be alright to assume (if it helps; and it often does) that $\varepsilon < 1$. The reason is that, if an N works for one ε, it will work for any bigger ε. In the same way we might want to reduce ε to below some value (if it is not already below it) if it helps to produce N.

An observation similar to that of the last paragraph: if N works for a given ε, then any number bigger than N would work as well (as an alternative N so to speak). This raises the question of whether we want to regard N as a function of ε; that is, should N be seen to be determined by ε? The answer is, that as far as the definition of limit is concerned, we do not have to display N as explicitly determined by ε. It is true that sometimes we may want this, for example if we wish to obtain explicit estimates of errors; but that is another issue, beyond that of proving that a sequence has the limit t. To produce N we may make some arbitrary choices along the way, without defining them explicitly. For example if we know that a certain set of numbers is not empty we can choose an element of it, without further explanation. And of course, the lowest N that works is uniquely determined by ε, though it may be impractical to give a formula for it, and, as we have seen, it is not needed to prove that a limit exists.

It is often a good idea to place some preliminary observations before introducing ε, but they should not refer to ε of course. For example one might want to construct a sequence b_n, such that $|a_n - t| \leq b_n$. Then in the body of the argument we might produce N, such that $b_n < \varepsilon$ for all $n \geq N$, if that should be simpler than dealing directly with $|a_n - t|$.

3.2.6 Denying That $\lim_{n \to \infty} a_n = t$

Proof by contradiction is an important mathematical tool, in fact one can nearly define mathematics as the domain of human discourse where proof by contradiction is completely accepted. One assumes that the conclusion of a proposition is false and deduces from this assumption a false statement, or a statement inconsistent with the proposition's premises.

Imagine that the conclusion is the statement $\lim_{n \to \infty} a_n = t$. To prove this by contradiction we have to begin by negating the statement $\lim_{n \to \infty} a_n = t$. What does this entail?

It is not simply equivalent to asserting that the limit of a_n is not t, for, literally, this asserts that a_n has a limit but the limit is not t. Rather we wish to assert the following: either the limit does not exist, or it exists and is not t.

To say that the limit exists and equals t means that for every ε that is positive we can accomplish a certain task. So to deny this means that there exists ε that is positive, and for which the task cannot be accomplished. This could be established by exhibiting just one value ε for which the task is impossible. But what is the task? It is to produce N that has a certain property. Therefore to say that the task is impossible means that for every N the property in question is not available. What is the property that N may or may not possess? It is that for all $n \geq N$ we have $|a_n - t| < \varepsilon$. So for N not to possess this property entails that there exists at least one n, such that $n \geq N$ and $|a_n - t| \geq \varepsilon$.

Finally we can set out what it means to negate the sentence $\lim_{n \to \infty} a_n = t$. It means the following: there exists $\varepsilon > 0$, such that for all natural numbers N there exists $n \geq N$, such that $|a_n - t| \geq \varepsilon$.

This can be written as a sentence of first-order logic (somewhat simplified to make it readable):

$$(\exists \varepsilon > 0)(\forall N \in \mathbb{N})(\exists n)(n \geq N \wedge |a_n - t| \geq \varepsilon).$$

3.2.7 Two Fundamental Limits

We now calculate two very important limits. The proofs are our first real use of the definition of limit.

Proposition 3.1

(1) $\lim_{n \to \infty} \dfrac{1}{n} = 0$.

(2) If $0 < x < 1$ then $\lim_{n \to \infty} x^n = 0$.

Proof of the First Limit Let $\varepsilon > 0$. By Proposition 2.4 there exists a natural number N such that $N > 1/\varepsilon$. If $n \geq N$ we have $1/n \leq 1/N < \varepsilon$ so that $|1/n| < \varepsilon$. $\square$

We do not have to produce the lowest N that works, although here that would be easy. Just take as N the lowest natural number greater that $1/\varepsilon$.

Proof of the Second Limit First some preparation. Let $0 < x < 1$. Then $1/x > 1$ and we let $1/x = 1 + h$ where $h > 0$. By the binomial rule (Sect. 2.2 Exercise 11)

$$\frac{1}{x^n} = (1 + h)^n > 1 + nh$$

(we drop all terms except the first and second). Therefore

$$0 < x^n < \frac{1}{1 + nh}.$$

Now we tackle the limit. Let $\varepsilon > 0$. To guarantee that $0 < x^n < \varepsilon$ it is enough to have $1/(1 + nh) < \varepsilon$, and that will be the case if $n \geq N$, where N is the least natural number above $((1/\varepsilon) - 1)/h$. $\square$

The N we found for the given ε was much bigger than was necessary since $(1 + h)^n$ is far above $1 + nh$. But we did not need to find the smallest N that works.

3.2.8 Bounded Sequences

Various notions of boundedness for sequences parallel the corresponding notions for sets of numbers, as defined in Sect. 2.5:

(a) A sequence $(a_n)_{n=1}^{\infty}$ is said to be *bounded above* if there exists K, such that $a_n < K$ for all n.
(b) A sequence $(a_n)_{n=1}^{\infty}$ is said to be *bounded below* if there exists K, such that $a_n > K$ for all n.
(c) A sequence $(a_n)_{n=1}^{\infty}$ is said to be *bounded* if there exists K such that $|a_n| \leq K$ for all n.

Obviously a sequence is bounded if and only if it is both bounded above and bounded below. As usual it is sometimes convenient to write the inequality as $-K \leq a_n \leq K$, or using set theory as $a_n \in [-K, K]$.

Proposition 3.2 *A convergent sequence is bounded.*

Proof Assume that $\lim_{n \to \infty} a_n = t$. Apply the definition of limit with $\varepsilon = 1$. There exists N, such that $|a_n - t| < 1$ for all $n \geq N$. But then $|a_n| < |t| + 1$ for all $n \geq N$. Choose $k = \max_{1 \leq n \leq N-1} |a_n|$ and we then have for all n that $|a_n| \leq K$ where $K = \max(k, |t| + 1)$. $\qquad\square$

3.2.9 The Limits ∞ and $-\infty$

Definition The sequence $(a_n)_{n=1}^{\infty}$ tends to infinity (or has the limit ∞), and we write $\lim_{n \to \infty} a_n = \infty$, if the following condition is satisfied:

For each real number K there exists a natural number N, such that $a_n > K$ for all $n \geq N$.

Definition The sequence $(a_n)_{n=1}^{\infty}$ tends to minus infinity (or has the limit $-\infty$), and we write $\lim_{n \to \infty} a_n = -\infty$, if the following condition is satisfied:

For each real number K there exists a natural number N, such that $a_n < K$ for all $n \geq N$.

A sequence with limit ∞ or $-\infty$ is not considered to be convergent. We sometimes say that it is divergent to ∞, or $-\infty$. The elements ∞ and $-\infty$ are not numbers, but they may be limits. We still say of such a sequence, though not convergent, that the limit exists.

Often, instead of saying "The sequence $(a_n)_{n=1}^{\infty}$ is convergent" we say "The limit $\lim_{n \to \infty} a_n$ exists and is a finite number". Here "finite number" means the same as "real number", or, "neither ∞ nor $-\infty$".

In general a sequence is said to be divergent when it has no finite limit. This includes oscillating sequences such as $a_n = (-1)^n$, as well as a vast assortment of wild behaviour, but also sequences that tend to ∞ or tend to $-\infty$. In this way "divergent" is genuinely the negation of "convergent".

3.2.10 Exercises (cont'd)

4. For each of the following limits find a suitable N for each positive ε:

   (a) $\displaystyle\lim_{n\to\infty} \frac{n+1}{n-1} = 1$

   (b) $\displaystyle\lim_{n\to\infty} \frac{n^2+1}{n^2-1} = 1$

   (c) $\displaystyle\lim_{n\to\infty} \frac{1}{\sqrt{n}} = 0$

   (d) $\displaystyle\lim_{n\to\infty} \frac{n^2-n+1}{n^2+n-1} = 1$

   (e) $\displaystyle\lim_{n\to\infty} \sqrt{n+1} - \sqrt{n} = 0.$

   Hint. There is a trick for doing this using the mathematics teacher's favourite identity: $(a+b)(a-b) = a^2 - b^2$. It can often be used in connection with square roots to avoid the use of continuity arguments. Write

   $$\sqrt{n+1} - \sqrt{n} = \frac{(\sqrt{n+1} - \sqrt{n})(\sqrt{n+1} + \sqrt{n})}{\sqrt{n+1} + \sqrt{n}}.$$

   (f) $\displaystyle\lim_{n\to\infty} \sqrt{1 + \frac{1}{n}} = 1$

   (g) $\displaystyle\lim_{n\to\infty} \sqrt{h_n} = \sqrt{a}$ where $\lim_{n\to\infty} h_n = a$ and $a > 0$.

5. Give a precise proof that the sequence $a_n = (-1)^n$ is divergent and has neither the limit ∞ nor $-\infty$. In brief, it has no limit.
6. Show that if $a > 1$ then $\lim_{n\to\infty} a^n = \infty$.
7. Show that if $a > 0$ then $\lim_{n\to\infty} a^{1/n} = 1$.
   Hint. If $a > 1$ write $a^{1/n} = 1 + b_n$ and estimate b_n.
8. Show that $\lim_{n\to\infty} n^{1/n} = 1$.
9. Let A be a closed interval (recall that there are five types of closed intervals; see Sect. 2.4). Let $(a_n)_{n=1}^{\infty}$ be a sequence in A that is convergent and let t be its limit. Show that t is in A. Prove also the converse of this: if A is an interval that is not closed, then there exists in A a convergent sequence whose limit is outside A.

   Note. This probably explains the appellation "closed". You cannot exit a closed interval by going to the limit of a convergent sequence lying in the interval.

3.3 Monotonic Sequences

Although a convergent sequence is always bounded, it is far from the case that a bounded sequence is always convergent. There is though one important case when a bounded sequence is convergent. It gives us a tool that is used over and over again; a simple instance of the structure of a sequence ensuring its convergence. Its proof is perhaps the first really important application of supremum and infimum.

Definition A sequence $(a_n)_{n=1}^{\infty}$ is said to be *increasing* if $a_n \leq a_{n+1}$ for all n. It is said to be *decreasing* if $a_n \geq a_{n+1}$ for all n. A sequence that is either increasing or decreasing is said to be *monotonic* (or *monotone*).

We shall sometimes refer to a sequence as being *strictly increasing*. This will mean rather obviously that $a_n < a_{n+1}$ for all n, similarly for *strictly decreasing*. A sequence can be both increasing and decreasing, if it is constant; a fact that we just have to live with. The terminology varies somewhat, but we want the simpler terminology for the more frequently encountered cases; and they are that $a_n \leq a_{n+1}$ for each n, or $a_n \geq a_{n+1}$ for each n.

Proposition 3.3 *Let $(a_n)_{n=1}^{\infty}$ be a monotonic sequence. Then exactly one of the following is the case:*

(1) It is convergent.
(2) It tends to $+\infty$.
(3) It tends to $-\infty$.

These three conclusions correspond respectively to the cases:

(1) It is bounded.
(2) It is unbounded and increasing.
(3) It is unbounded and decreasing.

Proof Consider the case where $(a_n)_{n=1}^{\infty}$ is bounded and increasing. The set

$$A = \{a_n : n \in \mathbb{N}\}$$

is then bounded. Note carefully that A is the set of values that occur in the sequence. This is shown by the use of curly brackets. Let $t = \sup A$. We shall show that $\lim_{n \to \infty} a_n = t$.

Let $\varepsilon > 0$. Since t is the supremum there exists N, such that $t - \varepsilon < a_N \leq t$. As the sequence is increasing and t is an upper bound, we have $t - \varepsilon < a_n \leq t$ for all $n \geq N$. We conclude that $\lim_{n \to \infty} a_n = t$.

Consider next the case when $(a_n)_{n=1}^{\infty}$ is unbounded but increasing. We shall show that $\lim_{n \to \infty} a_n = \infty$.

Let K be a real number. As the sequence is unbounded and increasing it is not bounded above. So there exists N, such that $a_N > K$. But then $a_n > K$ for all $n \geq N$. We conclude that $\lim_{n \to \infty} a_n = \infty$.

The case when the sequence is decreasing is similar and is left to the reader. □

Example We shall define a sequence by a process called iteration, which is extremely important and has many practical applications. Iteration is a simple case of inductive definition where there is a fixed function $f(x)$ and $a_{n+1} = f(a_n)$ for each n. The example is a very typical application of Proposition 3.3. We use the fact that the function $\sqrt{x}$ is strictly increasing, meaning that if $x < y$ then $\sqrt{x} < \sqrt{y}$.

Using induction we define the sequence $(a_n)_{n=0}^{\infty}$ by

$$a_0 = 1, \quad a_n = \sqrt{a_{n-1} + 2} \quad \text{for} \quad n \geq 1.$$

We shall show that $(a_n)_{n=0}^{\infty}$ is increasing and convergent.

We use induction to show that it is increasing, in fact strictly increasing. First of all we have $a_1 = \sqrt{3}$ so that $a_0 < a_1$. Assume that $a_{n-1} < a_n$ for a given $n \geq 1$. Since the function $\sqrt{x}$ is strictly increasing we have

$$a_n = \sqrt{a_{n-1} + 2} < \sqrt{a_n + 2} = a_{n+1}.$$

We deduce that $(a_n)_{n=0}^{\infty}$ is an increasing sequence.

We use induction also to show that the sequence is bounded, that in fact $a_n < 2$ for all n. In the first place $a_0 = 1$. Assume that $a_n < 2$ for a given $n \geq 0$. Then we find

$$a_{n+1} = \sqrt{a_n + 2} < \sqrt{2 + 2} = 2,$$

and conclude that $a_n < 2$ for all n.

By Proposition 3.3 the sequence is therefore convergent, and what is more $\lim_{n \to \infty} a_n = t = \sup_{n \geq 0} a_n$. Since $a_n < 2$ for all n we have $t \leq 2$. We cannot exclude the possibility $t = 2$ and later we shall see that this is indeed the case.

3.3.1   Limits and Inequalities

We saw in the example of the last section that $a_n < 2$ for all n, and concluded, without giving any justification, that $\lim_{n \to \infty} a_n \leq 2$. We have to assume that equality might hold for the limit, even though the inequalities are strict for the terms of the sequence.

Here is a general rule (not requiring the sequences to be monotonic):

Let $(a_n)_{n=1}^{\infty}$ and $(b_n)_{n=1}^{\infty}$ be convergent sequences such that $a_n \leq b_n$ for each n. Then $\lim_{n \to \infty} a_n \leq \lim_{n \to \infty} b_n$.

Exercise Prove this and also show that if $a_n < b_n$ for all n it is still possible for the limits to be equal.

The rule sometimes goes under the name of *preservation of inequalities*. Similar ideas are involved in the commonly used squeeze rule, covered in the section on limit rules.

3.3.2 Exercises

1. The following result is often used to establish a limit; in particular it is useful in some of the succeeding exercises, after making an informed guess about the limit.

   Suppose that for a given sequence $(a_n)_{n=1}^\infty$ there exist t and k, such that $0 < k < 1$ and $|a_n - t| \le k|a_{n-1} - t|$ for $n = 2, 3, \ldots$ Prove that $\lim_{n \to \infty} a_n = t$.

   Note. Often it happens in applications that $|a_n - t| \le k|a_{n-1} - t|$ only holds for all n from some index n_0. This makes no difference to the conclusion.

2. Show that the limit is 2 for the sequence $(a_n)_{n=0}^\infty$ defined above by the relation $a_n = \sqrt{a_{n-1} + 2}$, with starting value $a_0 = 1$.

3. A sequence $(a_n)_{n=1}^\infty$ is defined by $a_{n+1} = \frac{1}{4}(a_n^2 + 1)$, with starting value $a_1 = 0$. Show that a_n is increasing and is bounded above by $2 - \sqrt{3}$. Deduce that the sequence is convergent. Then prove that the limit is $2 - \sqrt{3}$.

   Hint. It may help to observe that $2 - \sqrt{3}$ is a root of $x = \frac{1}{4}(x^2 + 1)$.

4. A sequence is defined by $x_{n+1} = 2/(x_n + 1)$, with starting value $x_1 = 0$. Since the equation $x = 2/(x + 1)$ has only one positive root, and that is 1, the only reasonable candidate for a limit is 1. Prove that the limit is 1.

   Hint. It may help to observe that $x_n \ge \frac{2}{3}$ for all $n \ge 3$.

5. The ratio $x_n = a_{n+1}/a_n$ of successive Fibonacci numbers satisfies the relation

$$x_{n+1} = 1 + \frac{1}{x_n}.$$

As in the previous exercise a reasonable candidate for a limit is the positive root of the equation $x = 1 + 1/x$, or equivalently, $x^2 - x - 1 = 0$. This root is the famous number ϕ, or the Golden Ratio. Prove that the limit is ϕ without using the formula for the Fibonacci numbers obtained in Sect. 2.2 Exercise 9.

   Hint. Show that $|x_{n+1} - \phi| < |x_n - \phi|/\phi$.

6. Define a sequence inductively by

$$x_{n+1} = \frac{2x_n + 2}{x_n + 2},$$

with starting value $x_1 = 1$. Show that $\lim_{n \to \infty} x_n = \sqrt{2}$. Compare Sect. 2.2 Exercise 3.

3.4 Limit Rules

In this section we derive the most important limit rules, which enable us to find new limits from previously known ones, without the need to find N for each positive ε. They can be used to calculate limits involving rational functions of n, using as input only the two limits given in Proposition 3.1: $\lim_{n\to\infty} 1/n = 0$, and $\lim_{n\to\infty} x^n = 0$ given that $0 < x < 1$. Actually a third limit is needed, which one may think not worth stating:

$$\lim_{n\to\infty} C = C,$$

the limit of the constant sequence with all values equal to C.

Proposition 3.4 (Absolute value rule) *Let* $\lim_{n\to\infty} a_n = t$. *Then* $\lim_{n\to\infty} |a_n| = |t|$. *The rule holds also for* $t = \pm\infty$ *if we define* $|-\infty| = \infty$.

Proof We consider the case $t \neq \pm\infty$, leaving the remaining cases to the reader. We have the inequality (see Sect. 2.2.9)

$$\big||a_n| - |t|\big| \leq |a_n - t|.$$

Let $\varepsilon > 0$. Choose N, such that $|a_n - t| < \varepsilon$ for all $n \geq N$. Then for all $n \geq N$ we have $\big||a_n| - |t|\big| < \varepsilon$ and we are done. $\qquad\square$

Exercise Do the cases $t = \pm\infty$.

Proposition 3.5 (Sum and product rules) *Let* $(a_n)_{n=1}^{\infty}$ *and* $(b_n)_{n=1}^{\infty}$ *be convergent sequences, let* $\lim_{n\to\infty} a_n = s$ *and* $\lim_{n\to\infty} b_n = t$. *Then the sequences* $(a_n + b_n)_{n=1}^{\infty}$ *and* $(a_n \cdot b_n)_{n=1}^{\infty}$ *are convergent, and*

$$\lim_{n\to\infty} a_n + b_n = s + t, \quad \lim_{n\to\infty} a_n \cdot b_n = s \cdot t.$$

Note carefully that the sequences are supposed to be convergent, the limits $\pm\infty$ not allowed.

Proof for the Sum Let $\varepsilon > 0$. There exists N_1, such that $|a_n - s| < \varepsilon/2$ for all $n \geq N_1$; and there exists N_2, such that $|b_n - t| < \varepsilon/2$ for all $n \geq N_2$. Let $N = \max(N_1, N_2)$. For all $n \geq N$ we have

$$|(a_n + b_n) - (s + t)| \leq |a_n - s| + |b_n - t| < \frac{\varepsilon}{2} + \frac{\varepsilon}{2} = \varepsilon.$$

We conclude that $\lim_{n\to\infty} a_n + b_n = s + t$, thus completing the proof of the sum rule. $\qquad\square$

This argument has some features that occur quite often. Firstly, after introducing ε we produced an N, calling it N_1, that worked for $\varepsilon/2$. This was with hindsight. Other multiples of ε might be preferred in other contexts.

Secondly, we produced N that worked for two sequences simultaneously. It is clear that we could have any finite number of convergent sequences, and find N that works for a given ε for all the sequences at the same time. This is because if one N works for a given ε, then any integer bigger than N will also work. Moreover a finite set of natural numbers has a highest member. From now on, when working with a finite number of sequences we will usually take this trick for granted.

Proof for the Product   The convergent sequence $(a_n)_{n=1}^{\infty}$ is bounded (Proposition 3.2). Hence there exists $K > 0$, such that $|a_n| \le K$ for all n.

Let $\varepsilon > 0$. The most obvious way to begin is to use the definition of limit to find a natural number N, such that $|a_n - s| < \varepsilon$ and $|b_n - t| < \varepsilon$ for all $n \ge N$. For these values of n we have

$$
\begin{aligned}
|a_n \cdot b_n - s \cdot t| &= |a_n \cdot b_n - a_n \cdot t + a_n \cdot t - s \cdot t| \\
&\le |a_n \cdot b_n - a_n \cdot t| + |a_n \cdot t - s \cdot t| \\
&\le |a_n| \cdot |b_n - t| + |a_n - s| \cdot |t| \\
&< (K + |t|)\varepsilon.
\end{aligned}
$$

We interrupt the proof to interpose some discussion. We have arrived at the conclusion that for each $\varepsilon > 0$ there exists N, such that $|a_n \cdot b_n - s \cdot t| < C\varepsilon$ for all $n \ge N$; in this case $C = K + |t|$. It is clear that C does not depend on ε; nor does it depend on n or N. That is why we were careful to define the constant K before we introduced ε. It is always a good idea to introduce and define any constants, that one may want to use later, in the preamble, before the key phrase "Let $\varepsilon > 0$".

We restart the proof by backtracking to the point "Let $\varepsilon > 0$", and choose a slightly different N. There exists N, such that $|a_n - s| < \varepsilon/C$ and $|b_n - t| < \varepsilon/C$ for all $n \ge N$. For the same N we have $|a_n \cdot b_n - s \cdot t| < \varepsilon$ for all $n \ge N$. This concludes the proof. $\square$

Of course the backtracking requires the benefit of hindsight. Since we know that it can be done we can scrap it altogether in our proofs. From now on we will be content, in proving that $\lim_{n \to \infty} a_n = t$, to find N for each given $\varepsilon > 0$, such that for all $n \ge N$ we have $|a_n - t| < C\varepsilon$, provided we have made it clear that C is independent of ε, n and N.

Proposition 3.6 (Reciprocal rule) *Let $(a_n)_{n=1}^{\infty}$ be a convergent sequence, let $\lim_{n \to \infty} a_n = t$ and assume that $t \ne 0$. Then*

$$
\lim_{n \to \infty} \frac{1}{a_n} = \frac{1}{t}.
$$

Before we give the proof some explanatory discussion is needed. The terms $1/a_n$ form a sequence in the following sense: there exists n_0 such that $a_n \ne 0$ for all $n \ge n_0$ and the reciprocals form a sequence $(1/a_n)_{n=n_0}^{\infty}$. In fact we know that $\lim_{n \to \infty} |a_n| = |t|$ and $|t| > 0$. Hence there exists n_0, such that for all $n \ge n_0$ we have $|a_n| > \frac{1}{2}|t|$, and therefore also $a_n \neq 0$ (we are taking $\varepsilon = \frac{1}{2}|t|$ here). The reciprocal $1/a_n$ is then defined for all $n \geq n_0$. In this way we sidestep the possibility that a_n may be 0 for certain place numbers n.

Proof of Proposition 3.6 We have

$$\frac{1}{a_n} - \frac{1}{t} = \frac{t - a_n}{t \cdot a_n}.$$

Let $\varepsilon > 0$. There exists n_1 such that $|a_n - t| < \varepsilon$ for all $n \geq n_1$. We have seen that there exists n_0, such that $|a_n| > \frac{1}{2}|t|$ for all $n \geq n_0$. Let $N = \max(n_0, n_1)$. For all $n \geq N$ we have

$$\left| \frac{1}{a_n} - \frac{1}{t} \right| = \frac{|t - a_n|}{|t| \cdot |a_n|} < \frac{2}{|t|^2} \varepsilon.$$

The multiplier $2/|t|^2$ is independent of n, N and ε. We conclude (see the discussion following the proof for the product rule) that $\lim_{n \to \infty} 1/a_n = 1/t$. □

By using the rule for product and the rule for reciprocal we obtain the rule for quotient.

Proposition 3.7 (Quotient rule) *Let convergent sequences $(a_n)_{n=1}^{\infty}$ and $(b_n)_{n=1}^{\infty}$ be given. Let $\lim_{n \to \infty} a_n = s$ and $\lim_{n \to \infty} b_n = t$, and assume that $t \neq 0$. Then we have*

$$\lim_{n \to \infty} \frac{a_n}{b_n} = \frac{s}{t}.$$

Example Prove that $\lim_{n \to \infty} \dfrac{n^2 + 1}{2n^2 + 3}$ exists and find it.

We shall write the argument in excruciating detail referencing all rules. First we have

$$\frac{n^2 + 1}{2n^2 + 3} = \frac{1 + \dfrac{1}{n^2}}{2 + \dfrac{3}{n^2}}.$$

We know that $\lim_{n \to \infty} 1/n = 0$. By the product rule $\lim_{n \to \infty} 1/n^2 = 0$ and $\lim_{n \to \infty} 3/n^2 = 0$. By the sum rule

$$\lim_{n \to \infty} \left(1 + \frac{1}{n^2} \right) = 1 \quad \text{and} \quad \lim_{n \to \infty} \left(2 + \frac{3}{n^2} \right) = 2.$$

Finally by the quotient rule

$$\lim_{n \to \infty} \frac{1 + \dfrac{1}{n^2}}{2 + \dfrac{3}{n^2}} = \frac{1}{2}.$$

Actually the quotient rule tells you that the limit exists, as well as yielding its value. Usually in calculations using the rules of this section, we take the existence of the limit for granted, knowing that it is guaranteed by the rules being used. After a bit of practice most of the above steps can be carried out mentally.

Next we have the squeeze rule.[1]

Proposition 3.8   (Squeeze rule) *Suppose that* $(a_n)_{n=1}^\infty$, $(b_n)_{n=1}^\infty$, $(c_n)_{n=1}^\infty$ *are sequences and that*

$$a_n \le c_n \le b_n$$

for all n. Suppose that $(a_n)_{n=1}^\infty$ *and* $(b_n)_{n=1}^\infty$ *are convergent with the same limit,* $\lim_{n\to\infty} a_n = \lim_{n\to\infty} b_n = t$. *Then* $\lim_{n\to\infty} c_n = t$.

Proof Let $\varepsilon > 0$. There exists N, such that $|a_n - t| < \varepsilon$ and $|b_n - t| < \varepsilon$ for all $n \ge N$, (again, two sequences, same N). Hence, for all $n \ge N$, we have $a_n > t - \varepsilon$ and $b_n < t + \varepsilon$, from which we find that $t - \varepsilon < c_n < t + \varepsilon$. We conclude that $\lim_{n\to\infty} c_n = t$.   □

For the case of squeezing with infinite limits, the following easily proved rules can be used.

Suppose that $a_n \le b_n$ *for each* $n \ge 1$. *If* $\lim_{n\to\infty} a_n = \infty$ *then* $\lim_{n\to\infty} b_n = \infty$; *and if* $\lim_{n\to\infty} b_n = -\infty$ *then* $\lim_{n\to\infty} a_n = -\infty$.

Exercise Prove the squeeze rules with infinite limits.

Example We assume the reader is familiar with the function $\sin x$. All we need here is the fact that $|\sin x| \le 1$ for all x. Set $a_n = \sin n / n$. We have

$$0 \le |a_n| = \frac{|\sin n|}{n} \le \frac{1}{n}.$$

We conclude that $\lim_{n\to\infty} |a_n| = 0$, which is equivalent to $\lim_{n\to\infty} a_n = 0$.

3.4.1 Exercises

1. Prove some rules involving infinite limits:

   (a) Suppose that $\lim_{n\to\infty} a_n = t$, $\lim_{n\to\infty} b_n = \infty$ and t is a finite number. Then $\lim_{n\to\infty} a_n + b_n = \infty$.

   (b) Suppose that $\lim_{n\to\infty} a_n = t$, $\lim_{n\to\infty} b_n = \infty$ and t is a finite number, but is not 0. Then $\lim_{n\to\infty} a_n b_n$ equals ∞ if $t > 0$ and $-\infty$ if $t < 0$.

   (c) Suppose that $\lim_{n\to\infty} a_n = \infty$. Then $\lim_{n\to\infty} 1/a_n = 0$.

---

[1] Also known as the sandwich principle, or the two policemen and a drunk rule.

(d) Suppose that $\lim_{n\to\infty} a_n = 0$. Is it the case that $\lim_{n\to\infty} 1/a_n = \infty$? Explain
     what happens.

2. Find the limit

$$\lim_{n\to\infty} \frac{(6n + 1)(5n + 2)(4n + 3)(2n + 4)(2n + 5)(n + 6)}{(n - 1)(n + 7)(n - 11)(n + 15)(n - 21)(n + 101)}.$$

3. Let $(a_n)_{n=1}^\infty$ be the sequence of Fibonacci numbers (that is, $a_1 = a_2 = 1$, $a_n = a_{n-1} + a_{n-2}$ for $n \geq 3$). Compute the limit $\lim_{n\to\infty} a_{n+1}/a_n$ using the formula for a_n derived in Sect. 2.2 Exercise 9.

4. Let the sequence $(a_n)_{n=1}^\infty$ satisfy the recurrence relations

$$a_n = \beta a_{n-1} + \gamma a_{n-2}, \quad n = 3, 4, 5, \ldots$$

where a_1 and a_2 have prescribed values. Assume that the second-order equation $\lambda^2 - \beta\lambda - \gamma$ has two real roots r_1 and r_2, and that $0 < |r_1| < |r_2|$. Show that $\lim_{n\to\infty} a_{n+1}/a_n = r_2$ except in the case when $a_2 = r_1 a_1$.

5. Let a and b be positive constants. Find $\lim_{n\to\infty} (a^n + b^n)^{1/n}$.

6. Let a and b be positive constants. Find $\lim_{n\to\infty} \dfrac{a^n - b^n}{a^n + b^n}$.

7. Find $\lim_{n\to\infty} (n!)^{1/n}$.

    Hint. First show that if $1 \leq k < n$ then

$$(n!)^{\frac{1}{n}} > (k + 1)^{1 - \frac{k}{n}} (k!)^{\frac{1}{n}}.$$

    You might find Sect. 3.2 Exercise 7 useful.

8. Find $\lim_{n\to\infty} n - \sqrt{(n + a)(n + b)}$.

9. Let $(h_n)_{n=1}^\infty$ be a sequence of positive numbers such that $\lim_{n\to\infty} h_n = 1$. Let α be a non-zero rational. Prove that $\lim_{n\to\infty} h_n^\alpha = 1$.

    Hint. You will need the laws of exponents for rational powers (Sect. 2.2 Exercise 19). By the reciprocal rule one may assume that $\alpha > 0$. If α is an integer it's easy. If α is not an integer one may reduce it to the case $0 < \alpha < 1$ by subtracting an integer and using the product rule. Now use the squeeze rule, observing that if $0 < x < 1$ then $x < x^\alpha < 1$ whilst if $1 < x$ then $1 < x^\alpha < x$.

10. Recall the inequality of arithmetic and geometric means proved in elementary algebra. For positive real numbers a and b this states that

$$\sqrt{ab} \leq \frac{a + b}{2}$$

    with equality if and only if $a = b$.

Let a and b be positive and distinct. Define sequences $(a_n)_{n=0}^{\infty}$ and $(b_n)_{n=0}^{\infty}$ recursively by

$$a_0 = a, \qquad\qquad b_0 = b$$
$$a_{n+1} = \tfrac{1}{2}(a_n + b_n), \qquad b_{n+1} = \sqrt{a_n b_n}$$

for $n = 0, 1, 2, \ldots$.

(a) Show that

$$b_n < b_{n+1} < a_{n+1} < a_n$$

for all $n \geq 1$.

(b) Show that the limits $\lim_{n\to\infty} a_n$ and $\lim_{n\to\infty} b_n$ exist and are equal.

Note. The common limit is called the arithmetic-geometric mean of a and b and is sometimes denoted by $M(a, b)$. It was studied by Gauss and has some surprising applications in computation theory. It will be revisited, in Sect. 5.2 and again in Sect. 11.2.

3.5   Limit Points of Sets

In this section we shall study an important property that pertains to completely arbitrary subsets of $\mathbb{R}$. We shall also exhibit a second important application of supremum. The distinction between finite and infinite will play an important role in our considerations.

Definition Let A be a subset of $\mathbb{R}$. A number t is called a *limit point of the set A* if the following condition is satisfied:

For each $\varepsilon > 0$ there exists $x \in A$, such that $x \neq t$ but x lies in the interval $]t - \varepsilon, t + \varepsilon[$.

Do not confuse limit point of a set with limit of a sequence. A set may have many limit points, or none. Do note the following points listed here:

(a) A limit point t of A is not necessarily in A, though it may be in A.
(b) The condition that x lies in the interval $]t - \varepsilon, t + \varepsilon[$ and is not equal to t may be written as the inequalities $0 < |x - t| < \varepsilon$.
(c) We can express the definition using set theory: for each $\varepsilon > 0$ the set

$$A \cap \left( ]t - \varepsilon, t + \varepsilon[\setminus \{t\} \right)$$

is non-empty.
(d) It is easy to see that the set in the last item, if non-empty for every $\varepsilon > 0$, must be infinite for every $\varepsilon > 0$.

Exercise Prove this claim. By way of a hint: if the set is finite for a certain ε, what can you say about the set of numbers $|x - t|$ for which x is in the set?

(e) Informally, t is a limit point of the set A if we may approximate t by points in A, that are distinct from t if it happens that t is in A.

3.5.1   Weierstrass's Theorem on Limit Points

In this section the predicates "finite" and "infinite" will be used in proofs. We need to be a little clearer about their precise meaning without going too deeply into set theory.

As a working definition of finite set, we can use the following.

Definition A set A is finite if either it is empty, or, if not empty, there exists a natural number N and a sequence $(a_n)_{n=1}^N$, such that A is the set $\{a_n : 1 \le n \le N\}$.

The set $\{a_n : 1 \le n \le N\}$, as the curly brackets indicate, is just the set of values appearing as terms of the sequence $(a_n)_{n=1}^N$, ignoring all repetitions. To conclude that the set has N elements we must assume that the terms of the sequence are distinct. Children know that you have to be careful not to count the same sweet twice.

We could of course incorporate into the definition of finite set the requirement that the terms of the sequence are already distinct. However, it should be obvious that if A is finite according to our definition, and not empty, then there exists some natural number $L \le N$, such that A can be presented as a sequence with distinct terms and index set $\{1, ..., L\}$; just proceed from left to right throwing out repetitions. The number L, the cardinality of A, is uniquely defined by A, a fact that should be proved if this was a rigorous text on set theory, but which we shall simply accept as intuitively obvious.

There is then little mystery about the following notion.

Definition A set is infinite if it is not finite.

Thankfully, we have a plentiful supply of infinite sets: for example $\mathbb{N}$, $\mathbb{R}$, the intervals $]a, b[$ for $a < b$, and loads of sets formed from these using set-building operations. This is just the start.

A number of properties pertaining to the dichotomy of infinite set versus finite set will be frequently used. Maybe they are obvious, but it is useful to list them here. In a proper account of set theory they are theorems.

(a) A subset of a finite set is finite.
(b) If A is an infinite set and B is a set such that $A \subset B$ then B is infinite.
(c) The union of two finite sets is finite.

Some useful, and equally obvious (or otherwise) facts follow from these two.

(d) If A is an infinite set and B is a finite set then the set difference $A \setminus B$ (the set of all elements of A that are not in B) is infinite.
(e) The union of a finite number of finite sets is finite.

The following result is sometimes called Weierstrass's theorem on limit points. We will use it only once: to prove the Bolzano–Weierstrass theorem (Proposition 3.10), and, even so, another proof is suggested in the exercises that does not depend on Weierstrass's theorem. The dichotomy of finite set versus infinite set is essential here, and we apply it to completely arbitrary subsets of $\mathbb{R}$ and not only to sequences; Cantor showed us that these notions are different, there being sets of real numbers not expressible as sequences. Going beyond sequences might put Weierstrass's theorem outside fundamental analysis, except that there is a long tradition of including it, going back to Hardy's "A Course of Pure Mathematics".

Proposition 3.9 *Let A be an infinite but bounded set of real numbers. Then A has at least one limit point.*

Proof Define a subset $B \subset \mathbb{R}$ using the specification:

$$B = \{x \in \mathbb{R} : \text{ the set of all } y \in A, \text{ that satisfy } y < x, \text{ is finite}\}.$$

Now B is not empty; it contains all lower bounds of A and such points exist. And B is bounded above, since, A being infinite, all upper bounds of A (and such also exist) are also upper bounds of B. Hence the supremum $t = \sup B$ exists. We shall show that t is a limit point of A.

Let $\varepsilon > 0$. Then $t + \varepsilon$ is not in B so that *infinitely many elements of A are below* $t + \varepsilon$. On the other hand there must exist an element of B in the interval $]t - \varepsilon, t]$ (for otherwise it would not be true that $t = \sup B$). By rule (a) in the list preceding the proposition, we see that *at most finitely many elements of A are below* $t - \varepsilon$. By rule (d) we the see that infinitely many elements of A must lie in the interval $]t - \varepsilon, t + \varepsilon[$. At least one of them is not the same as t.   □

The proof was a typical use of supremum, and a particularly elegant one. We want to show that the set A has a limit point. The definition of limit point refers to a point t, so we need to conjure up a point on which to test the definition. It is supremum (or in other cases infimum) that does the conjuring, providing a likely candidate for limit point.

Compare this to the slightly simpler case of Proposition 3.3. A candidate for the limit of a sequence was needed; it was again provided by supremum. Ultimately it is axiom C1 that guarantees us that elements of $\mathbb{R}$, the existence of which we need, do in fact exist. As a matter of fact the set B in the proof of Proposition 3.9 is the left set of a Dedekind section, so we could have produced the point t by a direct appeal to axiom C1.

3.5.2   Exercises

1. Show that the limit points of an interval are all the points of the interval as well as the endpoints (whether or not the endpoints are in the interval); the only exception being the degenerate interval $[a, a] = \{a\}$, which has no limit points.

2. Find all limit points of the following subsets of $\mathbb{R}$:

   (a) $\mathbb{N}$
   (b) $\mathbb{Q}$
   (c) The set of all rationals of the form $a/10^n$, where a is an integer and n a natural number.
   (d) $\mathbb{R} \setminus \mathbb{Q}$ (the set of all irrationals)
   (e) The set A, where A is finite.
   (f) The set $\mathbb{R} \setminus A$, where A is finite.

3. Let k_1 and k_2 be real numbers, neither 0. Suppose that k_1/k_2 is irrational. Let A be the set of all real numbers that can be written as $mk_1 + nk_2$ for some integers m and n. Prove that every real number is a limit point of A.

   Hint. See Sect. 2.5 Exercise 3.

4. Show that if t is a limit point of a set A of real numbers, then there exists a sequence $(a_n)_{n=1}^{\infty}$ of points in A, all distinct from t (if t should be in A), such that $\lim_{n \to \infty} a_n = t$.

   Hint. Apply the definition of limit point to a sequence of ε's of the form $1/n$.

   Note. Actually a new axiom of set theory, the axiom of choice, is required to build the sequence for a set A in all generality. The use of this axiom goes beyond the scope of this book and the need for it has usually not bothered analysts. We shall rarely mention it. For most sets that we shall consider here, such as intervals, or finite unions of intervals, the sequence can be constructed more or less explicitly.

3.6   Subsequences

Let $(a_n)_{n=1}^{\infty}$ be a sequence of real numbers. Let $(k_n)_{n=1}^{\infty}$ be a strictly increasing sequence of natural numbers, that is, $k_n < k_{n+1}$ for each n. The sequence $(a_{k_n})_{n=1}^{\infty}$ is called a subsequence of the sequence $(a_n)_{n=1}^{\infty}$.

Thus starting with the sequence of natural numbers $(n)_{n=0}^{\infty}$, we can form the sequence of even natural numbers by taking $k_n = 2n$. We can form the sequence of primes by taking $k_n = \pi_n$, the latter being a common symbol for the nth prime. Clearly there is immense freedom to construct subsequences of a given sequence.

We have seen that a convergent sequence is always bounded, but that a bounded sequence is not always convergent, although it is if also monotonic. About an arbitrary bounded sequence we have the following proposition, the start of a story that extends far into analysis and topology.

Proposition 3.10 (Bolzano–Weierstrass theorem) *Every bounded sequence of real numbers has a convergent subsequence.*

Proof Let the real number sequence $(a_n)_{n=1}^\infty$ be bounded. If the same value t appears infinitely often in it, so that for example $a_{k_n} = t$ for $k_1 < k_2 < k_3 < \cdots$, then $\lim_{n\to\infty} a_{k_n} = t$ and we have a convergent subsequence.

Assume next that no real number appears infinitely often in the sequence. Then the set of all values in the sequence, let us call it A, is an infinite set of real numbers. (If that is not obvious, try to prove it using the properties of finite sets and infinite sets given in the last section.) The set A is bounded, and so has a limit point t (by Proposition 3.9). We construct a subsequence with limit t using induction. Find an index k_1 such that

$$0 < |t - a_{k_1}| < 1.$$

Suppose that we have found an increasing sequence of natural numbers $k_1, k_2, ..., k_n$ such that

$$0 < |t - a_{k_j}| < 1/j$$

for $j = 1, 2, ..., n$. There exists an integer, *higher than* k_n, which we can call k_{n+1}, such that

$$0 < |t - a_{k_{n+1}}| < 1/(n + 1).$$

The subsequence $(a_{k_n})_{n=1}^\infty$ thus constructed converges to t.   □

This proposition is immensely important. We do not want any mistakes in the proof. How could we be sure in the first step, when t occurred infinitely often in the sequence $(a_n)_{n=1}^\infty$, that the sequence $(k_n)_{n=1}^\infty$ really existed? In the induction argument of the second step, how could we be sure that k_{n+1} really existed?

This is one of those cases when there are hidden appeals to Proposition 2.1. In the first step there is an infinite set of natural numbers B, comprising the set of all k such that $a_k = t$. We need to arrange B as an increasing sequence $(k_n)_{n=1}^\infty$ of natural numbers. We take k_1 as the lowest member of B, then k_2 as the lowest member after removing k_1, then k_3 as the lowest member after removing k_1 and k_2, and so on. Because B is infinite we never empty it in a finite number of steps; there are always some numbers remaining and we can choose the lowest as the next term in the sequence. This can be expressed formally by the inductive definition:

$$k_{n+1} = \min(B \setminus \{k_1, ..., k_n\}).$$

In the second step, because t is a limit point of the set A, we know that for each $\varepsilon > 0$ there exists x in A, that satisfies $0 < |t - x| < \varepsilon$. In particular there exists x in A that satisfies $0 < |t - x| < 1/(n + 1)$. But we also want x to have an index higher than k_n. Consider the numbers $|t - a_m|$ as m ranges from 1 up to k_n. Some of these may be 0, whilst others are certainly non-zero, for example $|t - a_{k_n}|$. Choose $\varepsilon > 0$ smaller than $1/(n + 1)$ and smaller than all those numbers $|t - a_m|$ which are non-zero and for which $m \leq k_n$ (there are only finitely many of these). We could give a formula for ε using the min-function, but it would not be very readable. The set of natural numbers m such that $0 < |t - a_m| < \varepsilon$ is not empty (because t is a limit point of A) and all such m satisfy $m > k_n$ (because of the way we selected ε). We can choose the lowest of them as k_{n+1}.

Of course describing the proof in such fine detail might be thought rather pedantic, but as we defined a sequence as an assignment of terms to the natural numbers it may seem wise at least once to describe the assignment, especially where the stakes are so high. We can in future omit these details since for most purposes the first version we gave of the proof is quite sufficient. But it is worth reflecting on the mistakes that the great mathematicians of the past have made in analysis, so we should not allow ourselves to become too complacent.

Proposition 3.10 is often called the Bolzano–Weierstrass theorem. Another proof of this proposition depends on showing that every sequence has a monotonic subsequence. This bypasses the use of Weierstrass's theorem on limit points, but still depends on the dichotomy of finite set versus infinite set (see the exercises).

Another important use of subsequences arises in the negation of the statement $\lim_{n\to\infty} a_n = t$, often needed for constructing proofs by contradiction. Recall that the negation is equivalent to saying that there exists $\varepsilon > 0$, such that for every N there exists $n \geq N$, such that $|a_n - t| \geq \varepsilon$.

We can go further for the ε in question. We can produce a subsequence $(a_{k_n})_{n=1}^{\infty}$ such that $|a_{k_n} - t| \geq \varepsilon$. Consider the set of all n greater than N such that $|a_n - t| \geq \varepsilon$. The whole point is that this set is not empty. We can therefore assign to N the lowest number k (using here Proposition 2.1) greater than or equal to N for which $|a_k - t| \geq \varepsilon$. We use this assignment to define the subsequence $(a_{k_n})_{n=1}^{\infty}$ inductively. We start at $N = 1$ and assign k_1. Having assigned k_n we reset N to $k_n + 1$ and assign k_{n+1}, and so on. The result of this discussion is as follows:

Proposition 3.11 *The negation of the statement* $\lim_{n\to\infty} a_n = t$ *is equivalent to the following: there exists* $\varepsilon > 0$ *and a subsequence* $(a_{k_n})_{n=1}^{\infty}$, *such that* $|a_{k_n} - t| \geq \varepsilon$ *for all* n.

3.6.1 Exercises

1. Prove that every sequence has a monotonic subsequence. Use this to give another proof of the Bolzano–Weierstrass theorem.

    Hint. Consider the set of all integers n with the property that $a_m \leq a_n$ for all $m \geq n$, and reflect on the consequences of its being finite or infinite.
2. Show that if a sequence $(a_n)_{n=1}^{\infty}$ is not bounded above then there is a subsequence $(a_{k_n})_{n=1}^{\infty}$ such that $\lim_{n\to\infty} a_{k_n} = \infty$. A similar result holds if the sequence is not bounded below, but $-\infty$ replaces ∞.
3. Prove the following proposition. Suppose a sequence has the limit t (which may be $\pm\infty$). Show that every subsequence also has the limit t.
4. Prove a converse to the result of the previous exercise, with a twist. Assume that every subsequence of the sequence $(a_n)_{n=1}^{\infty}$ has a limit, but we do not assume that

they all have the same limit. We also allow ∞ and $-\infty$ here as limits. Prove that the sequence $(a_n)_{n=1}^{\infty}$ has a limit.

5. Let $(a_n)_{n=1}^{\infty}$ be a sequence of real numbers and suppose that the set of values $A = \{a_n : n = 1, 2, ...\}$ appearing in the sequence has a limit point t. Show that there exists a subsequence $(a_{k_n})_{n=1}^{\infty}$, possessing distinct terms, that converges to t.

3.7 Cauchy's Convergence Principle

We introduce a condition that is both necessary and sufficient for a sequence to be convergent, but does not mention a candidate for the limit. We can identify whether or not a sequence is convergent without going outside it; by studying in fact the terms alone. The condition has important, mainly theoretical, applications throughout analysis and allied subjects, and will reappear in this text in the context of limits of functions.

Proposition 3.12 (Cauchy's convergence principle) *A sequence $(a_n)_{n=1}^{\infty}$ of real numbers is convergent if and only if it satisfies the following condition (Cauchy's condition): for all $\varepsilon > 0$ there exist a natural number N, such that for all $m \geq N$ and $n \geq N$ we have $|a_m - a_n| < \varepsilon$.*

Proof As is often appropriate when proving that a condition is necessary and sufficient, we split the proof into two parts.

(a) *Cauchy's condition is necessary for convergence.* Assume $\lim_{n \to \infty} a_n = t$. Let $\varepsilon > 0$. Choose N, such that $|a_n - t| < \varepsilon/2$ for all $n \geq N$. If now $n \geq N$ and $m \geq N$ we have

$$|a_n - a_m| \leq |a_n - t| + |t - a_m| < \frac{1}{2}\varepsilon + \frac{1}{2}\varepsilon = \varepsilon.$$

(b) *Cauchy's condition is sufficient for convergence.* Assume that Cauchy's condition is satisfied. First we show that the sequence $(a_n)_{n=1}^{\infty}$ is bounded. We let $\varepsilon = 1$, find a corresponding N and let $m = N$ in Cauchy's condition. For all $n \geq N$ we have $|a_N - a_n| < 1$, which gives $|a_n| < |a_N| + 1$. That is to say, all terms of the sequence, except perhaps a finite number, satisfy $|a_n| < K$ where $K = |a_N| + 1$.

By the Bolzano–Weierstrass theorem (Proposition 3.10) there exists a convergent subsequence, $(a_{k_n})_{n=1}^{\infty}$, and we let $t = \lim_{n \to \infty} a_{k_n}$. Combined with Cauchy's condition this forces a_n to converge to t. For let $\varepsilon > 0$. Choose N, such that $|a_m - a_n| < \varepsilon/2$ for all $n \geq N$ and $m \geq N$. We can find a term in the convergent subsequence, for example a_{k_j}, with index $k_j \geq N$, and which satisfies $|a_{k_j} - t| < \varepsilon/2$. But then for all $n \geq N$ we have

$$|a_n - t| \leq |a_n - a_{k_j}| + |a_{k_j} - t| < \frac{1}{2}\varepsilon + \frac{1}{2}\varepsilon = \varepsilon.$$

This ends the proof. $\square$

2

The importance of Cauchy's principle for analysis and its further developments is such, that it is desirable to exhibit Cauchy's condition, separately from Proposition 3.12; for the reader's closer perusal:

Cauchy's condition. *For all $\varepsilon > 0$ there exist a natural number N, such that for all $m \geq N$ and $n \geq N$ we have $|a_m - a_n| < \varepsilon$.*

Cauchy's condition requires that $|a_m - a_n|$ should be smaller than ε, only in virtue of $m \geq N$ and $n \geq N$. The separation of m and n can be vast, in fact there is no upper limit on it. It is a common beginner's error to think that Cauchy's condition is satisfied if, for each $\varepsilon > 0$, there exists N, such that $|a_n - a_{n+1}| < \varepsilon$ for all $n \geq N$. One can rephrase the condition throwing more emphasis on the arbitrariness of $m - n$. This leads to a useful alternative formulation:

Cauchy's condition, second version. *For every $\varepsilon > 0$ there exists a natural number N, such that for all $n \geq N$, and for all natural numbers p, we have $|a_{n+p} - a_n| < \varepsilon$.*

3.8 Convergence of Series

In all mathematics there is not a single infinite series whose convergence has been established by rigorous methods. (Letter from N. H. Abel (1828))

The correct definition of convergence of the infinite series $\sum_{k=1}^{\infty} a_k$ is one of the main achievements of analysis and it dispersed a great deal of nonsense that had beset mathematics.

Let $(a_n)_{n=1}^{\infty}$ be a real number sequence and let $s_n = \sum_{k=1}^{n} a_k$ for $n = 1, 2, 3, \ldots$.

Definition If the sequence $(s_n)_{n=1}^{\infty}$ is convergent and $\lim_{n \to \infty} s_n = t$, we say that the infinite series $\sum_{k=1}^{\infty} a_k$ is convergent, and write

$$\sum_{k=1}^{\infty} a_k = t.$$

We call t *the sum of the series.* The numbers s_n are called *partial sums.* A series that is not convergent is said to be *divergent.*

Note that t is supposed to be a finite number when the series is convergent. If $\lim_{n \to \infty} s_n = \infty$ or $\lim_{n \to \infty} s_n = -\infty$ we write $\sum_{k=1}^{\infty} a_k = \infty$ or $\sum_{k=1}^{\infty} a_k = -\infty$, but the series in both these cases is divergent.

A series can be formed by starting at other indices than $k = 1$, for example $k = N$. Then to say $\sum_{k=N}^{\infty} a_k = t$ means that the partial sums $s_n = \sum_{k=N}^{n} a_k$ have the limit t. It is then easy to see that

$$\sum_{k=1}^{\infty} a_k = \sum_{k=1}^{N-1} a_k + \sum_{k=N}^{\infty} a_k,$$

given that either of the two infinite series appearing in this equation is convergent.

3.8.1 Rules for Series

Some simple facts about series follow easily from the limit rules for sequences given in Sect. 3.4. The reader is invited to supply the proofs. We obtain important and useful rules for manipulating series.

(i) (*Multiplication by a constant*) Let $\sum_{k=1}^{\infty} a_k$ be a convergent series and let its sum be t. Let α be a real number. Then the series $\sum_{k=1}^{\infty} \alpha a_k$ is convergent and

$$\sum_{k=1}^{\infty} \alpha a_k = \alpha t.$$

(ii) (*Sum of two series*) Let $\sum_{k=1}^{\infty} b_k$ be a second convergent series, and let its sum be s. Then the series $\sum_{k=1}^{\infty} (a_k + b_k)$ is convergent and

$$\sum_{k=1}^{\infty} (a_k + b_k) = t + s.$$

3.8.2 Convergence Tests

A big part of the theory of infinite series consists of the so-called convergence tests. These enable us to establish that a series is convergent (or in some cases divergent) by examining the sequence of terms, but without proposing a candidate for the sum.

Proposition 3.13 If $\sum_{k=1}^{\infty} a_k$ is convergent then $\lim_{k\to\infty} a_k = 0$.

Proof Let $s = \sum_{k=1}^{\infty} a_k$ and let $s_n = \sum_{k=1}^{n} a_k$ for each n. Then $\lim_{n\to\infty} s_n = s$ but we also have $\lim_{n\to\infty} s_{n-1} = s$. It follows that

$$\lim_{n\to\infty} (s_n - s_{n-1}) = s - s = 0.$$

But $s_n - s_{n-1} = a_n$, so that $\lim_{n\to\infty} a_n = 0$. □

The condition $\lim_{k\to\infty} a_k = 0$ is, in view of Proposition 3.13, a necessary condition for convergence of the series. It is far from being sufficient. Proposition 3.13 is thus a *divergence test* and a useful one. If $\lim_{n\to\infty} a_n$ does not exist, or if it exists but is not 0, then the series $\sum_{k=1}^{\infty} a_k$ is divergent. But if $\lim_{n \to \infty} a_n = 0$, one cannot conclude from that alone that $\sum_{k=1}^{\infty} a_k$ is convergent (it is a common beginner's error to think otherwise).

3.8.3   The Simplest Convergence Tests: Positive Series

A series $\sum_{k=1}^{\infty} a_k$ is said to be a *positive series* if $a_k \geq 0$ for each k. The partial sums s_n then form an increasing sequence. By Proposition 3.3 an increasing sequence is either convergent or it tends to ∞. So we have the following basic results.

Proposition 3.14 *Let $\sum_{k=1}^{\infty} a_k$ be a positive series. Then it is convergent if and only if there exists $K > 0$, such that $\sum_{k=1}^{n} a_k \leq K$ for each n.*

The proposition can be paraphrased loosely by saying that a positive series is convergent if and only if its partial sums are bounded.

Proof The sequence of partial sums is increasing and so is convergent if and only if it is bounded above.   □

Proposition 3.15 (The comparison test) *Assume that $\sum_{k=1}^{\infty} a_k$ and $\sum_{k=1}^{\infty} b_k$ are positive series and $a_k \leq b_k$ for each k. Then we have*

(1) If $\sum_{k=1}^{\infty} b_k$ is convergent then $\sum_{k=1}^{\infty} a_k$ is also convergent.
(2) If $\sum_{k=1}^{\infty} a_k$ is divergent then $\sum_{k=1}^{\infty} b_k$ is also divergent.

Proof We use the inequalities

$$\sum_{k=1}^{n} a_k \leq \sum_{k=1}^{n} b_k \leq \sum_{k=1}^{\infty} b_k,$$

in which the third member could be ∞. If $\sum_{k=1}^{\infty} b_k$ is convergent then the sums $\sum_{k=1}^{n} a_k$ are bounded above by the finite number $\sum_{k=1}^{\infty} b_k$, and the series $\sum_{k=1}^{\infty} a_k$ is therefore convergent. If $\sum_{k=1}^{\infty} a_k$ is divergent then the sums $\sum_{k=1}^{n} a_k$ are not bounded above (they tend to ∞), and so the sums $\sum_{k=1}^{n} b_k$ also tend to ∞.   □

Proposition 3.16 (Limit comparison test) *Given positive series $\sum_{k=1}^{\infty} a_k$ and $\sum_{k=1}^{\infty} b_k$, in which no terms are zero, we assume that the limit $\ell = \lim_{k \to \infty} a_k/b_k$ exists and satisfies $0 < \ell < \infty$. Then either the series $\sum_{k=1}^{\infty} a_k$ and $\sum_{k=1}^{\infty} b_k$ are both convergent or they are both divergent.*

Proof Assume that $\sum_{k=1}^{\infty} b_k$ is convergent. There exists N, such that for all $k \geq N$ we have $a_k/b_k < \ell + 1$, and therefore also $a_k < (\ell + 1)b_k$. But the series $\sum_{k=1}^{\infty} (\ell + 1)b_k$ is convergent so that, by the comparison test, the series $\sum_{k=1}^{\infty} a_k$ is also convergent.

Next assume that $\sum_{k=1}^{\infty} a_k$ is convergent. We note that $\lim_{k \to \infty} b_k/a_k = 1/\ell$ and use the same argument.   □

3.8.4 Geometric Series and D'Alembert's Test

In order to use the comparison test we need some series, whose convergence status is already known, to use as a yardstick.

A series $\sum_{k=1}^{\infty} a_k$ is called a geometric series, if there exists r independent of k, such that $a_k = r a_{k-1}$ for each k. In short, the ratio a_k/a_{k-1} is constant, provided no term is zero. It is convenient to write a geometric series with the starting index $k = 0$. Then $a_k = r^k a_0$ and the series has the form

$$\sum_{k=0}^{\infty} a_0 r^k.$$

From algebra we know that

$$\sum_{k=0}^{n} r^k = \begin{cases} \dfrac{r^{n+1} - 1}{r - 1} & \text{if } r \neq 1 \\ n + 1 & \text{if } r = 1. \end{cases}$$

Exercise Prove this formula.

The formula for the sum of a convergent geometric series is a basic result generally taught in school mathematics.

Proposition 3.17 *Assume that $a_0 \neq 0$. Then the geometric series $\sum_{k=0}^{\infty} a_0 r^k$ is convergent if and only if $|r| < 1$. In this case its sum is*

$$\sum_{k=0}^{\infty} a_0 r^k = \frac{a_0}{1 - r} = \frac{\text{First term}}{1 - \text{ratio}}.$$

Proof If $|r| < 1$ we have that

$$\sum_{k=0}^{n} a_0 r^k = a_0 \frac{r^{n+1} - 1}{r - 1}$$

and the limit is $a_0/(1 - r)$. If $|r| \geq 1$ the term $a_0 r^n$ does not converge to 0; hence the series diverges. $\qquad \square$

Proposition 3.18 (The ratio test or D'Alembert's test) *Let $\sum_{k=1}^{\infty} a_k$ be a positive series in which no term is 0. Assume that the limit*

$$t := \lim_{k \to \infty} \frac{a_{k+1}}{a_k}$$

exists. Then

(1) The series is convergent if $t < 1$.

(2) The series is divergent if $t > 1$.
(3) There is no conclusion if $t = 1$.

Proof (1) Assume that $t < 1$. Choose a number r such that $t < r < 1$. There exists N, such that $a_{k+1}/a_k < r$ for all $k \geq N$. For such k we have that $a_{k+1} < ra_k$, from which we find $a_k \leq a_N r^{k-N}$ for $k = N, N+1, N+2$ and so on (a sound proof can be made by induction from $k = N$). But now the series $\sum_{k=N}^{\infty} a_k$ is convergent as we see by comparing it to the convergent geometric series $\sum_{k=N}^{\infty} a_N r^{k-N}$. We restore the missing terms $a_1,...,a_{N-1}$ and conclude that $\sum_{k=1}^{\infty} a_k$ is convergent.
(2) Assume next that $t > 1$. Then there exists N, such that $a_{k+1}/a_k > 1$ for all $k \geq N$, and so a_k is increasing for $k \geq N$ and cannot have the limit 0.
(3) See below.   □

If $\lim_{k \to \infty} a_{k+1}/a_k = 1$ no conclusion can be obtained from the ratio test. We included this claim in the statement of the proposition, to provide some necessary emphasis, for it is intended to be used as a test and to be referenced as such. Although item 3 may appear to be an exceptional case, we are forced to come to grips with it. This will be abundantly clear from the material later in this chapter, and more especially in the study of power series in Chap. 11.

If the problem is that the limit does not exist, then it is sometimes possible to use a more delicate version of the ratio test that does not require the limit. Or else it may be possible to use *Cauchy's root test*, which involves calculating the usually rather hard limit $\lim_{n \to \infty} a_n^{1/n}$. But if $\lim_{k \to \infty} a_{k+1}/a_k = 1$ some other test is needed. There are many such tests known, thanks to the labours of nineteenth century mathematicians; for example *Raabe's test* or *Gauss's test*. These topics will be touched upon in Chap. 10.

Generally the ratio test is the first thing to try when faced with testing a series for convergence.

3.8.5   Exercises

1. Test the following series for convergence:

(a) $\displaystyle\sum_{n=0}^{\infty} n2^{-n}$

(b) $\displaystyle\sum_{n=0}^{\infty} n^{100}2^{-n}$

(c) $\displaystyle\sum_{n=0}^{\infty} n^{-100}2^{n}$

(d) $\displaystyle\sum_{n=0}^{\infty} \frac{1}{n!}.$

2. Let

$$a_n = \left(1 + \frac{1}{n}\right)^n$$

for $n = 1, 2, 3, \ldots$. Show that a_n is increasing and that for each n we have

$$a_n < \sum_{k=0}^{\infty} \frac{1}{k!}.$$

Deduce that the limit

$$\lim_{n \to \infty} \left(1 + \frac{1}{n}\right)^n$$

exists and is a finite number.

3. Continuation of the previous exercise. Show that for each $m \le n$ we have

$$\left(1 + \frac{1}{n}\right)^n \ge 1 + \sum_{k=1}^{m} \frac{\left(1 - \frac{1}{n}\right)\left(1 - \frac{2}{n}\right)\cdots\left(1 - \frac{k-1}{n}\right)}{k!}.$$

Deduce that

$$\lim_{n \to \infty} \left(1 + \frac{1}{n}\right)^n = \sum_{k=0}^{\infty} \frac{1}{k!}.$$

4. Draw conclusions for the following series using the ratio test. The conclusions may depend on the number x. You may assume that x is positive.

(a) $\displaystyle\sum_{n=0}^{\infty} \frac{(2n+1)! x^n}{(n!)^2}$

(b) $\displaystyle\sum_{n=0}^{\infty} \frac{n! x^n}{n^n}$

(c) $\displaystyle\sum_{n=0}^{\infty} \frac{(2n+1)!(3n)! x^n}{(n!)^5}$

(d) $\displaystyle\sum_{n=0}^{\infty} \sqrt{2^n + 3^n}\, x^n$

(e) $\displaystyle\sum_{n=0}^{\infty} \frac{a(a+c)\ldots(a+nc)}{b(b+c)\ldots(b+nc)} x^n$, $\quad (a, b, c$ positive constants$)$.

5. Test for convergence the series

$$\sum_{k=0}^{\infty} \frac{(4k)!}{(k!)^4} \frac{26390k + 1103}{396^{4k}}.$$

It is a result of S. Ramanujan that the sum of this series is

$$\frac{9801}{2\sqrt{2}} \frac{1}{\pi}.$$

Using a "hand-held" calculator approximate π using first one term, then two terms, of the series. The results are astonishing.

3.8.6   The Series $\sum_{n=1}^{\infty} 1/n^p$

Let $p > 0$. The ratio test gives no conclusion for the series $\sum_{n=1}^{\infty} 1/n^p$ since the quotient tends to 1. This type of series is sometimes called rather quaintly a p-series. We are going to estimate the sum directly.

We could straight away suppose that p is rational. We defined rational powers in 2.2, without giving the details, and we need the laws of exponents $(y^a)^b = y^{ab}$ and $y^a y^b = y^{a+b}$. The fact that the quotient tends to 1, that is,

$$\lim_{n \to \infty} \frac{n^p}{(n+1)^p} = 1,$$

follows from Sect. 3.4 Exercise 9.

The conclusions also hold for real p but proving them requires a definition of real powers, which will be fully covered together with the laws of exponents in Chap. 7.

Consider the terms from $n = 2^k$ to $2^{k+1} - 1$; there are 2^k of them and they decrease with increasing n. Therefore we have

$$\sum_{n=2^k}^{2^{k+1}-1} \frac{1}{n^p} \le 2^k \cdot \frac{1}{(2^k)^p} = \frac{1}{2^{k(p-1)}}.$$

It follows that

$$\sum_{n=1}^{2^{N+1}-1} \frac{1}{n^p} \le \sum_{k=0}^{N} \frac{1}{2^{k(p-1)}}.$$

On the right is a geometric series with ratio 2^{1-p}. For $p > 1$ it is convergent and this implies that the sums on the left-hand side are bounded above independently of N. Since the terms are positive it follows that the partial sums $\sum_{n=1}^{N} 1/n^p$ are also bounded above independently of N. We conclude that the series $\sum_{n=1}^{\infty} 1/n^p$ converges if $p > 1$.

We also obtain the estimate

$$\sum_{n=1}^{\infty} \frac{1}{n^p} < \frac{1}{1 - 2^{1-p}}.$$

Exercise Prove the estimate.

In particular we find that $\sum_{n=1}^{\infty} 1/n^2 < 2$. Euler solved the Basel problem in 1735 by showing that

$$\sum_{n=1}^{\infty} \frac{1}{n^2} = \frac{\pi^2}{6} = 1.6449...$$

Some very crafty proofs of this are known using only fundamental analysis, but it is best proved using Fourier series, which also yield formulas for $\sum_{n=1}^{\infty} 1/n^p$ in the cases when p is an even number.

When $p = 1$ we have the *harmonic series* $\sum_{n=1}^{\infty} 1/n$. The nth term tends to 0, but we cannot deduce convergence from this. We estimate the terms for $n = 2^k$ to $2^{k+1} - 1$ from below and obtain

$$\sum_{n=2^k}^{2^{k+1}-1} \frac{1}{n} \geq 2^k \cdot \frac{1}{2^{k+1}} = \frac{1}{2}$$

and so

$$\sum_{n=1}^{2^{N+1}} \frac{1}{n} \geq \frac{N+1}{2}.$$

The sums on the left therefore tend to infinity with increasing N and we conclude that the harmonic series $\sum_{n=1}^{\infty} 1/n$ is divergent. We even have an estimate of the size of the partial sums, though it greatly underestimates the rate of growth, which is, even so, rather small.

In the case $0 < p < 1$ the series $\sum_{n=1}^{\infty} n^{-p}$ diverges by the comparison test, since $n^{-p} > n^{-1}$.

3.8.7 Telescoping Series

This method can be used for series that are not necessarily positive. Given the sequence $(a_k)_{k=1}^{\infty}$ one can sometimes find another sequence $(b_k)_{k=1}^{\infty}$ such that

$$a_k = b_k - b_{k+1}, \quad k = 1, 2, 3, \dots.$$

Then we have

$$\sum_{k=1}^{n} a_k = \sum_{k=1}^{n} (b_k - b_{k+1}) = b_1 - b_{n+1}.$$

The series $\sum_{k=1}^{\infty} a_k$ is therefore convergent if and only if the limit $\lim_{n\to\infty} b_n$ exists, and if so then

$$\sum_{k=1}^{\infty} a_k = b_1 - \lim_{n\to\infty} b_n.$$

3.8.8 Exercises (cont'd)

6. Examine the following series for convergence:

   (a) $\displaystyle\sum_{n=1}^{\infty} \frac{n(n+1)}{(n+2)^2}$

   (b) $\displaystyle\sum_{n=1}^{\infty} \frac{n(n+1)}{(n+2)^3}$

   (c) $\displaystyle\sum_{n=1}^{\infty} \frac{n(n+1)}{(n+2)^4}$

   (d) $\displaystyle\sum_{n=1}^{\infty} \frac{1}{\sqrt{n}}.$

7. Let p be a natural number. Using the telescoping series

$$\sum_{n=1}^{\infty} \left(\frac{1}{n^p} - \frac{1}{(n+1)^p} \right)$$

   as a comparison series give another proof that the series $\sum_{n=1}^{\infty} 1/n^{p+1}$ converges.
8. Using the method of the previous exercise, but taking $p = \frac{1}{2}$, give another proof that the series $\sum_{n=1}^{\infty} n^{-3/2}$ converges.
9. Examine for convergence the series

$$\sum_{n=1}^{\infty} \frac{(n+1)(n+2)}{n^3 \sqrt{n}}.$$

10. Find N such that $\sum_{n=1}^{N} 1/n > 100$. It does not have to be the smallest N that works; that can be found by methods explained in the final chapter.
11. Let

$$s = \sum_{n=1}^{\infty} \frac{1}{n^2}.$$

   Express in terms of s the sums of the series

$$\sum_{n=1}^{\infty} \frac{1}{(2n)^2}, \quad \sum_{n=1}^{\infty} \frac{1}{(2n-1)^2}, \quad \sum_{n=1}^{\infty} \frac{(-1)^n}{n^2}.$$

The calculation would include a proof that the third series is convergent.

12. Prove Cauchy's condensation test. Let $\sum_{n=1}^{\infty} a_n$ be a positive series such that the terms a_n form a decreasing sequence. Then the series $\sum_{n=1}^{\infty} a_n$ is convergent if and only if the series $\sum_{n=1}^{\infty} 2^n a_{2^n}$ is convergent.

Hint. The method we used to study the series $\sum_{n=1}^{\infty} n^{-p}$ was essentially Cauchy's condensation test.

13. Use Cauchy's condensation test to study the series $\sum_{n=1}^{\infty} n^{-p} \ln n$.

Note. The function $\ln x$ is the natural logarithm of x and will be properly defined in a later chapter. Many readers will be familiar with it from school algebra. The only thing you need to know here is the formula $\ln(2^n) = n \ln 2$.

14. What can be said about the series $\sum_{n=1}^{\infty} n^{-p} (\ln n)^q$?
15. Prove the following theorem of Abel. Let the positive series $\sum_{n=1}^{\infty} a_n$ be convergent and assume that the sequence a_n is decreasing. Then $\lim_{n \to \infty} n a_n = 0$.

Note. This is a necessary condition for convergence that can stand beside Proposition 3.13 and settles the harmonic series.

16. As we have seen, the harmonic series $\sum_{n=1}^{\infty} 1/n$ diverges. Suppose we remove all terms for which the decimal representation of n includes the digit 9. Show that the resulting series converges.

3.9 Decimals Reprised

We now give an exact treatment of decimals based on infinite series. We consider a real number x in the interval $0 \le x < 1$ and study its decimal representation. By this is meant a representation as the sum of an infinite series

$$x = \frac{d_1}{10} + \frac{d_2}{10^2} + \frac{d_3}{10^3} + \cdots = \sum_{k=1}^{\infty} \frac{d_k}{10^k} \tag{3.1}$$

where each coefficient d_k is one of the natural numbers 0, 1, 2, 3, 4, 5, 6, 7, 8, 9. The usual notation for the series is

$$0.d_1 d_2 d_3 ...,$$

an expression that we shall call a decimal fraction,[2] or in short, a decimal. The coefficients are called decimal digits. Usage varies between countries as to whether a full-stop, a comma, or a centred dot is used.

We begin by showing that every decimal fraction represents a real number.

Proposition 3.19 *Let* $(d_k)_{k=1}^{\infty}$ *be a sequence in which each* d_k *is one of the natural numbers* 0, 1, ..., 9. *Then the series* $\sum_{k=1}^{\infty} d_k/10^k$ *is convergent and its sum is in the interval* [0, 1].

Proof It is enough to point out that

$$0 \le \frac{d_k}{10^k} \le \frac{9}{10^k}$$

and the geometric series $\sum_{k=1}^{\infty} 9/10^k$ is convergent with sum 1. $\square$

We have to make an irritating but necessary distinction between the real number x and the decimal fraction $0.d_1 d_2 d_3...$ that represents it, since two distinct decimal fractions can represent the same real number. This follows from the fact, just used, that

$$\sum_{k=1}^{\infty} \frac{9}{10^k} = 1,$$

or in the usual notation

$$0.\bar{9} = 1.$$

This is why we included 1 in the set of real numbers under consideration in Proposition 3.19. From this it follows that if $d_k < 9$ then the decimals

$$0.d_1...d_{k-1}d_k\bar{9} \quad \text{and} \quad 0.d_1...d_{k-1}(d_k+1)\bar{0}$$

represent the same real number.

Exercise Prove the claims made in the previous paragraph.

Let us call a decimal, of the kind that appears here on the left, a decimal that is *eventually* 9. We shall see that it is only in such cases that two distinct decimals represent the same real number.

We recall the algorithm for determining the decimal digits. For a given number x in the interval [0, 1[we define, by induction, sequences $(x_k)_{k=1}^{\infty}$ (the remainders) and $(d_k)_{k=1}^{\infty}$ (the digits), where $0 \le x_k < 1$ and d_k is one of the natural numbers in the range 0, ..., 9. Firstly we set $x_1 = x$. When x_k (a real number in the interval

---

[2]The terminology here is unconventional. Usually by a decimal fraction is meant a rational number whose denominator is a power of 10.

$0 \leq x_k < 1$) has been defined, we let d_k be the highest natural number less than or equal to $10x_k$, and set $x_{k+1} = 10x_k - d_k$. At each step the kth remainder determines the digits d_j for $j \geq k$ and the remainders x_j for $j \geq k + 1$.

Proposition 3.20 *Let $0 \leq x < 1$ and let the sequences $(x_k)_{k=1}^{\infty}$ and $(d_k)_{k=1}^{\infty}$ of remainders and digits be defined by the decimal algorithm. Then for each natural number n we have*

$$x - \left(\frac{d_1}{10} + \frac{d_2}{10^2} + \cdots + \frac{d_n}{10^n} \right) = \frac{x_{n+1}}{10^n}.$$

Proof We use induction. The result holds for $n = 0$ by the definition of x_1 (and note that the sum within parentheses, being empty, is 0). Suppose that it holds for a given n. Then $x_{n+2} = 10x_{n+1} - d_{n+1}$, so that

$$\frac{x_{n+2}}{10^{n+1}} = \frac{x_{n+1}}{10^n} - \frac{d_{n+1}}{10^{n+1}} = x - \left(\frac{d_1}{10} + \frac{d_2}{10^2} + \cdots + \frac{d_{n+1}}{10^{n+1}} \right).$$

$\square$

An obvious consequence is that the decimal algorithm accomplishes what it is intended to do.

Proposition 3.21 *Let $0 \leq x < 1$ and let the sequences $(x_k)_{k=1}^{\infty}$ and $(d_k)_{k=1}^{\infty}$ be defined by the decimal algorithm. Then*

$$x = 0.d_1 d_2 d_3 \ldots = \sum_{k=1}^{\infty} \frac{d_k}{10^k}.$$

Proof Since $0 \leq x_n < 1$ for all n, it is clear by Proposition 3.19 that

$$0 \leq x - \left(\frac{d_1}{10} + \frac{d_2}{10^2} + \cdots + \frac{d_n}{10^n} \right) < \frac{1}{10^n}$$

and the conclusion follows since $\lim_{n \to \infty} 10^{-n} = 0$. $\square$

We conclude that every real number in the interval $[0, 1[$ can be represented as a decimal fraction. We even have the error estimate that using n digits gives an error less than 10^{-n}. Moreover every real number in the interval $[0, 1[$ can be represented by a decimal *that is not eventually 9*. This is because a decimal that is eventually 9 can be replaced by one that is not eventually 9, and represents the same real number, as we have seen.

We come to the main conclusion regarding decimal fractions.

Proposition 3.22 *The decimal algorithm sets up a one-to-one correspondence between real numbers in the interval $0 \leq x < 1$ and decimal fractions that are not eventually 9.*

Proof Consider a decimal $0.d_1d_2d_3\ldots$ that is not eventually 9, and suppose it represents the real number x. Then $0 \le x < 1$. For we know that

$$\frac{9}{10} + \frac{9}{10^2} + \frac{9}{10^3} + \cdots = 1$$

and at least one of the coefficients d_k is not 9. Hence we have

$$\frac{d_1}{10} + \frac{d_2}{10^2} + \frac{d_3}{10^3} + \cdots < 1.$$

Next consider a real number x in the interval $0 \le x < 1$. We saw that x can be represented by a decimal *that is not eventually* 9. Let

$$x = \frac{d_1}{10} + \frac{d_2}{10^2} + \frac{d_3}{10^3} + \cdots$$

be such a representation. Now we can show that the decimal algorithm, applied to x, produces for the kth digit the displayed coefficient d_k, and the kth remainder is

$$x_k = \frac{d_k}{10} + \frac{d_{k+1}}{10^2} + \frac{d_{k+2}}{10^3} + \cdots \tag{3.2}$$

The proof of this claim is by induction. By definition

$$x_1 = x = \frac{d_1}{10} + \frac{d_2}{10^2} + \frac{d_3}{10^3} + \cdots,$$

so (3.2) holds for the case $k = 1$. Suppose that (3.2) is known to hold for a given k. Then

$$10x_k = d_k + \frac{d_{k+1}}{10} + \frac{d_{k+2}}{10^2} + \cdots$$

and the highest natural number less than or equal to this is d_k; it cannot be $d_k + 1$ since at least one of the succeeding digits is not 9, which implies

$$\frac{d_{k+1}}{10} + \frac{d_{k+3}}{10^2} + \frac{d_{k+4}}{10^3} + \cdots < 1.$$

Hence

$$x_{k+1} = \frac{d_{k+1}}{10} + \frac{d_{k+2}}{10^2} + \frac{d_{k+3}}{10^3} + \cdots$$

and the next digit is d_k.

These arguments show that the decimal algorithm produces *the unique* representation of each x in the interval $[0, 1[$ as a decimal that is not eventually 9. □

3.9.1 Exercises

1. The duodecimal system uses the base 12. There are various suggestions for writing the digits denoting 10 and 11. We shall simply use A and B. In the rest of this exercise we use the duodecimal system and express numbers using the base 12 (written "10" in the duodecimal system). Now express the following fractions as duodecimal expansions:

$$\frac{1}{3}, \quad \frac{1}{7}, \quad \frac{1}{A}, \quad \frac{1}{B}, \quad \frac{A}{B}.$$

2. The study of decimals has close links to number theory and algebra. This is apparent in the decimal expansion of $1/p$ where p is a prime. You might like to compute the decimal expansion of $1/p$ for prime p, up to, say, $p = 31$. You will see that, excepting the cases $p = 2$ and $p = 5$ (the prime divisors of 10), there is no initial string, and the period length divides $p - 1$.

You can check the following if you are patient. For $1/7$ the period is 6, the maximum possible. The maximum period occurs again for the prime denominators 17, 19, 23, 29, 47, 59, 61, 97 (these are the only ones under 100 for which the period is the maximum possible).

Note. The explanation, which requires some very basic group theory, is briefly as follows. The sequence $(x_n)_{n=1}^{\infty}$ generated by the decimal algorithm is given by $x_n = a_n/p$, where a_n is the remainder obtained on dividing p into 10^{n-1}. The numbers a_n are certain elements of the set $\{1, 2, \dots p - 1\}$. It is known from number theory that this set forms a cyclic group G (usually denoted by $(\mathbb{Z}/p\mathbb{Z})^*$) of order $p - 1$, under the operation of multiplication modulo p. The elements a_n are the remainders of the successive powers of 10; they form a subgroup of G, the one generated by 10 (or 10 reduced modulo p; that only makes a difference for $p < 10$). Starting at $a_1 = 1$, no repetition can occur until we reach 1 again. There is therefore no initial string. The order of the subgroup of G generated by 10 is therefore the length of the period in the decimal expansion of $1/p$, and, by Lagrange's theorem of group theory, it divides the order of G, that is, it divides $p - 1$. If 10 actually generates G we get a period of length $p - 1$, the maximum possible. It is an unsolved problem whether or not this happens for infinitely many primes (Artin's conjecture in number theory would imply that it does).

3.10 ($\lozenge$) Philosophical Implications of Decimals

Everyone knows the practical importance of decimals for doing calculations with real numbers. In this nugget we will consider an importance of a quite different kind.

We have not constructed the set of real numbers, instead we posited the existence of a set with certain properties. The indefinite article is important. In older treatises on analysis, it was common to construct such a set and prove that it satisfied the axioms A, B and C. The axioms are then theorems, as we might say. To construct such a set some raw materials are needed. These are commonly the natural numbers, and a good dose of set theory allowing the building of some sets. One quickly builds the rational numbers, as the set of fractions m/n, with the usual caveat about cancellation of factors. So let us assume that we already have a set representing $\mathbb{Q}$. We will look at a couple of constructions of $\mathbb{R}$ from $\mathbb{Q}$ that are historically important. It will be seen that quite a lot of set theory is needed to build the required sets, but we will not explain it in any detail.

Dedekind's construction. We have seen the notion of Dedekind section. Now we can apply the same notion to $\mathbb{Q}$. A Dedekind section of $\mathbb{Q}$ is a partition of $\mathbb{Q}$ into two subsets, D_l and D_r such that neither is empty, every rational belongs to D_l or D_r but not to both, and for all rationals x and y, if $x \in D_l$ and $y \in D_r$ then $x < y$. It is common to add the requirement, and we do so, that D_l has no highest member.

According to Dedekind a real number is a Dedekind section of the rationals. The rationals are viewed as particular reals by embedding the set of rationals into the set of reals as follows. The rational q is identified with the Dedekind section $\{D_l, D_r\}$ for which $D_l = \{s \in \mathbb{Q} : s < q\}$.

This leaves the daunting task of defining algebraic operations with, and ordering of, the Dedekind real numbers and proving the axioms A, B and C as theorems. Actually it turns out that ordering and the completeness axiom are really easy. If we have two Dedekind real numbers x and x', being the sections $\{D_l, D_r\}$ and $\{D'_l, D'_r\}$, then $x \leq y$ shall mean $D_l \subset D'_l$. The supremum of a set A of Dedekind real numbers that is bounded above always exists. It is the Dedekind section whose left set is the union of the left sets of the members of A.

Cantor's construction. Cantor had a completely different view of the real numbers. He saw sequences of rationals as the key to defining them. We have seen that every real number is the limit of a sequence of rationals. So we can think of a real number as a sequence of rationals, that either converges to a rational, or else wants to converge but has no rational to converge to.

This can be made precise. We single out those sequences of rationals that satisfy Cauchy's condition, in a form that mentions only rationals. Thus a sequence $(a_k)_{k=1}^{\infty}$ of rationals can be called a Cauchy sequence if it satisfies the following condition. For every rational $\varepsilon > 0$ there exists a natural number N, such that $|a_m - a_n| < \varepsilon$ for all $m \geq N$ and $n \geq N$.

Now we could say: a real number is a Cauchy sequence of rationals. However there is a problem. Different sequences of rationals could have the same limit when that limit is rational, which for starters makes it impossible to embed the rationals into the reals. Which Cauchy sequence of rationals are we to identify $\frac{1}{2}$ with? This is overcome by bunching Cauchy sequences of rationals, that we think should converge to the same limit, into sets, so-called equivalence classes. Two Cauchy sequences $(a_k)_{k=1}^{\infty}$ and $(b_k)_{k=1}^{\infty}$ belong to the same equivalence class if $\lim_{k \to \infty} a_k - b_k = 0$ (the limit being interpreted in a way that mentions only rationals).

According to Cantor's point of view, a real number is an equivalence class of Cauchy sequences of rationals.

Dedekind and Cantor might have had a most interesting argument over who had the nicer version of real numbers. But we can also imagine that they meet and have the following conversation (in German presumably, but a translation has most helpfully been provided). Cantor says "I'm thinking of a real number", (that is, an equivalence class of Cauchy sequences of rationals) "and it lies between the natural numbers 0 and 1". Dedekind says "I too am thinking of a real number" (that is, a Dedekind section of the rationals) "and it too lies between 0 and 1". Cantor asks "What are the decimal digits of your number"? Dedekind replies "They are 0, 1, 0, 2, 0, 3, 0, 4, and so on". "Interesting" says Cantor, "mine has the same digits. We are thinking of the same number".

We now see the force of Proposition 3.21. All versions of the real numbers are really the same, though they may look very different. We can identify a real number in Alice's version with a real number in Bill's version if they have the same decimal digits. This is a big comfort for we want the real numbers to be in some sense unique. In contrast, a field (a set with binary operations that satisfy axioms A) is not in any sense unique. There exists a field with 2 elements and another with 4. They are clearly in no way the same.

Only one thing can spoil this beautiful uniqueness of the real numbers. As decimal digits are nothing but a sequence of natural numbers in the range 0 to 9, Alice and Bill have to agree about what a sequence is. More precisely, although they may agree that a sequence is an assignment of terms to the natural numbers, they may disagree as to what constitutes an admissible assignment. For example it is possible that Bill requires the terms of a sequence to be in some sense computable for the assignment to be admissible. Alice may say "For each n the nth digit is 0 if the twin prime conjecture is true and it is 1 if it is untrue". Alice's number is either 0 or $\frac{1}{9}$. Presumably Alice does not know which it is (if she did she might qualify for a Fields Medal as the twin prime conjecture is unsolved at the time of writing), but has, according to standard thinking about sets, successfully defined a sequence of digits, and a very simple one at that, being entirely constant. Over this Bill may disagree.

Logicians say that in each model of set theory there is a unique model of the real numbers. This is just a way of saying that if Alice and Bill agree over how to assign terms in a sequence, they will have essentially the same real numbers. This is the philosophical importance of decimals.

3.10.1 Pointers to Further Study

→ Mathematical logic
→ Models of the real numbers
→ Axiomatic set theory

3.11 ($\Diamond$) Limit Inferior and Limit Superior

A sequence $(a_n)_{n=1}^{\infty}$ that is bounded is not necessarily convergent. Logically it is not correct to write the expression "$\lim_{n\to\infty} a_n$" without having first shown that the limit exists (though we often do so without coming to harm). In this uncomfortable situation we can use $\limsup_{n\to\infty} a_n$ (limit superior) and $\liminf_{n\to\infty} a_n$ (limit inferior), quantities that always exist if the sequence is bounded but not necessarily convergent. If we allow the values ∞ and $-\infty$ then they exist for all sequences whether bounded or not.

Limit inferior and limit superior are often used to prove that a limit exists and to calculate it. On the other hand most, if not all, calculations that use these notions can be carried out without them, and are not thereby appreciably longer. Use of these operations is very much a matter of personal preference, and one can usually get on quite well without them. However, it is right to mention that limit inferior and limit superior do appear in certain important formulas (such as that for the radius of convergence of a power series), and certain theorems (such as in Fatou's lemma of integration theory).

Let $(a_n)_{n=1}^{\infty}$ be a bounded sequence. We define the limit superior of the sequence $(a_n)_{n=1}^{\infty}$ by

$$\limsup_{n\to\infty} a_n := \lim_{n\to\infty} (\sup_{k\geq n} a_k).$$

To explain this better we let

$$h_n := \sup_{k\geq n} a_k = \sup\{a_n, a_{n+1}, a_{n+2}, \ldots\}.$$

As the sequence $(a_n)_{n=1}^{\infty}$ is bounded above, the number h_n is certainly finite. In fact we have $h_n \leq \sup_{k\geq 1} a_k$. As the sequence $(a_n)_{n=1}^{\infty}$ is bounded below, the sequence $(h_n)_{n=1}^{\infty}$ is bounded below; in fact $h_n \geq \inf_{k\geq 1} a_k$. Moreover h_n is decreasing, being the supremum of a set that shrinks with increasing n. The limit $\lim_{n\to\infty} h_n$ therefore exists and is a finite number.

In a similar way we define the limit inferior by

$$\liminf_{n\to\infty} a_n := \lim_{n\to\infty} (\inf_{k\geq n} a_k).$$

Now it is easy to obtain the following rules. For the moment all the sequences are supposed to be bounded.

(i)
$$\inf_{n\geq 1} a_n \leq \liminf_{n\to\infty} a_n \leq \limsup_{n\to\infty} a_n \leq \sup_{n\geq 1} a_n.$$

(ii) A sequence $(a_n)_{n=1}^{\infty}$ is convergent if and only if

$$\liminf_{n\to\infty} a_n = \limsup_{n\to\infty} a_n.$$

Given this equality the common value is $\lim_{n\to\infty} a_n$.

(iii) If $a_n \le b_n \le c_n$ we have

$$\limsup_{n\to\infty} a_n \le \limsup_{n\to\infty} b_n \le \limsup_{n\to\infty} c_n$$

and

$$\liminf_{n\to\infty} a_n \le \liminf_{n\to\infty} b_n \le \liminf_{n\to\infty} c_n.$$

Exercise Prove these rules.

We can also allow the values ∞ and $-\infty$. If a_n is not bounded above we set $\limsup_{n\to\infty} a_n = \infty$. If a_n is bounded above, but $\sup_{k\ge n} a_n$ tends to $-\infty$, we set $\limsup_{n\to\infty} a_n = -\infty$. Similarly we can assign infinite values to $\liminf_{n\to\infty} a_n$. By this means we can define limit inferior and limit superior for arbitrary sequences.

Examples

(a) $(1, -1, 1, -1, 1, -1, \ldots)$.

Limit inferior is -1, limit superior 1.

(b) $(\frac{1}{2}, 2, \frac{2}{3}, 2, \frac{3}{4}, 2, \frac{4}{5}, 2, \frac{5}{6}, 2, \frac{6}{7}, \ldots)$.

Limit inferior is 1, limit superior 2.

(c) $(\frac{1}{2}, -\frac{1}{2}, \frac{2}{3}, -\frac{2}{3}, \frac{3}{4}, -\frac{3}{4}, \frac{4}{5}, -\frac{4}{5}, \frac{5}{6}, -\frac{5}{6}, \ldots)$.

Limit inferior is -1, limit superior 1.

(d) $(1, 3, 2, 4, 3, 5, 4, 6, 5, 7, \ldots)$.

Limit superior and limit inferior are both ∞, which is also the limit.

(e) $a_n = \sin n$, $n = 1, 2, 3, \ldots$.

Limit inferior is -1, limit superior 1.

Exercise Check the above claims. For example (e) you will need to know that $\sin x$ is continuous, periodic, has maximum value 1, minimum -1, and its period 2π is irrational. So you might like to wait until these concepts have been properly treated in later chapters. You might also find useful Sect. 2.5 Exercise 3 and Sect. 3.5 Exercise 3.

We can show limit superior in action by proving the following result (though actually a proof avoiding it is not longer).

Proposition 3.23 *Let $(a_n)_{n=1}^{\infty}$ be a real sequence such that $\lim_{n\to\infty} a_n = t$. Let $\sigma_n = \left(\sum_{k=1}^{n} a_k\right)/n$. Then $\lim_{n\to\infty} \sigma_n = t$. The conclusion also holds if $t = \infty$ or $t = -\infty$.*

Proof We write the proof for the case when t is a finite number, leaving the cases $t = \pm\infty$ as an exercise. We first write

$$|\sigma_n - t| = \left|\left(\frac{1}{n}\sum_{k=1}^{n} a_k\right) - t\right| = \left|\frac{1}{n}\sum_{k=1}^{n}(a_k - t)\right| \leq \frac{1}{n}\sum_{k=1}^{n}|a_k - t|.$$

Let $\varepsilon > 0$. Choose N, such that $|a_n - t| < \varepsilon$ for all $n \geq N$. For a given $n \geq N$ we split up the sum into terms with $k \leq N - 1$ and terms with $k \geq N$. We find

$$|\sigma_n - t| \leq \frac{1}{n}\sum_{k=1}^{N-1}|a_k - t| + \frac{n - N + 1}{n}\varepsilon. \tag{3.3}$$

This holds for all $n \geq N$. Let $n \to \infty$ (but hold N and ε fixed). The right-hand side has the limit ε and we conclude (without knowing whether the left-hand side has a limit)

$$\limsup_{n\to\infty} |\sigma_n - t| \leq \varepsilon.$$

Now this must hold for all $\varepsilon > 0$ so in fact $\limsup_{n\to\infty} |\sigma_n - t|$ must be 0, that is, $\lim_{n\to\infty} |\sigma_n - t|$ exists and is 0. This gives $\lim_{n\to\infty} \sigma_n = t$. $\square$

3.11.1 Exercises

1. Prove the cases $t = \pm\infty$ of Proposition 3.23.
2. Finish the proof of Proposition 3.23 from Eq. (3.3) without using limit superior (it shouldn't be longer).
3. Let $(a_n)_{n=1}^{\infty}$ and $(b_n)_{n=1}^{\infty}$ be convergent sequences and let their limits be s and t respectively. Let the sequence $(c_n)_{n=1}^{\infty}$ be defined by $c_n = \frac{1}{n}\sum_{k=1}^{n} a_k b_{n-k+1}$. Prove that $\lim_{n\to\infty} c_n = st$.
4. Prove the following generalisation of Proposition 3.23. Suppose that all terms of the sequence $(c_n)_{n=1}^{\infty}$ are positive and that the series $\sum_{n=1}^{\infty} c_n$ diverges. Let $(a_n)_{n=1}^{\infty}$ be a sequence with limit t (may be $\pm\infty$). Define

$$\sigma_n = \frac{\sum_{k=1}^{n} c_k a_k}{\sum_{k=1}^{n} c_k}.$$

Show that $\lim_{n\to\infty} \sigma_n = t$.

3.11.2   Uses of Limit Inferior and Limit Superior

The dichotomy of finite versus infinite often lurks behind the appearance of limit inferior or limit superior. Given a sequence $(a_n)_{n=1}^{\infty}$, and a predicate $P(x)$ applicable to real numbers, we shall say that $P(a_n)$ is *eventually true* if there exists N, such that $P(a_n)$ is true for all $n \geq N$. This is the same as saying that $P(a_n)$ is false for, at most, finitely many place numbers n. We shall say that $P(a_n)$ is *infinitely often true* if $P(a_n)$ holds for infinitely many place numbers n.

   Now let $(a_n)_{n=1}^{\infty}$ be a bounded sequence. The reader is invited to prove the following characterisations of limit inferior and limit superior:

(i) $\liminf_{n\to\infty} a_n = t$ if and only if the following condition holds: for each $\varepsilon > 0$ the inequality $a_n > t - \varepsilon$ is eventually true and the inequality $a_n < t + \varepsilon$ is infinitely often true.

(ii) $\limsup_{n\to\infty} a_n = t$ if and only if the following condition holds: for each $\varepsilon > 0$ the inequality $a_n < t + \varepsilon$ is eventually true and the inequality $a_n > t - \varepsilon$ is infinitely often true.

   Certain frequently cited properties that a sequence may possess can be expressed succinctly with limit superior or limit inferior. In the following table we exhibit four such properties opposite their equivalent, and less wordy, formulations using limit superior. The abbreviation 'i.o.' stands for 'infinitely often'.

(a)	$\limsup_{n\to\infty} < t$	There exists $t' < t$, such that $a_n < t'$ eventually.
(b)	$\limsup_{n\to\infty} \leq t$	For all $\varepsilon > 0$, $a_n < t + \varepsilon$ eventually.
(c)	$\limsup_{n\to\infty} > t$	There exists $t' > t$, such that $a_n > t'$ i.o.
(d)	$\limsup_{n\to\infty} \geq t$	For all $\varepsilon > 0$, $a_n > t + \varepsilon$ i.o.

The reader is invited to prove these claims, and formulate similar ones using limit inferior.

   The ratio test can be generalised using limit inferior and limit superior, in a form that does not require that a_{n+1}/a_n converges.

Proposition 3.24 *Let $\sum_{n=1}^{\infty} a_n$ be a positive series in which no term is 0. The following conclusions hold:*

(1) If $\limsup_{n\to\infty} \dfrac{a_{n+1}}{a_n} < 1$ the series is convergent.

(2) If $\liminf_{n\to\infty} \dfrac{a_{n+1}}{a_n} > 1$ the series is divergent.

Proof In case 1 there exists $t < 1$, such that $a_{n+1}/a_n < t$ eventually. This implies (as the reader should check) that there exist n_0 and C, such that $a_n < Ct^n$ for all $n \geq n_0$. We obtain convergence of the series $\sum_{n=1}^{\infty} a_n$ by comparison with the series $\sum_{n=1}^{\infty} t^n$.

   In case 2 there exist $t > 1$, such that $a_{n+1}/a_n > t$ eventually. This implies divergence since a_n cannot tend to 0.   □

One sometimes encounters limits of the form $\lim_{n\to\infty} a_n^{1/n}$ and their treatment can be puzzling. The following result is often useful. It is another case where limit inferior or limit superior can be used optionally in the proof. We will need the known limit, for a given positive constant b, that $\lim_{n\to\infty} b^{1/n} = 1$. About the nth root function $x^{1/n}$: we will show in detail later that every positive real number has a unique positive n^{th} root. Moreover the nth root function is increasing.

Proposition 3.25 Let $(a_n)_{n=1}^{\infty}$ be a real sequence such that $a_n > 0$ for all n. Assume that $\lim_{n\to\infty} a_n/a_{n-1} = t$. Then $\lim_{n\to\infty} a_n^{1/n} = t$. The conclusion also holds if $t = \infty$.

Proof Obviously $t \geq 0$. We shall write out the proof in the case that t is a finite, positive number. The cases $t = 0$ and $t = \infty$ are left to the exercises.

Let $\varepsilon > 0$. Reduce ε, if necessary, so that $\varepsilon < t$ (it is here that we want $t > 0$). There exists N, such that

$$t - \varepsilon < \frac{a_n}{a_{n-1}} < t + \varepsilon$$

for all $n \geq N$. Then, for all $n \geq N$, we have

$$(t - \varepsilon)a_{n-1} < a_n < (t + \varepsilon)a_{n-1}$$

and by induction we find

$$(t - \varepsilon)^{n-N} a_N < a_n < (t + \varepsilon)^{n-N} a_N$$

for all $n \geq N + 1$. Taking the nth root, and using the fact that the nth root function is increasing, we find

$$(t - \varepsilon)^{1-\frac{N}{n}} a_N^{\frac{1}{n}} < a_n^{\frac{1}{n}} < (t + \varepsilon)^{1-\frac{N}{n}} a_N^{\frac{1}{n}}$$

for all $n \geq N + 1$. Let $n \to \infty$. Now

$$(t - \varepsilon)^{1-\frac{N}{n}} \to t - \varepsilon, \quad (t + \varepsilon)^{1-\frac{N}{n}} \to t + \varepsilon, \quad a_N^{\frac{1}{n}} \to 1$$

(all three follow from the limit $\lim_{n\to\infty} b^{1/n} = 1$). We conclude that

$$t - \varepsilon \leq \liminf_{n\to\infty} a_n^{\frac{1}{n}} \leq \limsup_{n\to\infty} a_n^{\frac{1}{n}} \leq t + \varepsilon.$$

This holds for all $\varepsilon > 0$. We conclude that $\liminf_{n\to\infty} a_n^{1/n}$ and $\limsup_{n\to\infty} a_n^{1/n}$ are equal to t. $\square$

3.11.3 Exercises (cont'd)

5. What minor modification is needed in the proof of Proposition 3.25 for the case $t = 0$?
6. Finish the proof of Proposition 3.25 by considering the case $t = \infty$.
7. Let $\limsup_{n\to\infty} a_n = t$. Show that there exists a subsequence $(a_{k_n})_{n=1}^{\infty}$, such that $\lim_{n\to\infty} a_{k_n} = t$. Here we may have $t = \pm\infty$. A similar result holds for $\liminf_{n\to\infty} a_n$.
8. Give another (actually the third of this text) proof of the Bolzano–Weierstrass theorem (Proposition 3.10) using the previous exercise.
9. Give another proof that a sequence $(a_n)_{n=1}^{\infty}$ that satisfies Cauchy's condition (Sect. 3.7) is convergent, by showing that $\liminf_{n\to\infty} a_n = \limsup_{n\to\infty} a_n$.

3.11.4 Pointers to Further Study

→ Semi-continuous functions
→ Radius of convergence
→ Fatou's lemma

3.12 (◊) Continued Fractions

Decimals provide the best known, and perhaps the most practical, way to approximate a real number by rational numbers. But they come at a price. Consider the following example:
$$x = 0.797997999799997999997....$$

The sequence of digits consists of isolated instances of the digit "7" interspersed with lengthening strings of the digit "9". This ensures that the number is irrational. If we truncate at the nth digit we obtain a rational approximation of the form $a/10^n$, and an error between $7/10^{n+1}$ and $1/10^n$.

The price of this error is the size of the denominator. To ensure an error less than $1/10^n$ we have to use a fraction with denominator 10^n. This is an expensive error, but it is possible to do much better. Approximations are possible with fractions a/b for which the error is less than $1/b^2$. They can be obtained through the *continued fraction algorithm*.

Whereas a decimal representation of a number x uses a sequence of digits from the range 0, 1, ..., 9, a continued fraction uses a sequence of integers, which can be arbitrarily large. This sequence is finite if and only if x is rational, unlike a decimal expansion, which can have infinitely many non-zero digits whilst representing a rational. The integers of the continued fraction are generated from the number x by a simple algorithm, which we describe in the next paragraph.

Let x be a real number. Using the notation $[x]$ for the highest integer less than or equal to x, we set

$$a_0 = [x] \quad \text{and} \quad x_1 = x - a_0.$$

If $x_1 \neq 0$ we set

$$a_1 = \left[\frac{1}{x_1}\right] \quad \text{and} \quad x_2 = \frac{1}{x_1} - a_1.$$

We continue inductively. Having reached a_{n-1} and x_n, and assuming that $x_n \neq 0$, we set

$$a_n = \left[\frac{1}{x_n}\right], \quad x_{n+1} = \frac{1}{x_n} - a_n.$$

The process terminates if, for some n, we have $x_n = 0$. Since x_n lies in the interval $0 \leq x_n < 1$ for every $n \geq 1$, it is clear that all the integers a_n, except possibly for a_0, are positive.

If the sequence terminates with a_n (because $x_{n+1} = 0$) then x must be rational. In fact, unravelling the reciprocals we find that

$$x = a_0 + \cfrac{1}{a_1 + \cfrac{1}{a_2 + \cfrac{1}{a_3 + \cfrac{1}{\ddots \cfrac{}{\;} \cfrac{1}{a_n}}}}}$$

This expression is known as a continued fraction. The study of them is *really old* with hints of them in ancient mathematics; for example they are closely related to the Euclidean algorithm.

There is a short notation for a continued fraction. We denote the above expression, whether or not the entries are integers, by

$$[a_0, a_1, a_2, \dots, a_n].$$

If x is irrational then the sequence of integers cannot terminate. We would then like to write

$$x = [a_0, a_1, a_2, \dots].$$

The right-hand side can be interpreted as the limit

$$\lim_{n \to \infty} [a_0, a_1, a_2, \dots, a_n]$$

if the limit exists. In fact the limit does exist, and it really does equal x. The proof is a bit lengthy and substantial parts of it will be left to the exercises.

We begin by writing

$$\frac{p_n}{q_n} = [a_0, a_1, a_2, ..., a_n]$$

where p_n and q_n are coprime integers (that is, integers with highest common divisor 1). The fraction $[a_0, a_1, a_2, ..., a_n]$ is called a convergent (anticipating the result; but it is convenient already to have a name for it).

There is a simple way to calculate the sequences $(p_k)_{k=0}^n$ and $(q_k)_{k=0}^n$ from the sequence $(a_k)_{k=0}^n$ without having to pick one's way through a pile of nested reciprocals. Both sequences satisfy the same recurrence relations, namely

$$p_k = a_k p_{k-1} + p_{k-2}, \quad q_k = a_k q_{k-1} + q_{k-2}, \quad k = 2, 3, ... \tag{3.4}$$

with the initial values $p_0 = a_0$, $p_1 = a_0 a_1 + 1$, $q_0 = 1$, $q_1 = a_1$.

Proof of the Recurrence Relations We prove that p_k and q_k satisfy the recurrences (3.4) by induction. As induction hypothesis we assume that for any continued fraction of length less than or equal to m, the corresponding numerators and denominators satisfy the recurrence relations (3.4).

For our given x we let

$$\frac{p_k'}{q_k'} = [a_1, a_2, a_3, ..., a_k], \quad k = 1, 2, 3, ...$$

where the integers p_k' and q_k' are coprime. The induction hypothesis is supposed to hold for the fraction $[a_1, a_2, a_3, ..., a_m]$, which is of length m, so that we have

$$p_m' = a_m p_{m-1}' + p_{m-2}', \quad q_m' = a_m q_{m-1}' + q_{m-2}'.$$

Furthermore, by the definition of continued fraction, the relation

$$\frac{p_k}{q_k} = a_0 + \frac{q_k'}{p_k'}$$

holds for all k and, recalling that p_k' and q_k' are coprime, we see that

$$p_k = a_0 p_k' + q_k', \quad q_k = p_k'.$$

Therefore, after some algebraic manipulation, we find

$$p_m = a_m p_{m-1} + p_{m-2}, \quad q_m = a_m q_{m-1} + q_{m-2}$$

and the proof of (3.4) is complete. □

The rest of the proof that $\lim_{n\to\infty} p_n/q_n = x$ is given in steps in Exercise 1 and builds almost entirely on the relations (3.4).

3.12.1 Exercises

1. Complete the proof that $\lim_{n \to \infty} p_n/q_n = x$ in the following steps:

   (a) Show that
   $$p_n q_{n+1} - p_{n+1} q_n = (-1)^{n+1} \quad n = 0, 1, 2, \dots$$

   (b) Show that
   $$\left| \frac{p_n}{q_n} - \frac{p_{n+1}}{q_{n+1}} \right| = \frac{1}{q_n q_{n+1}}.$$

   (c) Show that if the sequence a_n does not terminate then the integer sequences p_n and q_n are strictly increasing and satisfy $\lim_{n \to \infty} p_n = \lim_{n \to \infty} q_n = \infty$.
   (d) Assume that the fraction does not terminate and set $y_n = 1/x_n$. Show that, for each n, we have
   $$x = [a_0, a_1, \dots, a_{n-1}, y_n].$$

   Here a_n is replaced by y_n.
   (e) Assume that the fraction does not terminate. Then for each n the number x lies between $[a_0, a_1, \dots a_n]$ and $[a_0, a_1, \dots, a_n, a_{n+1}]$.
   (f) Show that
   $$\left| \frac{p_n}{q_n} - x \right| \le \frac{1}{q_n^2}, \quad n = 0, 1, 2, \dots$$

   This completes the proof that $p_n/q_n \to x$. It gives much more. We get an infinite sequence of rational approximations a/b to x such that the error is at most $1/b^2$. The price of an error less than ε is a denominator at most $1/\sqrt{\varepsilon}$. This is much better than what can be achieved by decimal expansions.

2. Show that the continued fraction of a rational number terminates.

   Hint. Show that if x is rational there can be at most finitely many rational approximations a/b that satisfy
   $$\left| x - \frac{a}{b} \right| \le \frac{1}{b^2}.$$

   From a non-terminating fraction we get infinitely many such approximations.
3. Show that
   $$\sqrt{2} = [1, \overline{2}].$$

   The overline means that the entry "2" repeats indefinitely. Tabulate values of p_k and q_k and observe that $p_5/q_5 = 99/70$, which gives $\sqrt{2}$ with an error less than 10^{-4}.
4. With y_k as defined in Exercise 1(d), show that for all k we have

$$x = \frac{p_{k-1}y_k + p_{k-2}}{q_{k-1}y_k + q_{k-2}}.$$

5. With y_k as defined in Exercise 1(d), show that $y_k = [a_k, a_{k+1}, a_{k+2}, ...]$, where $x = [a_0, a_1, a_2, ...]$.

6. Suppose that the continued fraction of x is periodic, that is to say, it is of the form $[\overline{a_0, a_1, ...a_n}]$, the overline indicating that the string of entries repeats indefinitely. Show that x satisfies a quadratic equation with integer coefficients.

7. Calculate the values of the following continued fractions.

   (a) $[\overline{1}]$, that is, $[1, 1, 1, ...]$. This is surely the simplest continued fraction that represents an irrational number.
   (b) $[\overline{1, 2}]$
   (c) $[\overline{1, 2, 1}]$.

8. Show that if the continued fraction of x has the form

$$[a_0, a_1, ...a_m, \overline{a_{m+1}, ...a_{m+n}}],$$

consisting of a string that repeats indefinitely after an initial string, then x is an irrational root of a quadratic equation with integer coefficients.

Hint. Use Exercises 4, 5 and 6, and observe that if y_k is the root of a quadratic equation then so is x.

Note. The converse is also true. Every quadratic irrational has a continued fraction of this form. This was shown by Lagrange. The proof is not hard but requires a little number theory.

9. The continued fraction algorithm gives a handy way of finding integers x and y that satisfy $ax - by = 1$ for a given pair of coprime integers a and b.
   Hint. Stare at the result of Exercise 1(a). This also shows that continued fractions are closely related to the Euclidean algorithm; the latter is often used to find x and y.

3.12.2 *Pointers to Further Study*

→ Number theory
→ Irrationality theory
→ Diophantine analysis

Chapter 4
Functions and Continuity

> *A function of a variable quantity is an analytic expression composed in any way whatsoever of the variable quantity and numbers or constant quantities*
>
> *L. Euler*

4.1 How Do We Talk About Functions?

A function or mapping assigns to each point in a set A a point in a set B. We shall allow a rather wide scope for the understanding of "assigns". It does not have to be an assignment using a formula in the ordinary sense, though that is very often the case in analysis.

An exact definition of the concept of function can be based on set theory. It seems to dispel all mysteries connected with the meaning of assignment, identifying a function with a certain set (its graph in fact), but the clarity thus gained is a little misleading for it raises the question of what sets are to be allowed. It is questionable whether the set-theoretical definition of function is needed for fundamental analysis.

As with any other object of interest in mathematics, we use letters to symbolise functions; "f" is often the first choice, if available, followed by "g". The set A is called the domain of the function whilst the set B is called its codomain. We write

$$f : A \to B$$

which is read "f is a function with domain A and codomain B" or "f maps A to B". If $x \in A$ (that is, if x is an element of A), we denote the element of B that f assigns to x by $f(x)$ and call it the value of f at x.

© The Editor(s) (if applicable) and The Author(s), under exclusive
license to Springer Nature Switzerland AG 2020
R. Magnus, *Fundamental Mathematical Analysis*, Springer Undergraduate
Mathematics Series, https://doi.org/10.1007/978-3-030-46321-2_4

Although it should be clear from the definition, it is worth emphasising that a function assigns only one value to each point in its domain. If we wish to say that the square root of 4 is plus 2 or minus 2, then we are not using a function.

The words "function" and "mapping" mean the same but "function" is mostly used when the codomain consists of numbers, rather than vectors or more exotic objects.

4.1.1 Examples of Specifying Functions

Here are some examples of how we specify a function. They show some acceptable ways to assign values of varying degrees of formality.

   (a) $f : \mathbb{R} \to \mathbb{R}, \quad f(x) = x^2.$

This says that f maps $\mathbb{R}$ to $\mathbb{R}$ and assigns to each x its square x^2. We say informally "f is the function x^2".

   (b) $f : [0, \infty[\to \mathbb{R}, \quad f(x) = \sqrt{x}.$

This says that f maps the set of positive real numbers, together with 0, to $\mathbb{R}$, and assigns to each such number x its square root $\sqrt{x}$. We say informally "f is the function $\sqrt{x}$".

   (c) $f :]0, 500] \to \mathbb{R}, \quad f(x) = \begin{cases} 10, & \text{if } 0 < x < 100 \\ 20, & \text{if } 100 \le x \le 500 \end{cases}$

The presentation here is called specifying a function by cases. This function could be a list of postal charges.

   (d)   The function $f : [0, 1[\to \mathbb{R}$, where $f(x) = 1$ if the digit 9 appears in the decimal representation of x and $f(x) = 0$ otherwise. We use a decimal representation that does not end in repeating 9's.

There seems to be no practical general way to compute $f(x)$ in this example, although we do have an algorithm for the decimal digits, so things are not as bad as they might be.

Another convenient, and less formal, way to specify functions is typified by the example:

   (e) $f(x) = \sqrt{x}, \quad (x > 0).$

The codomain is not given (it is not always important), but the domain is indicated, although sets are not mentioned.

Fig. 4.1 Two views of a function

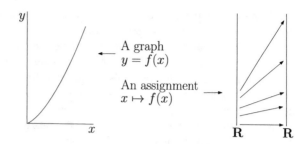

Often the domain is not mentioned at all when we specify a function informally, but it is inferred, as in the example:

(e) $\dfrac{(x-1)(x-2)}{(x-3)(x-4)}.$

Here we mean the function f with domain $A = \mathbb{R} \setminus \{3, 4\}$ and codomain $\mathbb{R}$, such that

$$f(x) = \frac{(x-1)(x-2)}{(x-3)(x-4)}, \quad (x \in A).$$

The graph of a function f with domain $A \subset \mathbb{R}$ and codomain $\mathbb{R}$ is the set of all pairs (x, y) such that $x \in A$ and $y = f(x)$. We can view the graph as a curve (of some sort) in the plane with coordinates x and y, and this is a useful way to visualise f. An example of Weierstrass of a continuous function that is nowhere differentiable, and the space filling curves of Peano, show the limitations of such a picture. Figure 4.1 illustrates the two ways to view a function: as a graph or as an assignment.

The domains of functions considered here will be subsets of $\mathbb{R}$. It is clearly too limiting to consider only functions with domain $\mathbb{R}$; the above examples illustrate this. However the typical domains for calculus are intervals, or finite unions of intervals.

Sometimes we speak of a function $y = f(x)$, instead of just f, as if we have the graph in mind. Or else we are thinking of x and y as *variables* and expressing a relation between them, a point of view common in physics (think of pressure P and volume V, and Boyle's Law of ideal gases). We may even say "the function $f(x)$", although strictly speaking $f(x)$ would be the value that f assigns to the number x. It offers a visual cue that a function is referred to, rather than a number that might be denoted by f. One should bear in mind that mathematics is not only a mode of thinking, but also a mode of communication.

4.2 Continuous Functions

If a parcel weighs 100 grammes, and the postal charges are as in example (c) of the last section, we will not be happy to pay 20 pounds in postage. We might object that it really weighs 99.99 grammes and the post office should have their scales checked.

The problem here is that the function in question is discontinuous. The definition of continuity of a function f at a point x_0 seems at first glance to be designed to exclude a jump in the graph of the function at x_0. This is a bit of an oversimplification; we will see later that it does more than this and excludes some other types of undesirable behaviour as well. The implications of the definition are not very obvious and will only gradually become clear. Until then the reader is asked to take on trust that the definition of continuity presented here is appropriate.

In the definitions below, the domain of the function is a subset A of $\mathbb{R}$. In most practical cases A is an interval or a finite union of intervals.

Definition Let $f : A \to \mathbb{R}$ where A is a subset of $\mathbb{R}$. Let $x_0 \in A$. We say that the function f is continuous at x_0, or that x_0 is a point of continuity of f, if the following condition is satisfied:

For each $\varepsilon > 0$ there exists $\delta > 0$, such that $|f(x) - f(x_0)| < \varepsilon$ for all x in A that satisfy $|x - x_0| < \delta$.

If f is not continuous at x_0 we say that f is discontinuous at x_0, or that x_0 is a point of discontinuity of f.

Definition Let $f : A \to \mathbb{R}$ where A is a subset of $\mathbb{R}$. We say that the function f is continuous if it is continuous at every point of A.

As in the definition of limit of a sequence, we have made a small concession to natural English. It would be more precise, but less natural, to define the condition of continuity as follows: for each $\varepsilon > 0$ there exists $\delta > 0$, such that for all x in A that satisfy $|x - x_0| < \delta$ we have $|f(x) - f(x_0)| < \varepsilon$. In a simplified first-order logic notation we can lay bare the logical structure of this condition:

$$(\forall \varepsilon > 0)(\exists \delta > 0)(\forall x \in A)\big(|x - x_0| < \delta \Rightarrow |f(x) - f(x_0)| < \varepsilon\big).$$

In most cases A is an interval. Even so, this is not quite general enough for our purposes. In the following pages, when we write $f : A \to \mathbb{R}$ we shall mean that A is a subset of $\mathbb{R}$ (not necessarily an interval).

The definition of continuity, like the definition of limit and the nature of the real numbers, took a long time to crystalise into its present form. It seems to have been thought that continuity must be seen as a property of the function as a whole, akin to saying that its graph hangs together in one piece; or even more loosely, that the graph can be drawn without lifting the pencil from the paper. It finally became clearer that the way forward was to define continuity at a point first, and only then to define continuity as a whole to mean that the function was continuous at each point. None of this was originally at all obvious.

4.2.1 Exercises

1. Because the set A is quite arbitrary the definition of continuity has a perhaps unexpected consequence. We say that a point $x_0 \in A$ is an isolated point of A if there exists $\delta > 0$ such that x_0 is the only point both in A and in the interval $]x_0 - \delta, x_0 + \delta[$. This means that if $x \in A$ and $|x - x_0| < \delta$ then $x = x_0$.

   Show that if $f : A \to \mathbb{R}$ and x_0 is an isolated point of A then f is automatically continuous at x_0.

2. Is 0 a point of discontinuity for the function $1/x$?

3. For each x_0 in its domain, and for each $\varepsilon > 0$, find a suitable δ, thus proving that the following functions are continuous:

   (a) $f(x) = 1$
   (b) $f(x) = x$
   (c) $f(x) = x^2$
   (d) $f(x) = \dfrac{1}{x}$.

4. Let $f : [0, 1[\to \mathbb{R}$ be the function defined in Sect. 4.1 Example (d), that assigns 1 to x if the decimal expansion of x contains the digit 9, using if possible the terminating expansion, and assigns 0 otherwise.
   Show that f is continuous at x if and only if $f(x) = 1$.

5. Suppose that the function $f : A \to \mathbb{R}$ is continuous at the point c and that $f(c) < d$ [respectively $f(c) > d$]. Show that there exists $\delta > 0$, such that $f(x) < d$ [respectively $f(x) > d$] for all x in A that satisfy $|x - c| < \delta$.

6. A function f with domain A is said to be upper semi-continuous [respectively, lower semi-continuous] at a point x_0 in A if the following condition is satisfied: for all $\varepsilon > 0$ there exists $\delta > 0$, such that $f(x) < f(x_0) + \varepsilon$ [respectively, $f(x) > f(x_0) - \varepsilon$] for all x in A that satisfy $|x - x_0| < \delta$.

   (a) Show that f is continuous at x_0 if and only if it is both upper semi-continuous and lower semi-continuous at x_0.
   (b) Suppose that f is upper semi-continuous [respectively lower semi-continuous] at a point c. Suppose that $f(c) < d$ [respectively $f(c) > d$]. Show that there exists $\delta > 0$, such that $f(x) < d$ [respectively $f(x) > d$] for all x in A that satisfy $|x - c| < \delta$.

4.2.2 Limits of Functions

Let $f : A \to \mathbb{R}$ where A is a subset $\mathbb{R}$. In practice A is often an interval, or an interval minus a finite set of points. The reader should recall the definition of limit point of a set (see Sect. 3.5).

Definition Let c be a limit point of the set A and let t be a real number. We say that the limit of f at c exists and equals t, and we write

$$\lim_{x \to c} f(x) = t,$$

if the following condition is satisfied:

For every $\varepsilon > 0$ there exists $\delta > 0$, such that $|f(x) - t| < \varepsilon$ for all x in A that satisfy $0 < |x - c| < \delta$.

We sometimes write more informally

$$f(x) \to t \quad \text{as} \quad x \to c$$

to express that the limit of f at c equals t.

Note the following points:

(a) Because c is required to be a limit point, there exists x in A with $0 < |x - c| < \delta$. This means that the situation, that the limit exists and equals t, cannot arise by default, which would happen with any number t whatsoever if there were no points x to be tested.

(b) If A is an interval (commonly the case), then c is either in A or else c is an endpoint (or both).

(c) If c is in A then the value $f(c)$ has no influence on the limit.

(d) The variable x in the expression "$\lim_{x \to c} f(x) = t$" is a bound variable. Any other letter may be used instead of "x", for example "$\lim_{q \to c} f(q) = t$" has the same meaning as "$\lim_{x \to c} f(x) = t$".

(e) The limit t may, or may not, be a value of the function f at some x not equal to c. It is quite possible for f to take the value t at points in the interval $]c - h, c + h[$, excluding c, for every $h > 0$. Confusion over this caused problems in early thinking about limits (as was also pointed out in connection with the limit of a sequence).

Although it might be thought nice to display δ as a function of ε, for example to get explicit error estimates, this is not necessary to verify the definition of limit, nor is it always helpful. To produce a δ that works for a given ε some arbitrary choices may have to be made, such as that of selecting in a non-explicit fashion a number from a non-empty set. It is not hard to see that the set of possible δ's for a given ε, if bounded above, is an interval of the form $]0, \delta_{\max}]$. In such a case we could, if we wished, define the function $\delta(\varepsilon) = \delta_{\max}$, but this is not necessarily useful.

Exercise Check the claim made at the end of the last paragraph about the set of possible δ's forming an interval.

The following result is often needed:

Proposition 4.1 *Let c be a limit point of the set A and suppose that the limit* $\lim_{x \to c} f(x)$ *exists (and is a finite number). Then there exists* $h > 0$, *such that* f *is bounded on the set* $]c - h, c + h[\cap A$.

Proof Using $\varepsilon = 1$ we see that there exists $h > 0$, such that $|f(x) - t| < 1$ for all x in A that satisfy $0 < |x - c| < h$. For such x we have $|f(x)| < 1 + |t|$. Let $K = 1 + |t|$ if $c \notin A$ and $K = \max(1 + |t|, |f(c)|)$ if $c \in A$. Then $|f(x)| < K$ for all x in $]c - h, c + h[\cap A$. □

The limit $\lim_{x \to c} f(x)$, if it exists, is unique. This was anticipated in our use of the definite article. It is impossible for distinct real numbers s and t, that both $\lim_{x \to c} f(x) = s$ and $\lim_{x \to c} f(x) = t$. If it was so we could choose ε, such that $0 < \varepsilon < \frac{1}{2}|s - t|$, and find $\delta > 0$, such that $|f(x) - s| < \varepsilon$ and also $|f(x) - t| < \varepsilon$ for all x in A that satisfy $0 < |x - c| < \delta$. Such points x exist since c is a limit point of A. But then we would have

$$|s - t| \leq |s - f(x)| + |f(x) - t| < 2\varepsilon < |s - t|,$$

which is impossible.

4.2.3 Connection Between Continuity and Limit

Arguments about continuity can often be rephrased as arguments about limits. This is due to the following result.

Proposition 4.2 *Let* $f : A \to \mathbb{R}$ *and let c be a point in A that is also a limit point of A. Then f is continuous at c if and only if* $f(c) = \lim_{x \to c} f(x)$.

Proof Assume first that f is continuous at c. Let $\varepsilon > 0$. There exists $\delta > 0$, such that $|f(x) - f(c)| < \varepsilon$ if $x \in A$ and $|x - c| < \delta$, and therefore in particular if $x \in A$ and $0 < |x - c| < \delta$. This says that $f(c) = \lim_{x \to c} f(x)$.

Next assume that $f(c) = \lim_{x \to c} f(x)$. Let $\varepsilon > 0$. There exists $\delta > 0$, such that $|f(x) - f(c)| < \varepsilon$ if $x \in A$ and $0 < |x - c| < \delta$. But then we also have $|f(x) - f(c)| < \varepsilon$ if $x \in A$ and $|x - c| < \delta$, since it obviously holds when $x = c$. □

If c is in A but is not a limit point of A, then f is automatically continuous at c. We can choose $\delta > 0$ so small that the conditions $|x - c| < \delta$ and $x \in A$ are only satisfied when $x = c$, and then $f(x) - f(c) = 0$.

4.2.4 Limit Rules

Limit rules allow us to establish new limits from old ones, usually without having to use the definition of limit. The limit rules for functions are similar to those for sequences, and the similarity extends to their proofs.

Proposition 4.3 *Let* $f : A \to \mathbb{R}$, $g : A \to \mathbb{R}$, *let* c *be a limit point of* A *and let* $\lim_{x \to c} f(x) = s$, $\lim_{x \to c} g(x) = t$. *We have the following rules:*

(1) (*Sum*) $\lim_{x \to c} f(x) + g(x) = s + t$
(2) (*Product*) $\lim_{x \to c} f(x) \cdot g(x) = s \cdot t$
(3) (*Absolute value*) $\lim_{x \to c} |f(x)| = |s|$
(4) (*Reciprocal*) *If* $s \neq 0$ *then* $\displaystyle \lim_{x \to c} \frac{1}{f(x)} = \frac{1}{s}$.

Note. In rule 4 the function $1/f(x)$ is possibly not defined for all $x \in A$ because f can have zeros. However, if $\lim_{x \to c} f(x) \neq 0$ the zeros of f, if any, other than possibly c itself, are a safe distance from c. More precisely there is an interval $I =]c - h, c + h[$, such that f has no zero in $I \cap A \setminus \{c\}$, on which domain we may define $1/f$.

Proof of the Limit Rules (1) Let $\varepsilon > 0$. We choose $\delta > 0$, such that $|f(x) - s| < \varepsilon/2$ and $|g(x) - t| < \varepsilon/2$ for all $x \in A$ that satisfy $0 < |x - c| < \delta$ (the same δ for both f and g). For such x we have

$$|f(x) + g(x) - (s + t)| \leq |f(x) - s| + |g(x) - t| < \frac{\varepsilon}{2} + \frac{\varepsilon}{2} = \varepsilon.$$

(2) Since the limit $\lim_{x \to c} f(x)$ exists, there exist $K > 0$ and $h > 0$, such that for all x in $]c - h, c + h[\cap A$ we have $|f(x)| < K$.

Let $\varepsilon > 0$. Choose $\delta_1 > 0$, such that $|f(x) - s| < \varepsilon$ and $|g(x) - t| < \varepsilon$ for all $x \in A$ that satisfy $0 < |x - c| < \delta_1$. Set $\delta = \min(\delta_1, h)$. If $0 < |x - c| < \delta$ we have

$$
\begin{aligned}
|f(x)g(x) - st| &= |f(x)g(x) - f(x)t + f(x)t - st| \\
&\leq |f(x)||g(x) - t| + |t||f(x) - s| \\
&< K\varepsilon + |t|\varepsilon \\
&< (K + |t|)\varepsilon.
\end{aligned}
$$

We conclude that $\lim_{x \to c} f(x) \cdot g(x) = s \cdot t$. It may help to reread the discussion in the proof of Proposition 3.5 in connection with the product of sequences.

(3) The proof is almost identical to the corresponding one for sequences.

(4) We have

$$\left| \frac{1}{f(x)} - \frac{1}{s} \right| = \frac{|s - f(x)|}{|s||f(x)|}.$$

By assumption $\lim_{x \to c} |f(x)| = |s|$ and $s \neq 0$. There therefore exists $h > 0$, such that $|f(x)| > \frac{1}{2}|s|$ for all x that satisfy $0 < |x - c| < h$. For such x we have

$$\left| \frac{1}{f(x)} - \frac{1}{s} \right| \leq \frac{2}{|s|^2} |s - f(x)|.$$

The conclusion follows from that fact that the right-hand side has the limit 0. $\square$

The justification for the last line of the proof of rule 4 is left to the reader. In fact there is a choice. One can give an argument starting "Let $\varepsilon > 0$" as in rules 1 and 2. Alternatively one can avoid mentioning ε at all by using a version of the *squeeze rule* for limits of functions; compare the corresponding rule for sequences, Proposition 3.8. The rule for functions reads as follows:

Let f, g and h be functions with domain A, and let c be a limit point of A. Assume that there exists $\delta > 0$, such that $g(x) \le f(x) \le h(x)$ for all x in A that satisfy $0 < |x - c| < \delta$, and that $\lim_{x \to c} g(x) = \lim_{x \to c} h(x) = t$. Then $\lim_{x \to c} f(x) = t$.

The proof of the squeeze rule is also left to the reader.

4.2.5 Continuity Rules

The limit rules give rise to continuity rules. Let $f : A \to \mathbb{R}$ and $g : A \to \mathbb{R}$ be continuous at c. Then

(i) The sum $f + g$ is continuous at c.
(ii) The product $f \cdot g$ is continuous at c.
(iii) The absolute value $|f|$ is continuous at c.
(iv) If $f(c) \ne 0$ then the reciprocal $1/f$ is continuous at c (where $1/f$ is defined sufficiently close to c to avoid zeros of f).

To begin the wholesale production of continuous functions we need to settle two initial cases, left to the reader to verify:

(a) The constant function $f : \mathbb{R} \to \mathbb{R}$, $f(x) = C$ for all x is everywhere continuous.
(b) The function $f(x) = x$ (identity function) is everywhere continuous.

From these and the continuity rules (i–iv), we immediately obtain a large number of continuous functions:

(c) The function x^n (where n is a fixed natural number) is continuous.
(d) The polynomial $f(x) = a_n x^n + a_{n-1} x^{n-1} + \cdots + a_0$ is continuous.
(e) If f and g are polynomials then the rational function $f(x)/g(x)$ is continuous (on its domain naturally, which excludes the zeros of g).

4.2.6 Left and Right Limits

Let $f :]a, b[\to \mathbb{R}$ and let $a < c < b$. We can consider separately the two functions

$$f_1 :]a, c[\to \mathbb{R}, \quad f_1(x) = f(x), \quad (a < x < c)$$
$$f_2 :]c, b[\to \mathbb{R}, \quad f_2(x) = f(x), \quad (c < x < b).$$

Fig. 4.2 The simplest
discontinuities

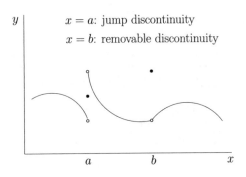

They are called the *restrictions* of f to the intervals $]a, c[$ and $]c, b[$.

The limits $\lim_{x \to c} f_1(x)$ and $\lim_{x \to c} f_2(x)$ are called the left-hand and right-hand limits of f at c. They are jointly called *one-sided limits*. The usual notation, which does not mention f_1 or f_2, refers to them by

$$\lim_{x \to c-} f(x) \quad \text{and} \quad \lim_{x \to c+} f(x),$$

or else even more simply, by

$$f(c-) \quad \text{and} \quad f(c+).$$

Furthermore when f has the domain $]a, b[$ it is quite common to denote the limits $\lim_{x \to a} f(x)$ and $\lim_{x \to b} f(x)$, quite unnecessarily, by $\lim_{x \to a+} f(x)$ and $\lim_{x \to b-} f(x)$, as a notational reminder that x can only approach a from the right and b from the left.

Proposition 4.4 *The function $f :]a, b[\to \mathbb{R}$ is continuous at $c \in]a, b[$ if and only if the one-sided limits $f(c-)$ and $f(c+)$ exist and are equal to $f(c)$.*

The proof of this is rather obvious, but the proposition is worth stating because it suggests somewhat graphically one of the characteristic ways we think about failure of continuity. If the left and right limits both exist but are unequal we say that f has a jump discontinuity at $x = c$. The difference $f(c+) - f(c-)$ is called the height of the jump, or simply the jump, at c. This being non-zero is the simplest way that a function f can be discontinuous, but it is by no means the only way.

It is possible for a function to be discontinuous at c because one or both of the one-sided limits $\lim_{x \to c-} f(x)$ and $\lim_{x \to c+} f(x)$ fail to exist. It is also possible that the one-sided limits are equal, so that there is no jump, but they are different from $f(c)$. Then f is discontinuous at c because, somehow, the "wrong value" is assigned to f at c. This is sometimes called a removable discontinuity; by changing the value at c the function can be made continuous there (Fig. 4.2).

4.2.7 The Limits $\lim_{x \to \infty} f(x)$ and $\lim_{x \to -\infty} f(x)$

Let us suppose that the domain of f includes an interval of the form $]a, \infty[$.

Definition We say that the limit $\lim_{x \to \infty} f(x)$ exists and equals t if the following condition is satisfied:

For each $\varepsilon > 0$ there exists K, such that $|f(x) - t| < \varepsilon$ for all x that satisfy the inequality $x > K$.

In a similar way, $\lim_{x \to -\infty} f(x) = t$ means that for each $\varepsilon > 0$, there exists K, such that $|f(x) - t| < \varepsilon$ for all x that satisfy $x < K$ (of course we assume that the domain of f includes an interval of the form $]-\infty, a[$).

The limit rules of Proposition 4.3 and the squeeze rule all hold with obvious modifications for limits of the kind $\lim_{x \to \infty} f(x)$ and $\lim_{x \to -\infty} f(x)$. The reader should write out the proofs.

Geometrically, saying that $\lim_{x \to \infty} f(x) = t$, or $\lim_{x \to -\infty} f(x) = t$, means that the line $y = t$ is a horizontal asymptote to the curve $y = f(x)$.

4.2.8 The Limits $\pm\infty$

Let $f : A \to \mathbb{R}$ and let c be a limit point of A. Most often A is an interval and c a point in A or an endpoint of A (or both).

Definition We say that f tends to ∞, or has the limit ∞ as x tends to c, and we write $\lim_{x \to c} f(x) = \infty$, when the following condition is satisfied:

For each K there exists $\delta > 0$, such that $f(x) > K$ for all $x \in A$ that satisfy $0 < |x - c| < \delta$.

To define $\lim_{x \to c} f(x) = -\infty$ we require $f(x) < K$ instead of $f(x) > K$.

Geometrically, saying that $\lim_{x \to c} f(x) = \infty$, or $\lim_{x \to c} f(x) = -\infty$, means that the line $x = c$ is a vertical asymptote to the curve $y = f(x)$, although in practice we usually speak of an asymptote when the limit is one-sided, for example, when $\lim_{x \to c-} f(x) = \infty$ (Fig. 4.3).

The reader should supply definitions for the notion $\lim_{x \to \infty} f(x) = \infty$, and three other similar ones obtained by inserting minus signs.

Although one says "tends to infinity" one never says "converges to infinity", the verb "converge" or the adjective "convergent" always implying a finite limit. Some say "diverges to infinity" in the cases $\lim_{n \to \infty} a_n = \infty$ or $\lim_{x \to c} f(x) = \infty$. The elements ∞ and $-\infty$ are not numbers, but they can be limits. So one has to be careful about saying "The limit $\lim_{x \to c} f(x)$ exists", always adding "and is a finite number." if that is necessary for clarity.

Fig. 4.3 Asymptotes as infinite limits

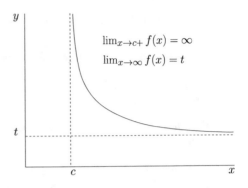

$$\lim_{x \to c+} f(x) = \infty$$
$$\lim_{x \to \infty} f(x) = t$$

4.2.9   Exercises (cont'd)

7. Find the following limits or explain why they do not exist:

   (a) $\lim\limits_{x \to 1} x^3 + x^2 + x + 1$

   (b) $\lim\limits_{x \to 0} \dfrac{1}{x}$

   (c) $\lim\limits_{x \to 0} \dfrac{1}{|x|}$

   (d) $\lim\limits_{x \to 1} \dfrac{x^3 - 1}{x - 1}$

   (e) $\lim\limits_{x \to 1} \dfrac{\sqrt{x} - 1}{x - 1}$

   (f) $\lim\limits_{x \to \infty} \sqrt{x}$

   (g) $\lim\limits_{x \to \infty} \sqrt{x + 1} - \sqrt{x}$

   (h) $\lim\limits_{x \to \infty} \sqrt{x(x + 1)} - x$

   (i) $\lim\limits_{x \to \infty} \sqrt[3]{x^2(x + 1)} - x$

   (j) $\lim\limits_{x \to 1} \min(x, 2 - x)$.

8. Prove that $\lim_{x \to \infty} x^3 - x^2 + x - 1 = \infty$ by showing how to find M, given K, such that $x^3 - x^2 + x - 1 > K$ for all $x > M$. It does not have to be the best M.

9. Let f be a function whose domain includes an interval of the form $]a, \infty[$. Show that $\lim_{x \to \infty} f(x) = \lim_{t \to 0+} f(1/t)$, in the sense that if either limit exists then so does the other and they are then equal; the limits $\pm\infty$ are allowed. This result, although simple, is used so much that it is worth pointing it out.

10. Another often used device is the following. Let f be a function defined on an interval A that contains 0, except possibly at 0 itself. Let $\lambda \neq 0$. Show that $\lim_{x \to 0} f(x) = \lim_{x \to 0} f(\lambda x)$. Again we imply that if one limit exists then so does the other.

Note. This is really an example of a composite function, where $f(x)$ is composed with the function λx, and is capable of countless variations.

11. Many readers may have been introduced at school to an important example of a limit, including a proof (of a sort) using the squeeze rule. We refer to the limit

$$\lim_{x \to 0} \frac{\sin x}{x} = 1.$$

It is essential here that the angle is measured by the arc of the unit circle that it sweeps out (that is, the angle is given in radians). The limit implies that x is an approximation to $\sin x$ when x is small. It is surprisingly good; for example $\sin 0.1 = 0.0998$ (0.1 rad, about 6°, is really not so small). The explanation for this unexpected accuracy is to be found in the power series expansion of $\sin x$, studied in Chap. 11.

The proof of this limit offered in school mathematics is in two steps. In the first the inequalities

$$\sin x < x < \frac{\sin x}{\cos x}$$

are established for $0 < x < \pi/2$, using a geometrical argument involving comparing the areas of three plane figures.

Complete the proof of the limit, assuming these inequalities and the continuity of $\cos x$.

4.2.10 Bounded Functions

Let $f : A \to \mathbb{R}$, where A is an arbitrary subset of $\mathbb{R}$. We define the following notions of boundedness for a function, paralleling those for sets and for sequences:

(a) The function f is said to be *bounded above*, if there exists K, such that $f(x) < K$ for all $x \in A$.
(b) The function f is said to be *bounded below*, if there exists K, such that $f(x) > K$ for all $x \in A$.
(c) The function f is said to be *bounded*, if it is both bounded above and bounded below. It is equivalent to saying that there exists $K > 0$, such that $|f(x)| < K$ for all $x \in A$.

If f is bounded above we set

$$\sup f := \sup\{f(x) : x \in A\}.$$

If f is bounded below we set

$$\inf f := \inf\{f(x) : x \in A\}.$$

More generally, if $B \subset A$ we denote the supremum (or infimum, with obvious changes) of f in B by

$$\sup_{B} f, \quad \text{or} \quad \sup_{x \in B} f(x), \quad \text{or} \quad \sup_{a \le x \le b} f(x);$$

the last if B is the interval $[a, b]$. There are many variations on these notations, more or less self-explanatory.

If f is continuous at x_0 then there exists $h > 0$, such that f is bounded in the set $]x_0 - h, x_0 + h[\cap A$. Because of this we say that a continuous function is *locally bounded*.

Exercise Prove this claim.

4.2.11 Monotonic Functions

The notions of increasing and decreasing for functions parallel those for sequences.

Definition A function $f :]a, b[\to \mathbb{R}$ is said to be increasing if, for all s and t such that $a < s < t < b$, we have $f(s) \le f(t)$. It is said to be decreasing if, for all s and t such that $a < s < t < b$, we have $f(s) \ge f(t)$. A function that is either increasing or decreasing is said to be monotonic.

As for sequences, we shall speak of a strictly increasing, or strictly decreasing, function when the inequalities are strict (that is, when equality is ruled out).

The terms "monotonic " and "monotone" are completely equivalent. It is a matter of taste, or even ease of speech, which one uses.

Proposition 4.5 *Let $f :]a, b[\to \mathbb{R}$ be an increasing function that is bounded above or a decreasing function that is bounded below. Then $\lim_{x \to b} f(x)$ exists and is a finite number (not $\pm\infty$). The conclusions also holds if $b = \infty$.*

Proof Suppose that f is an increasing function bounded above. There exists $K > 0$, such that $f(x) < K$ for all $x \in]a, b[$. The set M of all values taken by the function f is bounded above. Let $t = \sup M$. We shall show that $\lim_{x \to b} f(x) = t$.

Let $\varepsilon > 0$. By the definition of supremum, there exists $x_1 \in]a, b[$ such that $t - \varepsilon < f(x_1)$. Since f is increasing and bounded above by t, we must have $t - \varepsilon < f(x) \le t$ for all x in the interval $]x_1, b[$. We conclude that $\lim_{x \to b} f(x) = t$. For example, we can take $\delta = b - x_1$.

If $b = \infty$ a similar argument works. The case when f is decreasing is similar, using infimum instead of supremum. $\square$

Obviously similar conclusions hold for the limit $\lim_{x \to a} f(x)$. Furthermore it should be clear that if f is increasing, but not necessarily bounded above, then the limit $\lim_{x \to b} f(x)$ exists if we allow ∞ as a limit; and similarly a decreasing function not necessarily bounded below approaches a limit if we allow $-\infty$. With this understanding we can say that a monotonic function approaches limits at both ends of its interval of definition.

4.2.12 Discontinuities of Monotonic Functions

Let $f :]a, b[\to \mathbb{R}$ be an increasing function and let $a < c < b$. By Proposition 4.5 the one-sided limits $\lim_{x \to c-} f(x)$ and $\lim_{x \to c+} f(x)$ exist, and

$$f(c-) = \lim_{x \to c-} f(x) \le f(c) \le \lim_{x \to c+} f(x) = f(c+).$$

If $f(c-) = f(c+)$ then f is continuous at c, but otherwise f is discontinuous at c and has an upward jump discontinuity. The difference $f(c+) - f(c-)$ is called the height of the jump. Similar conclusions hold for decreasing functions.

It appears that a monotonic function can only fail to be continuous by having a jump, upwards for an increasing function, downwards for a decreasing one. As a corollary we can conclude that a monotonic function $f : [a, b] \to \mathbb{R}$, that takes all values between $f(a)$ and $f(b)$, or, as we might say, has no gaps in its range, is continuous. A rigorous proof of this is illustrated in Fig. 4.4. The converse is also true, a monotonic, continuous function has no gaps in its range. This is a simple consequence of the intermediate value theorem that we consider in detail later.

Most functions in practical applications are monotonic, or are increasing and decreasing piece-wise, switching between increasing and decreasing on successive intervals. So the commonest discontinuities are jumps. But it is easy to give an example of a function that is discontinuous without having a jump; the reader may consult the exercises.

Fig. 4.4 A monotonic function with no jumps is continuous at x_0

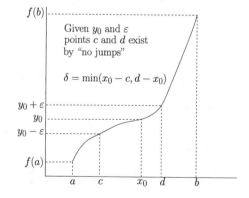

4.2.13 Continuity of $\sqrt{x}$

The function $f :]0, \infty[\to \mathbb{R}$, $f(x) = \sqrt{x}$ is defined in such a way that $\sqrt{x}$ is the unique positive real number y that satisfies $y^2 = x$.

Exercise Let $B = \{t \in \mathbb{R} : t > 0 \text{ and } t^2 < x\}$ and let $y = \sup B$. Show that y satisfies $y^2 = x$ and is the only positive solution.

It is easy to see that the function x^2 is increasing on the domain $]0, \infty[$ and the same is true for $\sqrt{x}$. We show next that $\sqrt{x}$ is continuous at the point c, where $c > 0$.

Let $\varepsilon > 0$ and reduce ε, if necessary, so that $\varepsilon < \sqrt{c}$. Because $\sqrt{x}$ is an increasing function we see that if

$$(\sqrt{c} - \varepsilon)^2 < x < (\sqrt{c} + \varepsilon)^2$$

then

$$\sqrt{c} - \varepsilon < \sqrt{x} < \sqrt{c} + \varepsilon.$$

We can therefore specify δ as follows:

$$\delta = \min \left((\sqrt{c} + \varepsilon)^2 - c, c - (\sqrt{c} - \varepsilon)^2 \right).$$

Actually the continuity of $\sqrt{x}$ can be viewed in the light of the fact that an increasing function without jumps is continuous, as was pointed out in the previous section. The function $\sqrt{x}$ has no jumps because every positive real number is the square root of another positive real number, namely that of its square.

4.2.14 Composite Functions

Let $f : A \to \mathbb{R}$ and $g : B \to \mathbb{R}$. If $f(A) \subset B$ we may compose the functions to obtain a function from A to $\mathbb{R}$:

$$g \circ f : A \to \mathbb{R}, \quad (g \circ f)(x) = g(f(x)), \quad x \in A.$$

Figure 4.5 illustrates a way to visualise the composition of functions, using the notion of a function as an assignment.

We have a new continuity rule.

Proposition 4.6 *If $x_0 \in A$, f is continuous at x_0 and g is continuous at $f(x_0)$, then $g \circ f$ is continuous at x_0.*

Proof Set $y_0 = f(x_0)$. Let $\varepsilon > 0$. Since g is continuous at y_0 there exists $\delta_1 > 0$, such that if $|y - y_0| < \delta_1$ and $y \in B$ then $|g(y) - g(y_0)| < \varepsilon$. But since f is continuous at x_0 there exists $\delta > 0$, such that if $|x - x_0| < \delta$ and $x \in A$ then $|f(x) - f(x_0)| < \delta_1$. This implies that $|g(f(x)) - g(f(x_0))| < \varepsilon$ for all $x \in A$ that satisfy $|x - x_0| < \delta$. $\square$

Fig. 4.5 A view of the composition $g \circ f$

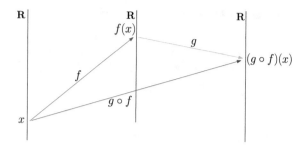

This continuity rule greatly increases our stock of continuous functions. As an example, $\sqrt{1 - x^2}$ is continuous on its domain $[-1, 1]$.

When taking a limit of a composite function, $\lim_{x \to a}(g \circ f)(x)$, given the two premises that $\lim_{x \to a} f(x) = b$ and $\lim_{y \to b} g(y) = t$, one has to be careful because for the second premise the value of $g(b)$ is immaterial. So in general the expected result, that the limit is t, can only be obtained on the assumption that there exists δ, such that f avoids the value b for all x that satisfy $0 < |x - a| < \delta$.

4.2.15 Limits of Functions and Limits of Sequences

We begin to explore the important role of sequences in considerations involving continuous functions.

Proposition 4.7 *Let $f : A \to \mathbb{R}$. Let c be a limit point of A and assume that $\lim_{x \to c} f(x) = t$. If $(x_n)_{n=1}^{\infty}$ is a sequence in $A \setminus \{c\}$ such that $\lim_{n \to \infty} x_n = c$, then $\lim_{n \to \infty} f(x_n) = t$.*

Proof Let $\varepsilon > 0$. There exists $\delta > 0$, such that $|f(x) - t| < \varepsilon$ for all x in A that satisfy $0 < |x - c| < \delta$. There exists a natural number N, such that $|x_n - c| < \delta$ for all $n \geq N$. But $x_n \neq c$, so that for all $n \geq N$ we also have $|f(x_n) - t| < \varepsilon$. □

It is interesting to ponder the question as to whether, given that c is a limit point of A, there must always exist a sequence $(x_n)_{n=1}^{\infty}$ in $A \setminus \{c\}$ such that $\lim_{n \to \infty} x_n = c$. In fact such a sequence always exists. It is a consequence of an axiom of set theory: the axiom of choice. This is a set-building axiom that is used to produce sequences in some cases when an assignment cannot be specified by any explicit procedure and requires an infinite number of *arbitrary choices*. In the cases which interest us, for example when A is an interval, or a finite union of intervals, it is not needed, as the existence of the sequence can be seen by an explicit procedure (though this may not be very obvious). For this reason we consider the axiom of choice to be beyond the scope of this text.

Proposition 4.7 has an important and much used consequence. We allow the reader to elucidate its proof.

Proposition 4.8 *Let f be continuous in its domain A, let $c \in A$ and let $(x_n)_{n=1}^{\infty}$ be a sequence in A such that $\lim_{n \to \infty} x_n = c$. Then $\lim_{n \to \infty} f(x_n) = f(c)$.*

Note that in the proposition we do not have to assume that x_n is never equal to c. Nor that c is a limit point of A, though if it is not, then the proposition has little practical value.

4.2.16   Iterations

Suppose that the function f is continuous in its domain A and that a sequence $(x_n)_{n=1}^{\infty}$ in A satisfies $x_{n+1} = f(x_n)$ for each n. Assume that the limit $\lim_{n \to \infty} x_n$ exists, is equal to t and that t lies in A. Then by Proposition 4.8 we have $f(t) = t$.

Sequences are often defined in this fashion, called iterating the function f, but it is really just a simple instance of inductive or recursive definition. Iterations are commonly used to solve the equation $f(x) = x$. For the method to succeed, two things are needed which can be tricky to check:

(a) The initial point x_1 must be chosen in such a way that each term of the sequence x_n is in A. We do not want x_n to land outside A for then we cannot continue the sequence with x_{n+1}. In short we must choose x_1 so that the whole sequence $(x_n)_{n=1}^{\infty}$ exists.

(b) The limit $\lim_{n \to \infty} x_n$ should exist and it should lie in A.

In the chapter on sequences we saw that a positive sequence can be defined recursively by

$$a_0 = 1, \quad a_{n+1} = \sqrt{2 + a_n},$$

and that a_n converges to a limit t. Now we see, thanks to the continuity of the function $\sqrt{x}$, that t must satisfy $t = \sqrt{2 + t}$, so that in fact $t = 2$, as expected.

Another example of such an iteration is

$$x_1 = 1, \quad x_{n+1} = 1 + \frac{1}{1 + x_n}.$$

The limit is $\sqrt{2}$, the unique positive root of $1 + 1/(1 + x) = x$.

Yet another, and quite important, example is furnished by *Newton approximations*. These are widely used to approximate solutions of equations that cannot be obtained in closed form. The scheme

$$x_1 = 1, \quad x_{n+1} = \frac{1}{2}\left(x_n + \frac{2}{x_n}\right),$$

converging to $\sqrt{2}$ (sometimes called the Babylonian method for calculating $\sqrt{2}$ and apparently known to the ancients) results from applying Newton's method to the equation $x^2 - 2 = 0$. Newton approximations will be studied in Chap. 5.

4.2.17 Exercises (cont'd)

12. Write out a nice proof of the claim made earlier in this section that a monotonic function with no gaps in its range is continuous.

13. Let f and g be continuous functions with the same domain A. Define the functions $\max(f, g)$ and $\min(f, g)$, with domain A by

$$\max(f, g)(x) = \max(f(x), g(x)), \quad \min(f, g)(x) = \min(f(x), g(x)).$$

    Show that $\max(f, g)$ and $\min(f, g)$ are continuous.

    Hint. You can use the definition of continuity, or else you can express these functions in terms of other functions and use limit-rules.

14. Let $f : \mathbb{R} \to \mathbb{R}$ be defined by $f(x) = \sin(1/x)$ if $x \neq 0$ and $f(0) = 0$. Show that f is discontinuous at $x = 0$, but the limits $\lim_{x \to 0+} \sin(1/x)$ and $\lim_{x \to 0-} \sin(1/x)$ do not exist.

    Note. The function $\sin x$ will be familiar to the reader from school trigonometry. It will be rigorously defined later but for now all we need to know is that $\sin x$ is continuous and periodic (with period 2π, but that is not important), its maximum is 1 and its minimum -1.

15. Let $f : \mathbb{R} \to \mathbb{R}$ be defined by $f(x) = 1$ if x is rational and by $f(x) = 0$ if x is irrational. Show that f is everywhere discontinuous.

16. What can you say about the continuity or otherwise of the function f, given by $f(x) = x$ for rational x and $f(x) = 0$ for irrational x?

17. Show that the function f defined by $f(x) = x \sin(1/x)$ for $x \neq 0$ and $f(0) = 0$ is everywhere continuous.

18. Let f be the function with domain $]0, 1[$ defined by the following prescription: if x is irrational then $f(x) = 0$; if x is rational then $f(x) = 1/b$, where b is the denominator of x when it is expressed as a fraction in lowest terms. Show that for all c in the domain we have $\lim_{x \to c} f(x) = 0$. Deduce from this that f is continuous at each irrational x, but discontinuous at each rational x.

19. For each real number x we denote by $[x]$ the highest integer n, such that $n \leq x$. Define the function $f : [0, \infty[\to \mathbb{R}$ by

$$f(0) = 0, \quad f(x) = \sum_{n=[1/x]}^{\infty} 2^{-n}, \quad (x > 0).$$

    (a) Show that f is increasing.

    (b) Show that f is continuous at all points $x \geq 0$ except at those of the form $x = 1/n$ for some $n \in \mathbb{N}_+$.

    (c) Calculate the jump of the function at $x = 1/n$.

    (d) Which of the following is true:
    - (i) $\lim_{x \to \frac{1}{n}-} f(x) = f\left(\frac{1}{n}\right)$?
    - (ii) $\lim_{x \to \frac{1}{n}+} f(x) = f\left(\frac{1}{n}\right)$?
    - (iii) Neither?

20. A function with domain all of $\mathbb{R}$ is called periodic if there exists a number $k \neq 0$, such that $f(x + k) = f(x)$ for all x. A non-zero number k with this property is called a period of f.

   (a) Let k_1 and k_2 be periods and let m and n be integers (not necessarily positive). Show that $mk_1 + nk_2$ is either 0 or else it is a period.
   (b) Suppose there exists a lowest positive period T. Show that the set of all periods is precisely the set of non-zero multiples nT as n ranges over the non-zero integers.
   (c) Suppose that f is periodic but there is no lowest positive period. Show that the set of periods has the following property: every open interval $]a, b[$, however small $b - a$ is, contains a period.
   (d) Suppose that f is periodic, continuous, and that there exists no lowest positive period. Show that f is constant.

   Note. A lowest positive period, if it exists, is called the fundamental period. Compare this exercise with Sect. 2.5 Exercise 3.

21. Let $a_n = \sin n$ for $n = 1, 2, \ldots$. Show that for every t in the interval $[-1, 1]$ there exists a subsequence $(a_{k_n})_{n=1}^\infty$ that converges to t.

   Hint. Use your knowledge of $\sin x$ from school mathematics. In particular it oscillates between -1 and 1 with period 2π. In addition you will need two facts of analysis, to be proved later: $\sin x$ is continuous; and π is irrational. See also Sect. 2.5 Exercise 3.

22. $(\diamondsuit)$ The notions of limit inferior and limit superior can be defined for functions. Suppose that f has domain A and c is a limit point of A. We define

$$\limsup_{x \to c} f(x) := \lim_{h \to 0+} \left(\sup_{x \in A,\, 0 < |x - c| < h} f(x) \right)$$

$$\liminf_{x \to c} f(x) := \lim_{h \to 0+} \left(\inf_{x \in A,\, 0 < |x - c| < h} f(x) \right).$$

   Draw up a list of properties analogous to those stated for sequences in Sect. 3.11. Look for opportunities to use them.

23. The notions of upper semi-continuity and lower semi-continuity were defined in Exercise 6.

   (a) Let $(f_k)_{k=1}^\infty$ be a sequence of functions and suppose that they are all upper semi-continuous at x_0. Define the function g by $g(x) = \inf_{1 \leq k < \infty} f_k(x)$. Show that g is upper semi-continuous at x_0. The same holds if "upper semi-continuous" is replaced by "lower semi-continuous" and "inf" by "sup".
   (b) Find an example of a sequence $(f_k)_{k=1}^\infty$ of continuous functions such that the function $g(x) := \inf_{1 \leq k < \infty} f_k(x)$ fails to be lower semi-continuous at at least one point.

4.3 Properties of Continuous Functions

In this section we prove a number of general propositions about continuous functions using the full weight of the completeness axiom C1. In all cases it is the existence of supremum and infimum of bounded non-empty sets that is exploited. These propositions flaunt the success of analysis and show that axiom C1 and the definition of continuity, which took so long to evolve to their present forms, are correct.

It might be a good idea for the reader to reprise the contents of Sect. 2.4, particularly the paragraphs under the heading "Using supremum and infimum to prove theorems", before reading on.

4.3.1 The Intermediate Value Theorem

If a continuous function is defined on an interval and among its values are the real numbers y_1 and y_2, then among its values is *every real number between y_1 and y_2*. Its *range* (the set of all its values) includes the interval with endpoints y_1 and y_2.

Proposition 4.9 *Let $f : [a, b] \to \mathbb{R}$ be continuous and suppose that η is a real number between $f(a)$ and $f(b)$; in other words we suppose that $f(a) < \eta < f(b)$ or $f(b) < \eta < f(a)$. Then there exists $t \in {]a, b[}$ such that $f(t) = \eta$.*

Proof We preface the proof by pointing out how continuity is used in it. Let $a < t < b$ and assume that $f(t) < \eta$. We can deduce from this that there exists $\delta > 0$, such that $f(x) < \eta$ for all x in the interval ${]t - \delta, t + \delta[}$. For there exists $\varepsilon > 0$ such that $f(t) + \varepsilon < \eta$. By the continuity of f there exists $\delta > 0$, such that

$$f(t) - \varepsilon < f(x) < f(t) + \varepsilon$$

for all x in the interval ${]t - \delta, t + \delta[}$. In particular for such x we have $f(x) < \eta$. In a similar way, if $f(t) > \eta$ there exists $\delta > 0$, such that $f(x) > \eta$ for all x in the interval ${]t - \delta, t + \delta[}$.

Similar considerations are valid for the endpoints. For example if $f(a) < \eta$ then there exists $\delta > 0$, such that $f(x) < \eta$ for all x in the interval $[a, a + \delta[$.

Let us prove the proposition on the assumption that $f(a) < \eta < f(b)$. Let A be the set of all x in $[a, b]$, such that $f(s) < \eta$ for all s in $[a, x]$. The set A is bounded (it is a subset of $[a, b]$) and is not empty (it contains a). Let $t = \sup A$ (the supremum exists by Proposition 2.3). We shall show that $f(t) = \eta$.

We show first that t lies in the open interval ${]a, b[}$. By the considerations of the first and second paragraphs, since f is continuous and $f(a) < \eta$ there exists $\delta_1 > 0$, such that $f(x) < \eta$ for all x in $[a, a + \delta_1]$. Hence $t \geq a + \delta_1$. Since f is continuous and $f(b) > \eta$ there exists $\delta_2 > 0$, such that $f(b - \delta_2) > \eta$. We deduce that $t \leq b - \delta_2$. For if t was strictly above $b - \delta_2$, there would be an element of A above $b - \delta_2$ and we would have $f(b - \delta_2) < \eta$, contrary to the definition of δ_2. From these arguments we conclude that $a + \delta_1 \le t \le b - \delta_2$, that is, t is in the open interval $]a, b[$.

We shall eliminate the possibilities $f(t) < \eta$ and $f(t) > \eta$, by showing that each leads to a contradiction. This will prove that $f(t) = \eta$, as required.

Assume first that $f(t) < \eta$. Then there exists $\delta > 0$, such that $f(x) < \eta$ for all x in the interval $]t - \delta, t + \delta[$. But there exists $s \in A$ such that $t - \delta < s < t$ (because $t = \sup A$) and therefore $f(x) < \eta$ for all x in $[a, s]$. We deduce that $f(x) < \eta$ for all x in $[a, t + \delta[$ so that A contains a number strictly higher than t. That is impossible since t is an upper bound of A.

Assume next that $f(t) > \eta$. Then there exists $\delta > 0$, such that $f(t - \delta) > \eta$. This implies that $t - \delta$ is an upper bound of A, for otherwise A would have an element strictly above $t - \delta$ and we would have $f(t - \delta) < \eta$, contrary to the definition of δ. But it is impossible that $t - \delta$ is an upper bound of A, since t is its lowest upper bound.

Both assumptions, that $f(t) < \eta$ and that $f(t) > \eta$, have led to contradictions. Finally we conclude that $f(t) = \eta$.

The case $f(a) > \eta > f(b)$ is handled in a similar fashion; or else the former conclusion can be applied to the function $-f$.   □

4.3.2 Thoughts About the Proof of the Intermediate Value Theorem

We cannot prove the intermediate value theorem without using axiom C1 (or something equivalent to it like the existence of the least upper bound). The theorem is not valid in $\mathbb{Q}$. The equation $x^2 = 2$ has no solution in $\mathbb{Q}$ although $1^2 < 2 < 2^2$.

The set A used in the proof can be replaced by the set B of all x in $[a, b]$, such that $f(x) \le \eta$. Then $t := \sup B$ is a solution of $f(x) = \eta$, by an argument very similar to that used to prove Proposition 4.9. The difference is that this solution is the highest one in the interval whilst the solution given in the proof is the lowest. Of course they can be the same solution.

Another proof can be given using what is called the method of bisection. This is based on the fact that an increasing sequence that is bounded above is convergent, itself a consequence of axiom C1. We shall meet this method in the next proposition.

The property of continuous functions encapsulated in the intermediate value theorem was long considered a possible defining feature of continuity. We shall say that a function defined on *an interval A* has the *intermediate value property* if it satisfies the following condition:

For all a and b in A such that a < b, if η is such that $f(a) < \eta < f(b)$ or $f(a) > \eta > f(b)$, then there exists x, such that $a < x < b$ and $f(x) = \eta$.

The intermediate value property seems to assert that the graph of the function is in one piece; it can be drawn without lifting the pencil from the paper. However, we have to give up any idea of using this property as an alternative definition of continuity, despite its appealing character. It turns out that there are discontinuous functions that have the property. See the exercises in this section.

Finally, do not confuse the intermediate value theorem with the mean value theorem of calculus.

4.3.3 The Importance of the Intermediate Value Theorem

The intermediate value theorem is very powerful. It gives convenient conditions for concluding that an equation $f(x) = \eta$ has a solution. It does not say that the solution is unique, and indeed, multiple solutions can exist. However, additional arguments can imply uniqueness.

Solving an equation is a common problem of applied mathematics, perhaps even the commonest, and so a theoretical proof that a solution exists is useful. Frequently one has to resort to an approximation method that in a sequence of steps produces more and more correct decimal digits of a solution. It is important to know that there is a number there that is being approximated, and it is this that the theorem guarantees.

As an example of the theorem in action, we can let n be a positive integer greater than or equal to 2, and deduce that every positive real number has a unique positive nth root. For let $f : [0, \infty[\to \mathbb{R}$ be the function $f(x) = x^n$, and let $c > 0$. A solution of $x^n = c$ exists in the interval $]0, \infty[$ because firstly, f is continuous; secondly, $f(0) = 0$; and thirdly, as $\lim_{x \to \infty} x^n = \infty$ there must exist b such that $f(b) > c$. The solution is unique because f is an increasing function, more precisely, if $0 \le x_1 < x_2$ then $x_1^n < x_2^n$.

This defines at a stroke the function $\sqrt[n]{y}$ that assigns to each positive y its unique positive nth root.

4.3.4 The Boundedness Theorem

Continuous functions defined on bounded and closed intervals are bounded. This is the content of the boundedness theorem.

Proposition 4.10 *Let $f : [a, b] \to \mathbb{R}$ be continuous. Then f is a bounded function.*

Proof We assume that f is unbounded on $[a, b]$ and derive a contradiction. The argument, based on the method of bisection, though intuitive, is rather long; so some patience may be required. For this reason we first give a rough description of the proof.

If we divide the interval $[a, b]$ into two equal parts the function f must be unbounded on at least one of them. We can therefore choose one of the two intervals, such that f unbounded on it. The length of the new interval is half that of the original interval. Now we repeat this procedure, divide the new interval into two equal parts and choose one part, such that f is unbounded on it. Repeating this procedure we obtain a sequence of intervals, with length that tends to 0, and such that f is unbounded on each. The intervals shrink down to a single point t. We obtain a contradiction because f is continuous at t and is therefore bounded on an interval of the form $]t - \delta, t + \delta[$.

In this rough description it may not be clear how the completeness axiom is needed; so we proceed to describe the proof more precisely. Recall that f is supposed to be unbounded on $[a, b]$ and from this we wish to obtain a contradiction.

Set $a_1 = a$ and $b_1 = b$. Let c_1 be the midpoint $\frac{1}{2}(a_1 + b_1)$. Then f is unbounded either on $[a_1, c_1]$ or on $[c_1, b_1]$ (or on both; that is not excluded here). Let $a_2 = a_1$, $b_2 = c_1$ if f is unbounded on $[a_1, c_1]$, and let $a_2 = c_1$, $b_2 = b_1$ if f is bounded on $[a_1, c_1]$ (because then f is unbounded on $[c_1, b_1]$). In both cases $a_1 \leq a_2 < b_2 \leq b_1$, the function f is unbounded on the new interval $[a_2, b_2]$, and the length of $[a_2, b_2]$ is half the length of $[a_1, b_1]$. We repeat this step and obtain an interval $[a_3, b_3]$ which is either $[a_2, c_2]$ or $[c_2, b_2]$, where $c_2 = \frac{1}{2}(a_2 + b_2)$, and f is unbounded on $[a_3, b_3]$. Furthermore $a_1 \leq a_2 \leq a_3 < b_3 \leq b_2 \leq b_1$.

From this procedure there result two sequences in the interval $[a, b]$, an increasing sequence $(a_n)_{n=1}^{\infty}$, and a decreasing sequence $(b_n)_{n=1}^{\infty}$. Moreover $a_n < b_n$, and because $b_{n+1} - a_{n+1} = \frac{1}{2}(b_n - a_n)$ we have $b_n - a_n \to 0$. Finally f is unbounded on $[a_n, b_n]$ for each n. By Proposition 3.3 concerning bounded, monotonic sequences, the proof of which depended on the completeness axiom, both sequences are convergent, and since $b_n - a_n \to 0$ they have the same limit t, which lies in the interval $[a, b]$ (the reader should check the last claim; see, for example Sect. 3.2 Exercise 9).

Now f is continuous at t. So there exists $\delta > 0$, such that $|f(x) - f(t)| < 1$ for all x in $[a, b]$ that satisfy $|x - t| < \delta$. For such x we have $|f(x)| < |f(t)| + 1$, so that f is bounded on the set $[a, b] \cap \,]t - \delta, t + \delta[$. But we know that a_n and b_n both converge to t. Hence there exists N, such that a_n and b_n both lie in the interval $]t - \delta, t + \delta[$ for $n \geq N$; and this is the same as saying that the interval $]t - \delta, t + \delta[$ includes the interval $[a_n, b_n]$ for all $n \geq N$. But then f is also bounded on $[a_n, b_n]$. This is a contradiction since we chose a_n and b_n so that f was unbounded on $[a_n, b_n]$.   □

It is essential for the general validity of the boundedness theorem that the domain is a bounded and closed interval. The boundedness theorem does not hold on intervals of other kinds.

Exercise For each type of interval A, except the bounded closed interval, and the empty interval, find an example of a unbounded, continuous function f with domain A.

4.3.5 Thoughts About the Proof of the Boundedness Theorem

Continuity was not used in its full strength in the proof of the boundedness theorem; it only mattered that f was *locally bounded* (which is a consequence of continuity). Local boundedness is the property that for each x_0, there exists $\delta > 0$, such that f is bounded in the set $]x_0 - \delta, x_0 + \delta[\cap [a, b]$.

It is also possible to prove the boundedness theorem using a method similar to that used for the intermediate value theorem. Let A be the set of all x in $[a, b]$, such that f is bounded on $[a, x]$. Then setting $t = \sup A$, one proceeds to show that $t = b$ and $t \in A$.

Exercise Write out the details of the proof of the boundedness theorem suggested in the previous paragraph.

By the same token it is possible to prove the intermediate value theorem by the method of bisection. Just divide the interval into two equal parts and choose the one for which $f(x) - \eta$ has a different sign at the endpoints and continue. This is a practical method for approximating a solution. If the interval is $[0, 1]$ we can divide into 10 equal parts, and repeating the process obtain a decimal expansion of one of the solutions.

Exercise Write out the details of the proof of the intermediate value theorem suggested in the previous paragraph.

A rather short proof of the boundedness theorem can be based on Proposition 3.10, the Bolzano–Weierstrass theorem, which states that a bounded sequence has a convergent subsequence. Here is a sketch of it, omitting some subtle set-theoretic details. Suppose f is unbounded. Then there must exist a sequence $(x_n)_{n=1}^\infty$ in $[a, b]$ such that the sequence $(f(x_n))_{n=1}^\infty$ is unbounded. By Bolzano–Weierstrass there is a convergent subsequence $(x_{k_n})_{n=1}^\infty$ of $(x_n)_{n=1}^\infty$, say with limit t. Next it is shown that t is in the interval $[a, b]$ (important here that the interval is closed; see Sect. 3.2 Exercise 9), so that $f(x_{k_n}) \to f(t)$ by continuity, whilst at the same time $f(x_{k_n})$ is unbounded, which is a contradiction.

The sketched proof just given is not just an academic curiosity. The Bolzano–Weierstrass theorem is capable of great generalisation, into the area of multivariate calculus, and even beyond, into the realm of infinite-dimensional spaces. It means that versions of the boundedness theorem, and the extreme value theorem of the next section, emerge repeatedly in advanced work.

4.3.6 The Extreme Value Theorem

A continuous function defined on a bounded and closed interval is not just bounded. It attains a maximum and a minimum. This is the extreme value theorem.

Proposition 4.11 *Let $f : [a, b] \to \mathbb{R}$ be continuous. Then there exist c_1 in $[a, b]$, such that $f(x) \le f(c_1)$ for all x in $[a, b]$; and c_2 in $[a, b]$, such that $f(x) \ge f(c_2)$ for all x in $[a, b]$.*

Proof We know that f is bounded in $[a, b]$. Let $M = \sup_{a \le x \le b} f(x)$ (that is, M is the supremum of the set of all values of f). Assume that f does not attain a maximum. Then $f(x) < M$ for all x in $[a, b]$. The function $g(x) := 1/(M - f(x))$ is then continuous in $[a, b]$ (since the denominator is nowhere 0). However g cannot be bounded; however large we choose K there exists $x \in [a, b]$ such that $f(x) > M - 1/K$ (because M is the supremum of f) and then $g(x) > K$. This contradicts the boundedness theorem because g, being continuous in $[a, b]$, must be bounded. We conclude that there exists x in $[a, b]$ such that $f(x) = M$; that is, f attains a maximum value in $[a, b]$.

Similar arguments show that f attains a minimum value.   □

We sketch a second proof of the extreme value theorem based on Bolzano–Weierstrass, ignoring again some subtle set theoretical details. We know that f is bounded; so let $M = \sup f$. For each positive integer n there exists x_n in $[a, b]$, such that $f(x_n) > M - 1/n$. The sequence $(x_n)_{n=1}^{\infty}$ has a convergent subsequence, say, $(x_{k_n})_{n=1}^{\infty}$. Let its limit be t. Then $t \in [a, b]$ and by continuity of f we have $f(t) = \lim_{n \to \infty} f(x_{k_n}) \ge M$. We conclude that $f(t) = M$. A similar argument shows that the infimum of f is also attained.

The argument of the last paragraph can even be modified to prove the boundedness theorem and the extreme value theorem simultaneously. We let $M = \sup f$ and rewrite the last paragraph to allow the a priori possibility of $M = \infty$. The conclusion, that $f(t) \ge M$, shows at once that M is finite and is attained. All of this is capable of much generalisation.

4.3.7 Using the Extreme Value Theorem

Seeking the maximum or minimum of a function is a common problem of applied mathematics. Just as the intermediate value theorem can justify that what is sought, a solution of an equation, actually exists, so also the extreme value theorem can guarantee that what is sought, a maximum value or a minimum value, actually exists.

The limitation of the extreme value theorem to a bounded closed interval can sometimes be overcome. It is often possible to gain some knowledge of the maximum or minimum of a continuous function $f(x)$ on an unbounded interval, if we can control the function as $x \to \pm\infty$.

As an example we suppose that f is continuous on all of $\mathbb{R}$, and that $\lim_{x \to \infty} f(x) = \lim_{x \to -\infty} f(x) = 0$. If now f takes a positive value somewhere, then it must attain a maximum. For suppose that $f(a) > 0$. We can find K, such that for $|x| > K$ we have $|f(x)| < f(a)$. But then the maximum value of f on the interval $[-K, K]$, which is attained by the extreme value theorem, is the maximum value of f on all of $\mathbb{R}$. This argument is easily adapted to different cases.

4.3.8 Exercises

1. Let $f(x) = x^n + a_{n-1}x^{n-1} + \cdots + a_1 x + a_0$ be a polynomial function with leading coefficient 1, and with *odd degree n*.

   (a) Show that $\lim_{x \to \infty} f(x) = \infty$ and $\lim_{x \to -\infty} f(x) = -\infty$.
   (b) Show that the equation $f(x) = y$ has at least one real root for every y.

2. Let $f(x) = x^n + a_{n-1}x^{n-1} + \cdots + a_1 x + a_0$ be a polynomial function with leading coefficient 1, and with *even degree n*.

   (a) Show that $\lim_{x \to \infty} f(x) = \lim_{x \to -\infty} f(x) = \infty$.
   (b) Show that the function $f(x)$ attains a minimum value m at some point.
   (c) Show that for $y > m$ the equation $f(x) = y$ has at least two solutions in $\mathbb{R}$, whereas for $y < m$ it has no solution.

3. Prove the following fact, used several times in this section: if A is a closed interval, then no sequence in A can converge to a point outside A.
   Note. The interval does not have to be bounded. This is really the same as Sect. 3.2 Exercise 9.

4. Show that the equation $x^5 - x^2 + 1 = 0$ has a root in the interval $-1 < x < 0$.

5. Let $f : [a, b] \to [a, b]$ be continuous. Show that there exists x in $[a, b]$ such that $f(x) = x$.

6. Let f be a continuous function defined in an interval A (or, more generally, f is a function that satisfies the intermediate value property). Let $t_1, t_2, ..., t_m$ be points in A, and let $c_1, c_2, ..., c_m$ be positive numbers. Set

   $$ w := \frac{\sum_{j=1}^{m} c_j f(t_j)}{\sum_{j=1}^{m} c_j}. $$

   Show that there exists ξ in A such that $f(\xi) = w$.

7. Consider the function f with domain $\mathbb{R}$ given by

   $$ f(x) = \sin\left(\frac{1}{x}\right), \quad (x > 0), \qquad f(x) = 0, \quad (x \le 0). $$

   Show that f is discontinuous but has the intermediate value property.
   Hint. You will need to know that $\sin x$ is continuous, periodic (the period is 2π but that is not needed) and that $\sin x$ oscillates between its maximum 1 and its minimum -1. These facts will be established properly in a later chapter.

8. Let A be an interval and let $f : A \to \mathbb{R}$ be continuous.

   (a) Show that the range of f (the set of all its values) is an interval.
       Hint. See Sect. 2.4 Exercise 6.
   (b) Suppose that A is closed and bounded. Show that the range of f is a closed and bounded interval.

9. Let f be continuous and defined in an interval A. A line segment joining two points $(a, f(a))$ and $(b, f(b))$ in the graph $y = f(x)$, where a and b are distinct points in A, is called a chord. As is usual in analytic geometry the slope of the chord is the number

$$p := \frac{f(a) - f(b)}{a - b}.$$

Let B be the set of all numbers p, such that there exists a chord with slope p. Show that B is an interval.

Hint. Given two chords, the first whose endpoints have x-coordinates a_0 and b_0, and the second whose endpoints have x-coordinates a_1 and b_1, consider the variable chord whose endpoints have x-coordinates a_t and b_t, where

$$a_t = (1 - t)a_0 + ta_1, \quad b_t = (1 - t)b_0 + tb_1, \quad 0 \le t \le 1.$$

10. Let A be an interval and $f : A \to \mathbb{R}$. Show that f has the intermediate value property if and only if, for every interval B such that $B \subset A$, the set $f(B)$ (the set of all y such that $y = f(x)$ for some $x \in B$) is an interval. In short, f has the intermediate value property if and only if it maps intervals to intervals.

11. ($\Diamond$) The notions of upper and lower semi-continuity were defined in Sect. 4.2 Exercise 6. Let $f : [a, b] \to \mathbb{R}$ be upper semi-continuous (that is, it is upper semi-continuous at all points of its domain).

   (a) Let $a \le c \le b$ and let $(x_n)_{n=1}^{\infty}$ be a sequence in $[a, b]$ such that $x_n \to c$. Show that

   $$\limsup_{n \to \infty} f(x_n) \le f(c).$$

   Note that the limit superior may have infinite absolute value.

   (b) Show that f attains a maximum value in $[a, b]$.
   Hint. Let $M = \sup_{[a,b]} f$ (allowing ∞ as a possible value). Let $(x_n)_{n=1}^{\infty}$ be a sequence in $[a, b]$, such that $f(x_n) \to M$. Revisit the last paragraph of the section "The extreme value theorem".

   (c) Obtain similar results in the case that f is lower semi-continuous, replacing limit superior with limit inferior, reversing the inequality sign, and concluding that f attains a minimum.

4.4 Inverses of Monotonic Functions

Let us run over some important concepts about functions. They all contribute to understanding the set of solutions of an equation $f(x) = y$.

Let $f : A \to B$ where A and B are quite arbitrary sets. One may think of f purely as an assignment of an element in B to each element in A. The concepts we shall define are set-theoretical in nature, although we shall apply them in this text almost entirely to functions of a real variable.

(a) The function f is said to be *injective* if it maps distinct elements in A to distinct elements in B. This is equivalent to saying that if x_1 and x_2 are elements of A, and $f(x_1) = f(x_2)$, then $x_1 = x_2$. Equivalently, if y is an element of B and the equation $f(x) = y$ has a solution, then the solution is unique.

(b) The function f is said to be *surjective*, if the equation $f(x) = y$ has a solution for all y in B. We also say that f maps A *on to* B, since every element of B appears as a value $f(x)$ for some x in A.

(c) The function f is said to be *bijective* if it is both injective and surjective. This is equivalent to saying that the equation $f(x) = y$ has a unique solution for every y in B.

(d) Even if f is not surjective we can make it so by restricting the codomain to those elements y such that the equation $f(x) = y$ has a solution. These form a subset of B called the *range* of f (already mentioned in Sect. 4.3). A function always maps its domain on to its range.

Given that the function $f : A \rightarrow B$ is bijective we can define the inverse function $f^{-1} : B \rightarrow A$, also bijective. For each $y \in B$ we simply let $f^{-1}(y)$ be the unique x in A such that $f(x) = y$. The important cases in this text are when A and B are sets of real numbers. The key observation is that a strictly monotonic function is injective.

Proposition 4.12 *Let $f :]a, b[\rightarrow \mathbb{R}$ be continuous and strictly increasing, allowing here the possibilities $a = -\infty$ and $b = \infty$. Let*

$$c = \lim_{x \to a+} f(x) \quad and \quad d = \lim_{x \to b-} f(x)$$

(again allowing $c = -\infty$ or $d = \infty$ so that the limits always exist). Then the range of f is the interval $]c, d[$ and f is a bijective function from $]a, b[$ to $]c, d[$. The inverse function

$$f^{-1} :]c, d[\rightarrow]a, b[$$

is strictly increasing and continuous.

Proof Let $y \in]c, d[$. Then the equation $f(x) = y$ has a solution by the intermediate value theorem. To see this let y_1 and y_2 satisfy

$$c < y_1 < y < y_2 < d.$$

We have

$$c = \lim_{x \to a+} f(x) \quad and \quad d = \lim_{x \to b-} f(x).$$

Therefore we can find x_1 and x_2, such that

$$a < x_1 < x_2 < b, \quad f(x_1) < y_1, \quad f(x_2) > y_2.$$

So the equation $f(x) = y$ has a solution between x_1 and x_2. The solution is unique because f is strictly increasing. Note how the phrasing of this paragraph works just as well for infinite values of a, b, c or d, as for finite ones.

The inverse function

$$f^{-1} :]c, d[\to]a, b[$$

is therefore defined and is bijective. It is easy to see that f^{-1} is strictly increasing. The proof that it is continuous is really the same as the one that we used to prove that $\sqrt{x}$ is continuous. Let $c < y_0 < d$ and $x_0 = f^{-1}(y_0)$. Then $y_0 = f(x_0)$. We shall prove that f^{-1} is continuous at y_0.

Let $\varepsilon > 0$. Let $y_1 = f(x_0 - \varepsilon)$ and $y_2 = f(x_0 + \varepsilon)$ (reduce ε if necessary so that $x_0 - \varepsilon$ and $x_0 + \varepsilon$ fall within the interval $]a, b[$). Now we set

$$\delta = \min(y_2 - y_0, y_0 - y_1).$$

Because of monotonicity, if

$$y_0 - \delta < y < y_0 + \delta$$

then

$$x_0 - \varepsilon < f^{-1}(y) < x_0 + \varepsilon. \qquad \square$$

Another proof, perhaps simpler, that f^{-1} is continuous, is based on the observation that f^{-1} is monotonic and its range, the interval $]a, b[$, is without gaps. Therefore f^{-1} has no jumps, and must be continuous as the discontinuities of a monotonic function consist only of jumps.

Proposition 4.12 provides us with the most commonly used tool to produce inverse functions in the calculus of functions of one variable. Here are some examples:

(a) We consider the function $f(x) = x^n$ on the domain $]0, \infty[$ (given that n is a fixed positive integer). Then f is continuous, strictly increasing and, as is easily seen, maps $]0, \infty[$ on to $]0, \infty[$. The inverse function, $f^{-1}(y) = y^{1/n}$, is continuous and maps $]0, \infty[$ on to itself. It is also written as $\sqrt[n]{y}$. This short paragraph could replace all our previous lengthy deliberations about the nth root function.

(b) The function $f(x) = \sin x$ is not monotonic but we can restrict it to an interval where it is strictly increasing, for example $]-\pi/2, \pi/2[$. It maps this interval on to the interval $]-1, 1[$; so we get an inverse function $\arcsin y$, that maps the interval $]-1, 1[$ on to the interval $]-\pi/2, \pi/2[$, and is also increasing.

(c) A result similar to Proposition 4.12 holds for strictly decreasing functions; the reader is invited to formulate it. It can be used to produce an inverse function $\arccos y$ for $\cos x$, using the interval $]0, \pi[$ on which cosine is decreasing.

4.4.1 Exercises

1. Let α be a rational number. Prove that the function

$$f(x) = x^\alpha, \quad (0 < x < \infty)$$

   is continuous.

2. Let A and B be sets and let $f : A \to B$ be bijective. Show that

$$f \circ f^{-1} = id_B, \quad f^{-1} \circ f = id_A$$

   where $id_A : A \to A$ and $id_B : B \to B$ are the *identity functions*: $id_A(x) = x$ and $id_B(y) = y$.

3. For each of the following functions, describe inverse functions, restricting the domain if necessary:

   (a) $f(x) = x + \dfrac{1}{x}, \quad (x > 0)$

   (b) $f(x) = x^3 + x, \quad (x \in \mathbb{R})$.
      You may not be able to give a formula here, but describe the domain and range of an inverse.

   (c) $f(x) = x^4 + x^2 + 1, \quad (x \in \mathbb{R})$.

4. Let A be an interval, let f be a function with domain A and suppose that f is continuous and injective. Prove that f is monotonic.
   Hint. Use the intermediate value theorem. It may be simpler to consider first the case $A = [a, b]$.

4.5 Two Important Technical Propositions

The contents of this section, Cauchy's principle for the limit of a function, and the small oscillation theorem, may be read later when they are required. We also meet the notion of uniform continuity.

Proposition 4.13 (Cauchy's principle) *Let $f : A \to \mathbb{R}$ and let c be a limit point of A. Then the limit $\lim_{x \to c} f(x)$ exists and is a finite number if and only if the following condition (Cauchy's condition) holds: for each $\varepsilon > 0$ there exists $\delta > 0$, such that $|f(x_1) - f(x_2)| < \varepsilon$ for all x_1 and x_2 in A that satisfy $0 < |x_1 - c| < \delta$ and $0 < |x_2 - c| < \delta$.*

Proof That Cauchy's condition is necessary for the existence of a finite limit follows by virtually the same argument as was used to prove Cauchy's principle for sequences, Proposition 3.12.

Let us prove that the condition is sufficient. Assume that Cauchy's condition is satisfied. Let $(a_n)_{n=1}^\infty$ be a sequence in A that converges to c and is such that no term a_n

is equal to c.[1] The sequence $(f(a_n))_{n=1}^\infty$ satisfies Cauchy's condition for sequences, and so the limit $\lim_{n\to\infty} f(a_n) = t$ exists and is finite. If $(b_n)_{n=1}^\infty$ is another sequence in $A \setminus \{c\}$ with limit c then the limit $\lim_{n\to\infty} f(b_n) = s$ exists by the same token. But now we must have $s = t$. For we can construct a third sequence with limit c whose terms, $(d_n)_{n=1}^\infty$, are taken from $(a_n)_{n=1}^\infty$ and $(b_n)_{n=1}^\infty$ alternately. Then the limit $\lim_{n\to\infty} f(d_n)$ exists, but this is only possible if $s = t$.

We see therefore that $\lim_{n\to\infty} f(a_n)$ exists, and is finite, for every sequence in $A \setminus \{c\}$ that converges to c, and the limit t is the same for every such sequence.

Now we prove that $\lim_{x\to c} f(x) = t$. Let $\varepsilon > 0$. There exists $\delta > 0$, such that $|f(x_1) - f(x_2)| < \varepsilon/2$ for all x_1 and x_2 in A that satisfy $0 < |x_1 - c| < \delta$ and $0 < |x_2 - c| < \delta$. There exists y in A, such that $0 < |y - c| < \delta$ and $|f(y) - t| < \varepsilon/2$ (simply choose a sequence in $A \setminus \{c\}$ with limit c, and choose for y a term in the sequence sufficiently near to c). If now x is in A and $0 < |x - c| < \delta$, we obtain

$$|f(x) - t| \le |f(x) - f(y)| + |f(y) - t| < \frac{\varepsilon}{2} + \frac{\varepsilon}{2} = \varepsilon$$

as needed.   □

Let us display Cauchy's condition for limits of functions so that the reader can better compare it to Cauchy's condition for limits of sequences, given in Sect. 3.7:

For each $\varepsilon > 0$ there exists $\delta > 0$, such that $|f(x_1) - f(x_2)| < \varepsilon$ for all x_1 and x_2 in A that satisfy $0 < |x_1 - c| < \delta$ and $0 < |x_2 - c| < \delta$.

There is a version of Cauchy's principle for the limit $\lim_{x\to\infty} f(x)$, and with obvious changes, for $\lim_{x\to-\infty} f(x)$. It is very useful for studying improper integrals (Chap. 12). In the following, it is reasonable to assume that f is defined in an interval of the form $]a, \infty[$.

The limit $\lim_{x\to\infty} f(x)$ exists and is finite if and only if the following condition is satisfied: for all $\varepsilon > 0$ there exists K, such that $|f(x) - f(y)| < \varepsilon$ for all x and y that satisfy $x > K$ and $y > K$.

Exercise Prove the last assertion.

4.5.1 The Oscillation of a Function

Let $f : [a, b] \to \mathbb{R}$ be a bounded function. The difference $\sup_{[a,b]} f - \inf_{[a,b]} f$ is called the oscillation of f on the interval $[a, b]$. Recall that $\sup_{[a,b]} f$ is the supremum and $\inf_{[a,b]} f$ the infimum of the set $\{f(x) : x \in [a, b]\}$.

---

[1]It is obvious that such sequences exist if A is an interval, and this is the case in all applications considered in this text. For general sets we must appeal to the so-called axiom of choice of set theory.

More generally we define the oscillation of f on a subset A of its domain as the difference:

$$\Omega_A f := \sup_A f - \inf_A f.$$

It is easy to see that f is continuous at the point c if and only if for each $\varepsilon > 0$ there exists $\delta > 0$, such that the oscillation of f on the set $]c - \delta, c + \delta[\cap A$ is less than ε. The proof is left to the exercises, but we shall use this fact in the proof of the small oscillation theorem.

The following proposition will be needed to prove that continuous functions are integrable.

Proposition 4.14 (The small oscillation theorem) *Let $f : [a, b] \to \mathbb{R}$ be continuous. Let $\varepsilon > 0$. Then it is possible to partition the interval $[a, b]$ with finitely many points*

$$a = t_0 < t_1 < t_2 < \cdots < t_m = b$$

such that the oscillation of f on each interval $[t_j, t_{j+1}]$, $j = 0, 1, 2, m - 1$, is less than ε.

Proof Let $\varepsilon > 0$. Partition the interval into two parts $[a, c]$ and $[c, b]$ using the midpoint $c = \frac{1}{2}(a + b)$. If, for the given ε, the conclusion of the proposition is true for both the intervals $[a, c]$ and $[c, b]$, then it is true for $[a, b]$.

Turn this around and use the method of bisection. We suppose, for the given ε, that the conclusion of the proposition is not true for the interval $[a, b]$. Then either it is not true for $[a, c]$ or it is not true for $[c, b]$, and we choose that interval for which it is not true (we can choose the left interval if it fails for both). We repeat this for the new interval, and so on.

We can therefore construct an increasing sequence $(a_n)_{n=1}^{\infty}$ and a decreasing sequence $(b_n)_{n=1}^{\infty}$, such that $a_1 = a$, $b_1 = b$, $a_n < b_n$, $b_n - a_n$ tends to 0, and for each n there is no partition of $[a_n, b_n]$ into finitely many intervals on each of which the oscillation of f is less than ε.

Now $(a_n)_{n=1}^{\infty}$ and $(b_n)_{n=1}^{\infty}$ converge to the same limit t in $[a, b]$. Suppose first that $a < t < b$. Since f is continuous at t there exists an interval $[t - h, t + h]$ on which the oscillation of f is less than ε. But when n is sufficiently large we have $t - h < a_n < b_n < t + h$. The interval $[t - h, t + h]$ then includes the interval $[a_n, b_n]$, so that the oscillation of f on $[a_n, b_n]$ must be less than ε also. The conclusion of the proposition holds for the interval $[a_n, b_n]$ and the given ε; without even partitioning it. This contradicts the definition of the sequences a_n and b_n.

If $t = a$ we use the interval $[a, a + h]$ instead of $[a - h, a + h]$, and if $t = b$ the interval $[b - h, b]$, in the argument of the last paragraph. $\square$

4.5.2   Uniform Continuity

There is another form of the small oscillation theorem that is useful. It concerns the notion of uniform continuity. Suppose the function f is continuous in its domain A. This means it is continuous at each point a in A. Suppose we consider its continuity at a. For each $\varepsilon > 0$ there exists $\delta > 0$, such that for all x in A that satisfy $|x - a| < \delta$ we have $|f(x) - f(a)| < \varepsilon$.

If we move to a different point b, instead of a, but keep the same ε, it is not certain that the same δ will work. But if it does not work we can make it work by decreasing it, since we know that *some* δ will work for b (and the same ε as at a). We can clearly look simultaneously at a finite number of points $a_1, a_2, ..., a_m$, and find one δ that works for them all (always for the same ε as before). We just take the lowest of the m values of δ that we found for the points individually. But to treat infinitely many points, for example all points of the domain A, and find one δ that works for them all, may be impossible. Tolkien's One Ring can rule the others only so long as the others are finitely many.

The upshot of the discussion of the last paragraph is to define the notion of *uniform continuity*. We say informally that a function is uniformly continuous on its domain A, if it is continuous in A, and for each $\varepsilon > 0$ a single δ can be found that works for all points of A.

This can be expressed in a different way, which is nicer, and more symmetrical. We have to make $|f(x) - f(a)|$ small only in virtue of $|x - a|$ being small, and without pinning down a in advance. We therefore say that a function f is uniformly continuous on the domain A if it satisfies the following condition:

For all $\varepsilon > 0$ there exists $\delta > 0$, such that for all x and y in A that satisfy $|x - y| < \delta$ we have $|f(x) - f(y)| < \varepsilon$.

Proposition 4.15 *Let $f : [a, b] \to \mathbb{R}$ be continuous. Then f is uniformly continuous.*

Proof Let $\varepsilon > 0$. There exists a partition of $[a, b]$, such that the oscillation of f is less than $\varepsilon/2$ on each subinterval. Let δ be smaller than the length of the shortest subinterval of the partition. If $|x - y| < \delta$ then either x and y belong to the same subinterval, so that $|f(x) - f(y)| < \varepsilon/2$; or else x and y belong to adjacent subintervals, in which case $|f(x) - f(y)| < \varepsilon$.   □

4.5.3   Exercises

1. Let f be a function on the domain $]a, b[$. We suppose that b is finite. Suppose that there exists a constant K, such that for all x and y in the domain we have $|f(x) - f(y)| \le K|x - y|$. Show that $\lim_{t \to b-} f(t)$ exists and is a finite number.

   Note. The condition on f is called a Lipschitz condition. It is a strong version of continuity.

2. Prove that if a function f satisfies a Lipschitz condition (see Exercise 1) on a domain A then it is uniformly continuous on A.
3. It is important for the validity of Propositions 4.14 and 4.15 that the domain of f is a closed and bounded interval. For each type of interval A, that is not closed and bounded, find an example of a continuous function with domain A that is not uniformly continuous.
4. Prove that $\Omega_A f = \sup_{x,y \in A} |f(x) - f(y)|$, where the supremum is taken over all pairs of points x and y in A. This characterisation of oscillation is often useful.

In the remaining exercises it is simplest and most useful to assume that the domain A is an interval.

5. Prove that a function f is continuous at a point c in A if and only if the following condition is satisfied: for all $\varepsilon > 0$ there exists $\delta > 0$, such that $\Omega_{]c-\delta,c+\delta[\cap A} f < \varepsilon$.
6. Given that f is a bounded function, show that for each c in A the limit

$$\omega_f(c) := \lim_{h \to 0+} \Omega_{]c-h,c+h[\cap A} f$$

exists and is a finite number. This enables us to define the function ω_f, called the point oscillation of f.

The point oscillation ω_f of a function f is studied in the next three exercises. Assume that the domain of f is an interval.

7. Show that f is continuous at c if and only if $\omega_f(c) = 0$.
8. Show that if the left limit $\lim_{x \to c-} f(x)$ exists and is a finite number then $\lim_{x \to c-} \omega_f(x) = 0$. Show that the converse is false. (Similar results hold for the right limit.)
9. Give an example to show that the function ω_f is not necessarily continuous. However, it is upper semi-continuous. This means that for each point c, and for each $\varepsilon > 0$, there exists $\delta > 0$, such that $\omega_f(x) < \omega_f(c) + \varepsilon$ for all x that satisfy $|x - c| < \delta$. Prove this.

4.6 ($\diamond$) Iterations of Monotonic Functions

We present some simple, practical and general conclusions about iterations. The main object is to exploit continuity and monotonicity to compute a solution of a fixed point problem $f(x) = x$, given that we know that the solution exists. Further developments (such as the study of convergence rates) require derivatives and will be taken up later.

Let $f : A \to \mathbb{R}$ be a continuous function, where A is a subset of $\mathbb{R}$. We assume that we can define an infinite sequence in A by the iteration scheme $a_{n+1} = f(a_n)$, using some initial point a_1. We know that if the sequence converges to t, and if t is

Fig. 4.6 Picture of the
proof. Iterating an increasing
function; the case $a_1 < a_2$

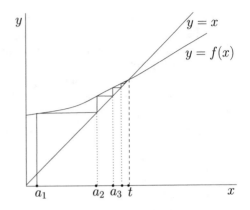

in A, then t is a solution to $f(x) = x$. In practice an iteration scheme such as this is used to approximate a solution to $f(x) = x$.

The problem is to see whether the sequence $(a_n)_{n=1}^{\infty}$ is convergent. Even if the equation $f(x) = x$ is known to have exactly one solution in A, it is not guaranteed that the iterations converge. Here are two conclusions that are sometimes useful. They are capable of some variation, which increases their practicality, but it is left to the reader to explore this. Both depend on monotonicity and continuity of the function iterated.

(i) Let $f :]0, \infty[\to \mathbb{R}$, where $f(x) > 0$ for all x in its domain and f is continuous and increasing. Suppose it is known that the equation $f(x) = x$ has exactly one solution t. Then the following conclusions hold. If $a_1 < t$ and $a_1 < a_2$, then a_n is increasing and converges to t; if $a_1 > t$ and $a_1 > a_2$, then a_n is decreasing and converges to t.

(ii) Let $f :]0, \infty[\to \mathbb{R}$, where $f(x) > 0$ for all x in its domain and f is continuous and decreasing. Assume that the equation $f(x) = x$ has exactly one solution t and that t is also the unique solution of $f(f(x)) = x$. Then the following conclusions hold. If $a_1 < t$ and $a_1 < a_3$, then a_{2n-1} (the subsequence with odd place numbers) is increasing and converges to t whilst a_{2n} (the subsequence with even place numbers) is decreasing and converges to t.

Proof of the First Rule In the case $a_1 < t$ and $a_1 < a_2$ it is seen by induction (left to the reader to verify) that $a_n < a_{n+1}$ for all n, and $a_n < t$ for all n. The sequence a_n is therefore increasing and bounded above; it therefore converges, to s say. But now $f(s) = s$, and since there is only one solution we must have $s = t$. The case $a_1 > t$ and $a_1 > a_2$ is similar. $\qquad \square$

Proof of the Second Rule The second rule follows from the first. Let $g = f \circ f$. Assume that $a_1 < t$. Now g is increasing and t is the sole solution of $g(x) = x$. The sequence $b_n = a_{2n-1}$ satisfies $b_{n+1} = g(b_n)$. We apply the first rule to g. If $b_1 < b_2$,

Fig. 4.7 Picture of the proof. Iterating a decreasing function; the case $a_1 < a_3$

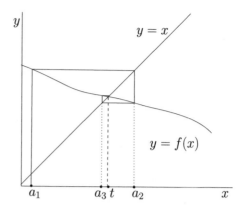

(that is $a_1 < a_3$) then $b_n = a_{2n-1}$ is increasing and tends to t. The sequence $c_n = a_{2n}$ also satisfies $c_{n+1} = g(c_n)$. We see that $c_1 = a_2 > t$ since f is decreasing and $f(t) = t$. Moreover $c_2 = a_4 = f(a_3) < f(a_1) = a_2 = c_1$. The sequence $c_n = a_{2n}$ is therefore decreasing and tends to the limit t.   □

The proofs of these rules are pictured in Figs. 4.6 and 4.7.

4.6.1 Exercises

1. Draw conclusions about the results of iterating the following functions, using a positive starting value.

   (a) $f(x) = \sqrt{2 + x}$
   (b) $f(x) = \sqrt{2x}$
   (c) $f(x) = \dfrac{2x + 2}{x + 2}$
   (d) $f(x) = 1 + \dfrac{1}{x}$
       (connected with the ratio of successive Fibonacci numbers).
   (e) $f(x) = 1 + \dfrac{1}{1 + x}$
       (connected with the continued fraction of the limit; see the nugget "Continued fractions").

4.6.2 Pointers to Further Study

→ Numerical analysis
→ Dynamical systems

Chapter 5
Derivatives and Differentiation

Big fleas have little fleas upon their backs to bite 'em, And little fleas have lesser fleas, and so, ad infinitum.

Augustus de Morgan

There's a problem with continuity. Suppose that f is continuous at a point x_0. Suppose we want to compute $f(x_0)$ with an error less than ε, for example $\varepsilon = 10^{-5}$, but we do not know x_0 exactly. We know that there exists δ, such that if $|x - x_0| < \delta$ then $|f(x) - f(x_0)| < 10^{-5}$. We do not therefore have to know x_0 exactly; a certain number of decimal places will suffice.

But what if δ needed to be uncomfortably small compared to ε in order to achieve the desired accuracy? What, for example, if δ was 10^{-10}, or 10^{-100} or even less....? The function may be continuous but continuity does not seem so useful here.

The problem is that f could be increasing or decreasing very rapidly at the point x_0. But what does that mean—the rate of increase or decrease of a function at a point?

The concept of the rate of growth of a function at a point is the key to the calculus of Newton and Leibniz and is what we call the derivative. As soon as it was introduced it became possible to solve important problems in geometry and physics with the new calculus, in spite of the fact that an acceptable definition of derivative was not given for some 200 years.

5.1 The Definition of Derivative

The average rate of growth of a function f between distinct points x_0 and x is the quotient

$$\frac{f(x) - f(x_0)}{x - x_0}.$$

R. Magnus, *Fundamental Mathematical Analysis*, Springer Undergraduate Mathematics Series, https://doi.org/10.1007/978-3-030-46321-2_5

The rate of growth at the point x_0 is defined as the limit.

Definition Let $f :]a, b[\to \mathbb{R}$ and $a < x_0 < b$. If the limit

$$\lim_{x \to x_0} \frac{f(x) - f(x_0)}{x - x_0} = A$$

exists and is a finite number we say that f is differentiable at x_0 and call A, *the derivative of f at x_0*.

We emphasise that A is a finite number; if the limit is ∞ or $-\infty$ we may sometimes say that the derivative is ∞ or $-\infty$ respectively, but we will never say that f is differentiable at x_0.

Another version of the definition of derivative, that arises by replacing $x - x_0$ by h, is

$$\lim_{h \to 0} \frac{f(x_0 + h) - f(x_0)}{h},$$

provided that the limit exists and is a finite number. The quotient appearing here is called a difference quotient. It is defined for both positive and negative values of h, though not for $h = 0$, but $|h|$ should not be so big that $x_0 + h$ falls outside the domain of f.

If f is differentiable at x_0 we denote its derivative at x_0 by $f'(x_0)$. We say that the function f is differentiable in the interval $]a, b[$ if f is differentiable at every point of the interval.

The definition of derivative follows a pattern that we have set in defining limit and the sum of an infinite series, and will continue in defining integral. The quantity in question that we wish to define does not necessarily exist. The definition of the quantity states when it exists, and given that it exists defines its value. Just as it is illogical to write $\lim_{x \to a} f(x)$ without first ascertaining whether the limit exists (though we often do this), we should not write $f'(c)$ without first ascertaining whether f is differentiable at c.

If f is differentiable in the interval $]a, b[$ we get a new function

$$f' :]a, b[\to \mathbb{R}, \quad f'(x) = \text{ derivative of } f \text{ at } x.$$

The operation of creating f' from f is called differentiation of the function f.

5.1.1 Differentiability and Continuity

Proposition 5.1 *Let the function f be differentiable at the point x_0. Then f is continuous at x_0.*

Proof We have, for $x \neq x_0$,

$$f(x) - f(x_0) = \left(\frac{f(x) - f(x_0)}{x - x_0} \right) \cdot (x - x_0)$$

so that, by the rule for the limit of a product,

$$\lim_{x \to x_0} (f(x) - f(x_0)) = \lim_{x \to x_0} \left(\frac{f(x) - f(x_0)}{x - x_0} \right) \cdot \lim_{x \to x_0} (x - x_0) = f'(x_0) \cdot 0 = 0.$$

In other words

$$\lim_{x \to x_0} f(x) = f(x_0)$$

which says that f is continuous at x_0. $\square$

Continuity is therefore necessary for differentiability, but it is far from being sufficient.

5.1.2 Derivatives of Some Basic Functions

Now we can begin to differentiate functions *from first principles*, that is, by applying the definition of derivative as the limit of the difference quotient.

(a) Let f be the constant function, $f(x) = C$ for all real numbers x. Then

$$\frac{f(x + h) - f(x)}{h} = \frac{C - C}{h} = 0$$

and so $f'(x) = 0$.

(b) Next let f be the so-called identity function, defined by $f(x) = x$ for all real numbers x. Then

$$\frac{f(x + h) - f(x)}{h} = \frac{x + h - x}{h} = 1$$

and so $f'(x) = 1$.

(c) Next let f be the function $f(x) = x^2$. Then

$$\frac{f(x + h) - f(x)}{h} = \frac{(x + h)^2 - x^2}{h} = \frac{2hx + h^2}{h} = 2x + h$$

and so, by the rule for the limit of a sum,

$$f'(x) = \lim_{h \to 0} (2x + h) = 2x.$$

We could go on, but it is far better to use the differentiation rules, as set out in the next paragraphs. These allow one to differentiate without considering difference quotients and limits. They make differentiation an almost mechanical procedure, and are of immense practical and historical importance. Without them there would be no calculus justifying the name.

5.1.3 Exercises

1. Differentiate from first principles, that is, using the definition of derivative as limit of the difference quotient:

   (a) $ax + b$, where a and b are constants.
   (b) x^3
   (c) x^n (n a natural number)
   (d) $\sqrt{x}$
   (e) $\sqrt[3]{x}$.
       Hint. Use algebraic properties of these functions. The only analytic input needed is their continuity.

2. Differentiate the circular functions $\sin x$ and $\cos x$ from first principles (that is, by calculating the limit of the difference quotient). You will need algebraic input in the form of the addition formulas

$$\sin(u + v) = \sin u \cos v + \cos u \sin v,$$
$$\cos(u + v) = \cos u \cos v - \sin u \sin v,$$

   and two facts of analysis to be proved later: the continuity of both functions, and the limit
$$\lim_{x \to 0} \frac{\sin x}{x} = 1.$$

   Note. The circular functions will be defined analytically in a later chapter. The reader has doubtlessly been introduced to them through school mathematics, in which it is usual to obtain the addition formulas by geometry and the limit of $\sin x/x$ by geometric intuition.

3. Differentiate the exponential function e^x from first principles. You will need the algebraic input that e^x satisfies the first law of exponents:

$$e^{x+y} = e^x . e^y,$$

   and the analytic input that
$$\lim_{x \to 0} \frac{e^x - 1}{x} = 1,$$

   equivalent to giving the derivative of e^x at $x = 0$; this essentially pins down the special base e.

Note. Just like the circular functions the exponential function and its inverse the natural logarithm will be defined analytically in a later chapter. We do not have to take for granted the existence of a function with these properties.

4. An exponential function a^x can be defined for any positive base a. It satisfies the law of exponents $a^{x+y} = a^x a^y$. For the sake of this exercise we shall adopt the notation $E_a(x) = a^x$. Assuming that E_a is differentiable derive the formula

$$E_a'(x) = k_a E_a(x)$$

where $k_a = E_a'(0)$.

Note. The special base e could be defined as the number that satisfies $k_e = 1$, though there would be difficulties involved—for example, why does such a number exist and why is it unique? Compare the previous exercise.

5. The natural logarithm $\ln x$ is the inverse function to the exponential function (Exercise 3) and from it $\ln x$ inherits the law of logarithms: $\ln(xy) = \ln x + \ln y$. Differentiate $\ln x$ from first principles.

Hint. You may need to figure out first why $\lim_{h \to 0} \ln(1+h)/h = 1$.

6. Let $f(x) = |x|$. Show that f is differentiable at all points except $x = 0$. Show that $f'(x) = x/|x|$ if $x \neq 0$.

7. Let $a_1, a_2, ..., a_n$ be a strictly increasing sequence of real numbers. Let $f(x) = \sum_{j=1}^n |x - a_j|$ for each real x.

   (a) Show that f is continuous at every point x, whereas it is differentiable everywhere except at the points a_j, $(j = 1, ..., n)$.
   (b) Show that the derivative is constant in each of the open intervals $]a_k, a_{k+1}[$, as well as in $]-\infty, a_1[$ and in $]a_n, \infty[$, and find a formula for it.
   (c) Sketch the graphs in the cases

   $$y = |x + 1| + |x| + |x - 1|$$

   and

   $$y = |x + 2| + |x + 1| + |x - 1| + |x - 2|.$$

8. Let f be the function with domain $\mathbb{R}$ defined by letting $f(x) = x$ if x is rational and $f(x) = 0$ if x is irrational.

   (a) Are there any points at which f is differentiable?
   (b) Are there any points at which the function $g(x) := xf(x)$ is differentiable?

9. Let $f :]0, 1[\to \mathbb{R}$ be the function defined in Sect. 4.2, Exercise 18. Recall that $f(x) = 0$ if x is irrational and $f(x) = 1/b$ if x is the fraction a/b expressed in lowest terms. Show that f is nowhere differentiable.

   Hint. Show that if x is irrational then there exist arbitrarily small h such that $\left| (f(x+h) - f(x))/h \right| > 1$.

5.2 Differentiation Rules

The elementary differentiation rules put the calculus into analysis. There are two groups of rules. The first deals with functions constructed by algebraic operations, addition, multiplication and division, from other functions. The second comprises the rule for differentiating composite functions (the chain rule) and the rule for differentiating inverse functions.

Let $f :]a, b[\to \mathbb{R}$, $g :]a, b[\to \mathbb{R}$. Sum, product and quotient of functions are defined pointwise:

$$(f + g)(x) = f(x) + g(x), \quad (fg)(x) = f(x)g(x), \quad \left(\frac{f}{g}\right)(x) = \frac{f(x)}{g(x)}.$$

Take care not to confuse product fg and composition $f \circ g$.

We state the first group of differentiation rules in the following lengthy proposition. In the proofs we rely entirely on the limit rules of Sect. 4.2; we never have to say "Let $\varepsilon > 0$".

Proposition 5.2 Let $f :]a, b[\to \mathbb{R}$, $g :]a, b[\to \mathbb{R}$. Let c be in the interval $]a, b[$ and assume that both f and g are differentiable at the point c. Let α be a numerical constant. Then αf, $f + g$ and fg are differentiable at c and we have

(1) $(\alpha f)'(c) = \alpha f'(c)$ (Multiplication by a constant)
(2) $(f + g)'(c) = f'(c) + g'(c)$ (Sum of functions)
(3) $(fg)'(c) = f'(c)g(c) + f(c)g'(c)$ (Product of functions; Leibniz's rule).

If moreover $g(c) \neq 0$, then $1/g$ and f/g are differentiable at c, and we have the further rules:

(4) $\left(\dfrac{f}{g}\right)'(c) = \dfrac{g(c)f'(c) - g'(c)f(c)}{(g(c))^2}$ (Quotient rule)

(5) $\left(\dfrac{1}{g}\right)'(c) = -\dfrac{g'(c)}{(g(c))^2}$ (Reciprocal rule).

Proof The rule for the derivative of αf is a special case of the rule for fg and that for $1/g$ a special case of that for f/g (left to the reader to see why).

Now for the proofs of rules 2, 3 and 4. Firstly the sum. We examine the difference quotient:

$$\frac{(f + g)(c + h) - (f + g)(c)}{h} = \frac{f(c + h) - f(c) + g(c + h) - g(c)}{h}$$
$$= \frac{f(c + h) - f(c)}{h} + \frac{g(c + h) - g(c)}{h}$$

and taking the limit we obtain the rule $(f + g)'(c) = f'(c) + g'(c)$.

Secondly the product. We transform the difference quotient:

$$\frac{(fg)(c+h) - (fg)(c)}{h} = \frac{f(c+h)g(c+h) - f(c)g(c)}{h}$$

$$= \frac{f(c+h)g(c+h) - f(c)g(c+h) + f(c)g(c+h) - f(c)g(c)}{h}$$

$$= \frac{f(c+h) - f(c)}{h}g(c+h) + f(c)\frac{g(c+h) - g(c)}{h}.$$

We take the limit as $h \to 0$, using the rules for the limits of sums and products, and remembering that $\lim_{h\to 0} g(c+h) = g(c)$ since g, being differentiable, is continuous at c. Thus we obtain the rule for the product, $(fg)'(c) = f'(c)g(c) + f(c)g'(c)$.

Next the quotient. Again we transform the difference quotient by algebra:

$$\frac{\left(\dfrac{f}{g}\right)(c+h) - \left(\dfrac{f}{g}\right)(c)}{h} = \frac{1}{h}\left(\frac{f(c+h)}{g(c+h)} - \frac{f(c)}{g(c)}\right)$$

$$= \frac{1}{h}\left(\frac{f(c+h)g(c) - f(c)g(c+h)}{g(c+h)g(c)}\right)$$

$$= \frac{\left(\dfrac{f(c+h) - f(c)}{h}\right)g(c) - f(c)\left(\dfrac{g(c+h) - g(c)}{h}\right)}{g(c+h)g(c)}.$$

We let $h \to 0$, use the rules for limits of sums, products and quotients, remember that $\lim_{h\to 0} g(c+h) = g(c)$ and obtain the limit

$$\frac{g(c)f'(c) - g'(c)f(c)}{(g(c))^2}.$$

$\square$

5.2.1 Differentiation of the Power Function

If n is a positive integer and f is the function x^n then we have

$$f'(x) = nx^{n-1}.$$

This is now easy to prove without considering the limit of a difference quotient. We use induction. The rule is known for $n = 1$. Let us assume it holds for a particular integer n and write $x^{n+1} = x \cdot x^n$. Using the rule for differentiating a product we obtain for the derivative of x^{n+1} the formula

$$1 \cdot x^n + x \cdot nx^{n-1} = (n+1)x^n.$$

This proves the rule generally for positive integers n.

Next we consider $f(x) = x^{-n} = 1/x^n$. The rule for the derivative of a quotient gives the formula

$$f'(x) = -\frac{nx^{n-1}}{x^{2n}} = -nx^{-n-1}.$$

So we have now shown that the derivative of x^a is ax^{a-1} in all cases when a is an integer (positive or negative).

What about the power $x^{1/n}$, which denotes the nth root $\sqrt[n]{x}$, or the fractional power $x^{m/n} = \sqrt[n]{x^m}$? For these we need the rule for differentiating inverse functions, and the celebrated chain rule, often referred to somewhat misleadingly as the rule for functions of a function. The latter rule, which we take first, is used to differentiate composite functions and is perhaps the most remarkable of the differentiation rules.

5.2.2   The Chain Rule

Proposition 5.3 *Let* $f : A \to \mathbb{R}$, $g : B \to \mathbb{R}$, *where* A *and* B *are open intervals and* $f(A) \subset B$. *Form the composition* $g \circ f : A \to \mathbb{R}$,

$$(g \circ f)(x) = g(f(x)), \quad (x \in A).$$

Let $x_0 \in A$, *assume that* f *is differentiable at* x_0 *and* g *is differentiable at* $f(x_0)$. *Then* $g \circ f$ *is differentiable at* x_0 *and*

$$(g \circ f)'(x_0) = g'((f(x_0))f'(x_0).$$

The chain rule is illustrated in Fig. 5.1

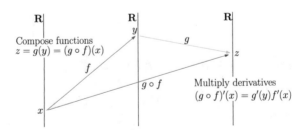

Fig. 5.1 A view of the chain rule

Proof of the Chain Rule Let $y_0 = f(x_0)$. As for the previous rules we start by applying some algebra to the difference quotient:

$$\frac{(g \circ f)(x_0 + h) - (g \circ f)(x_0)}{h}$$

$$= \frac{g(f(x_0 + h)) - g(f(x_0))}{f(x_0 + h) - f(x_0)} \cdot \frac{f(x_0 + h) - f(x_0)}{h}.$$

$$(5.1)$$

The second factor on the right-hand side has the limit $f'(x_0)$. As for the first factor it looks as if it should have the correct limit $g'(y_0)$. For we can think of $f(x_0 + h)$ as $y_0 + k$ (effectively defining the new quantity k) and then the first factor is the quotient

$$\frac{g(y_0 + k) - y_0}{k}.$$

As $h \to 0$ we have $k \to 0$ also and we seem to have a proof.

But there is a problem here. Although h is not 0 (as befits a correctly formed difference quotient) the denominator k, defined to be the difference $f(x_0 + h) - f(x_0)$, can be 0, and the first factor is then not defined for such values of h. There could even exist such values of h that are arbitrarily small which are then impossible to escape.

To save the proof we shall define a function R, the domain of which is a suitably small interval $]-\alpha, \alpha[$, in such a way that formula (5.1) for the difference quotient is correct if $R(f(x_0 + h) - f(x_0))$ replaces the first factor.

For $\alpha > 0$ and suitably small (the reader should try to figure out what "suitably small" means in this context and why we have to say it) we set

$$R(t) = \begin{cases} \dfrac{g(y_0 + t) - g(y_0)}{t} & \text{if } 0 < |t| < \alpha \\ g'(y_0) & \text{if } t = 0. \end{cases}$$

Note that R is continuous at the point $t = 0$ because

$$\lim_{t \to 0} \frac{g(y_0 + t) - g(y_0)}{t} = g'(y_0).$$

Moreover

$$g(y_0 + t) - g(y_0) = R(t)t$$

both when $t \neq 0$ and when $t = 0$. In this equation we replace t by the difference $f(x_0 + h) - f(x_0)$. This is allowed if $|h|$ is sufficiently small and then we have

$$g(f(x_0 + h)) - g(f(x_0)) = R(f(x_0 + h) - f(x_0))(f(x_0 + h) - f(x_0)).$$

Division by h when the latter is not 0, but still sufficiently small, gives

$$\frac{g\big(f(x_0 + h)\big) - g\big(f(x_0)\big)}{h} = R\big(f(x_0 + h) - f(x_0)\big)\left(\frac{f(x_0 + h) - f(x_0)}{h}\right).$$

Now we may let h tend to 0 and by the limit rules the right-hand side has the limit

$$\left(\lim_{h \to 0} R\big(f(x_0 + h) - f(x_0)\big)\right) f'(x_0) = R(0) f'(x_0) = g'\big(f(x_0)\big) f'(x_0).$$

In slightly more detail (we seriously want *this* proof to be correct) we can introduce the function $\phi(h) := f(x_0 + h) - f(x_0)$. Then ϕ is continuous at $h = 0$, and $\phi(0) = 0$. Moreover the function R is continuous at 0 as we saw. Hence the composition $R \circ \phi$ is continuous at 0 and so $\lim_{h \to 0} R(\phi(h)) = R(\phi(0)) = R(0)$, as we wrote above.   □

5.2.3 Differentiation of Inverse Functions

This is the last of the elementary differentiation rules. The lengthy preamble repeats the conditions (see Proposition 4.12) under which the inverse function exists and should not distract the reader from the extraordinary simplicity of the formula that is the conclusion.

Proposition 5.4

Preamble. *Let $f :]a, b[\to \mathbb{R}$ be continuous and strictly increasing (the point a may be $-\infty$ and b may be ∞). Let $c = \lim_{x \to a+} f(x)$ and $d = \lim_{x \to b-} f(x)$ (the limits exist if we allow $c = -\infty$ and $d = \infty$). The inverse function $g :]c, d[\to \mathbb{R}$ therefore exists, is continuous, and maps the interval $]c, d[$ on to the interval $]a, b[$.*
Conclusion. *Let $a < x_0 < b$ and assume that f is differentiable at x_0, and that $f'(x_0) \neq 0$. Then g is differentiable at $f(x_0)$ and*

$$g'(f(x_0)) = \frac{1}{f'(x_0)}.$$

A similar conclusion holds if f is strictly decreasing; the only difference is that $d < c$ and g has the domain $]d, c[$.

Proof Let $y_0 = f(x_0)$. We have to show that $g'(y_0) = f'(x_0)^{-1}$. Connect the variables h and k by the equation

$$y_0 + k = f(x_0 + h), \quad \text{equivalently} \quad h = g(y_0 + k) - g(y_0)$$

(recall that $g(y_0) = x_0$). The second equation here shows h as a function of k; it is a continuous, injective function of k, defined when k is sufficiently small. Moreover $h = 0$ when $k = 0$.
    We also have

$$\frac{g(y_0 + k) - g(y_0)}{k} = \frac{h}{k} = \frac{h}{f(x_0 + h) - f(x_0)}.$$

Let $k \to 0$ and think of h as a function of k, as defined above. Then h tends to 0, but is not 0 as long as $k \neq 0$. By the rule for the limit of a reciprocal, the right-hand side has the limit $f(x_0)^{-1}$. □

Assuming that $f'(x) \neq 0$ for all x in the interval $]a, b[$, we have the conclusion that

$$(f^{-1})'(y) = \frac{1}{f'(f^{-1}(y))} \tag{5.2}$$

for all y in the interval $]c, d[$ (or in $]d, c[$ if f is decreasing).

We shall see later (Sect. 5.6) that if $f'(x) > 0$ for all x in the open interval A, then f is strictly increasing in A, so that Proposition 5.4 is immediately applicable.

5.2.4 Differentiation of Fractional Powers

Let $f(x) = x^{1/n}$, where n is a positive natural number. We have here the inverse function of the function $g(x) = x^n$. The domain is the interval $]0, \infty[$. By the rule for differentiating an inverse function (that is, we apply (5.2) to the function g with x instead of y) we have

$$f'(x) = (g^{-1})'(x) = \frac{1}{g'(g^{-1}(x))} = \frac{1}{n(x^{\frac{1}{n}})^{n-1}} = n^{-1}x^{\frac{1-n}{n}} = \frac{1}{n}x^{\frac{1}{n}-1}.$$

Next we consider the function $f(x) = x^{m/n}$, where m is an integer, positive or negative. This is the composition $(x^m)^{1/n}$. By the chain rule we have

$$f'(x) = \frac{1}{n}(x^m)^{\frac{1}{n}-1}mx^{m-1} = \frac{m}{n}x^{\frac{m}{n}-1}.$$

The conclusion is striking. The derivative of the power function x^a is ax^{a-1} for every rational power a.

It is a further task to define the power function x^a for irrational powers and prove that the same differentiation formula continues to be valid.

5.2.5 Exercises

1. Differentiate the following functions. You may assume that the domain of each function is the set of all x for which the formula makes sense.

(a) $\dfrac{1}{x^2 + 2}$

(b) $\dfrac{x^2 - x + 1}{x^2 + x - 1}$

(c) $\sqrt{\dfrac{x^2 - x + 1}{x^2 + x - 1}}$

(d) $\sqrt[4]{\dfrac{x^2 - x + 1}{x^2 + x - 1}}$

(e) $\sqrt{1 + \sqrt{x}}$

(f) $\sqrt{1 + \sqrt{1 + \sqrt{x}}}$

(g) $\sqrt{1 + \sqrt{1 + \sqrt{1 + \sqrt{x}}}}.$

2. Define the function f on the whole real line by

$$
f(x) = \begin{cases} x, & (x \le 0) \\ 1 - 2\sqrt{1 - x}, & (0 < x < 1) \\ x, & (1 \le x \le 2) \\ \sqrt{x^2 + 5} - 1, & (2 < x). \end{cases}
$$

Determine where f is differentiable, and where it is, find its derivative.

Hint. In this, and in similar examples where a function is defined by cases, the differentiation rules are only useful in the open intervals between the partition points. At the partition points something else is required, such as arguing by examination of the difference quotient.

3. For this exercise we assume some knowledge of the circular functions $\sin x$ and $\cos x$, including their derivatives (see Sect. 5.1, Exercise 2). Determine where the following functions are differentiable and calculate the derivative when it exists:

(a) $f(x) = \begin{cases} \sin \frac{1}{x}, & (x > 0) \\ 0, & (x \le 0). \end{cases}$

(b) $f(x) = \begin{cases} x \sin \frac{1}{x}, & (x > 0) \\ 0, & (x \le 0). \end{cases}$

(c) $f(x) = \begin{cases} x^2 \sin \frac{1}{x}, & (x > 0) \\ 0, & (x \le 0). \end{cases}$

4. Show that the function $f(x) = x^5 + x$ is strictly increasing on the whole real line and calculate $(f^{-1})'(2)$.

5. Let f_k, $(k = 1, 2, ..., n)$ be differentiable functions. Let g be their product, $f_1 f_2 ... f_n$. Show that

$$
\frac{g'(x)}{g(x)} = \sum_{k=1}^{n} \frac{f_k'(x)}{f_k(x)}
$$

at every point x at which none of the denominators is 0.

5.3 Leibniz's Notation

There are several notational systems in use for derivatives. They reflect the differing views of Newton and of Leibniz. Newton used dots to signify the derivative, as in $\dot{x}$ and $\dot{y}$. One might say that the various dashes, as in f' and f'', popularised by Lagrange, reflect Newton's notation. Leibniz introduced the expressions dx and dy, signifying in his view infinitesimal changes in the variables x and y ("d" for Latin "*differentia*"), and leading to the differential quotient dy/dx. He also introduced the integral sign "$\int$", an elongated "S" (for Latin "*Summa*"). Each notation has its advantages and it is best to learn how to use both.

5.3.1 Tangent Lines

We often think of a function f as a curve in the (x, y)-plane. The curve in question is the set of all points (x, y) that satisfy $y = f(x)$, in other words the graph of f. Leibniz's notation reflects the geometric intuition behind the idea of a tangent line to a curve.

A line is a curve of the form $y = mx + c$ with constants m (the slope) and c (the intercept). The curve $x = a$ is also a line but it is not a graph of a function in the above sense. It is though a graph if we think of x as a function of y (in this case a constant function).

We could ask whether every curve in the plane can be described (perhaps locally; in small sections at a time) as a graph, in which y is a function of x, or x is a function of y. The question arises even for familiar everyday curves like the circle and shows the limitation of thinking of a curve simply as a graph. This gets us into the area of differential geometry. We would have to give a general definition of curve, a task that is not so straightforward.

The ancient Greek geometers tried to define a tangent line to a curve as a line that meets the curve in only one point. This works for circles (and more generally conic sections) but not for more complicated curves. Differential calculus allows us to give a correct definition of tangent line to a curve when that curve is a graph $y = f(x)$, and its extension to differential geometry does the job for more general curves. For this reason it is said that differentiation solved the problem of tangents.

Consider a differentiable function f. The tangent line to the curve $y = f(x)$, at a point (x_0, y_0) on the graph (that this point lies on the graph means that $y_0 = f(x_0)$), is the line through the point (x_0, y_0) that has the slope $f'(x_0)$. In other words it is the line

$$y - y_0 = f'(x_0)(x - x_0)$$

Fig. 5.2 Leibniz's
differential quotient

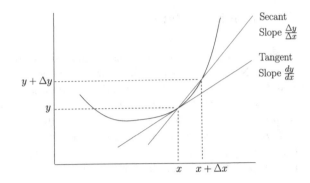

or equivalently
$$y = f'(x_0)x + (y_0 - f'(x_0)x_0).$$

The intuitive thinking behind this is that the tangent line at (x_0, y_0) is the limit of a secant line through the two points, (x_0, y_0) and $(x_0 + \Delta x, y_0 + \Delta y)$, both on the graph $y = f(x)$, the limit being taken as $\Delta x \to 0$. The slope of the secant line is $\Delta y / \Delta x$ and we want to make Δx, and as a result Δy, tend to 0.

In the view of seventeenth century mathematicians, who did not possess a definition of limit, the quantity Δx was actually supposed to become infinitely small, the tangent being thought to intersect the curve at two distinct points infinitely close together. For the slope of the tangent we obtain a quotient of infinitely small quantities, or infinitesimals. This intuition lies behind Leibniz's notation for derivatives (Fig. 5.2).

5.3.2 Differential Quotients

Leibniz proposed setting an infinitesimal dx in place of Δx, as the notion of limit was not available to him. He would have said that y underwent a corresponding change, which was also an infinitesimal dy, and the derivative was the quotient dy/dx. Although dx and dy are infinitesimals (whatever that means) the quotient is an ordinary real number. He called the infinitesimals dx and dy differentials. The derivative was then the *differential quotient*.

According to the prevailing modern view the derivative is not a quotient; it is though the limit of a quotient, namely the limit of the difference quotient. In spite of this it is possible to define differentials, expressed in the classical notation dx and dy, without resorting to the mysterious infinitesimals. This is very useful for calculus in several variables and differential geometry of surfaces and their generalisations, manifolds. It means, for example, that classical formulas, such as $dy = f'(x)\,dx$, remain valid with an appropriate interpretation of their symbols. However, that is a whole new topic.[1]

Here are some examples of statements written using Leibniz's notation. It will be seen that they have certain advantages, notably brevity and flexibility, over their equivalents using function symbols:

(a)   If $y = x^3$ then $\dfrac{dy}{dx} = 3x^2$.

That is, if f is the function $f(x) = x^3$ then $f'(x) = 3x^2$.

(b) $\dfrac{d}{dx} x^3 = 3x^2$.

Same meaning as the previous item. We again avoid using a symbol for the function, as well as mentioning the variable y.

(c) $\dfrac{d}{dx} x^3 \Big|_{x=1} = 3$.

In other words if f is the function $f(x) = x^3$ then $f'(1) = 3$. The vertical stroke with the subscript "$x = 1$" means evaluate the preceding expression at $x = 1$.

5.3.3   The Chain Rule and Inverse Functions in Leibniz's Notation

Many calculations using the chain rule or the inverse-function rule are easier to carry out using Leibniz's notation. This makes it particularly useful for effecting a change of variables in a differential equation, a subject not covered in the present text.

Functions f and g are given and we wish to differentiate the composed function $g \circ f$. We consider that the function f sets up a relation between variables x and y, namely $y = f(x)$, whilst g sets up a relation between variables y and z, namely $z = g(y)$. Then the composition $g \circ f$ sets up the relation $z = (g \circ f)(x)$.

We can differentiate the composition $g \circ f$ using the chain rule. In Leibniz's notation we are finding the differential quotient dz/dx and this is given by the striking formula

$$\frac{dz}{dx} = \frac{dz}{dy}\frac{dy}{dx}.$$

This is of course just the formula

$$(g \circ f)'(x) = g'(f(x))f'(x).$$

---

[1] This has nothing to do with what is known as non-standard analysis. In the latter the real number system is extended by including infinitely small quantities and infinitely large quantities.

The first factor on the right-hand side, that is dz/dy, must be interpreted with some care. We first differentiate z with respect to y, but then express this as a function of x, using the relationship between y and x.

We illustrate these steps by differentiating $\sqrt{1-x^2}$. We set $y = 1 - x^2$ and $z = \sqrt{y}$. Then

$$\frac{dz}{dx} = \frac{dz}{dy}\frac{dy}{dx} = \frac{1}{2\sqrt{y}}(-2x) = -\frac{x}{\sqrt{1-x^2}}.$$

Consider next inverse functions. If y is a function of x, namely $y = f(x)$, we can turn this round and look at x as a function of y, namely $x = f^{-1}(y)$. The rule for differentiating f^{-1} takes the memorable form

$$\frac{dx}{dy} = 1 \bigg/ \frac{dy}{dx}.$$

This is the same formula as the less intuitive

$$(f^{-1})'(y) = \frac{1}{f'(f^{-1}(y))}.$$

As an example we shall differentiate the function $x^{1/n}$. Let $y = x^{1/n}$ and turn it around giving $x = y^n$. Then

$$\frac{dx}{dy} = ny^{n-1}$$

so that by the rule we find

$$\frac{dy}{dx} = \frac{1}{ny^{n-1}} = \frac{1}{nx^{\frac{n-1}{n}}} = \frac{1}{n}x^{\frac{1}{n}-1}.$$

5.3.4 Tangents to Plane Curves

In analytic geometry, the simplest way to represent a circle with centre (a, b) and radius r is by means of the equation $(x - a)^2 + (y - b)^2 = r^2$. Here the curve is not seen as a graph; in order to do so we must solve for y as a function of x, or for x as a function of y. To represent a curve in analytic geometry as a graph, we usually have to break it into pieces.

A simple example is that of the unit circle $x^2 + y^2 = 1$. Solving for y we obtain two solutions, and two graphs:

$$y = \sqrt{1-x^2}, \quad (-1 \le x \le 1) \quad \text{the upper semicircle}$$

$$y = -\sqrt{1-x^2}, \quad (-1 \le x \le 1) \quad \text{the lower semicircle.}$$

Now we can differentiate these formulas in order to compute the tangents to the circle, using the appropriate formula for each semicircle.

However, there is another way to calculate the tangent at a point (x_0, y_0) on the curve without solving for y as a function of x. Suppose that we are looking at a part of the circle that can be represented as a graph $y = f(x)$, where f is differentiable, and contains the point (x_0, y_0). Then $y_0 = f(x_0)$ and the equation

$$x^2 + \left(f(x)\right)^2 = 1$$

holds for all x in some interval containing x_0. We may differentiate with respect to x, using the differentiation rules, and obtain

$$2x + 2f(x)f'(x) = 0.$$

In particular $f'(x_0) = -x_0/f(x_0) = -x_0/y_0$.

The calculation just given would normally be done without introducing a function symbol, using Leibniz's notation

$$x^2 + y^2 = 1 \ \Rightarrow \ 2x + 2y\frac{dy}{dx} = 0 \ \Rightarrow \ \frac{dy}{dx} = -\frac{x}{y}$$

or else a form of Newton's notation

$$x^2 + y^2 = 1 \ \Rightarrow \ 2x + 2yy' = 0 \ \Rightarrow \ y' = -\frac{x}{y}.$$

This procedure is called implicit differentiation. The differentiation proceeds with respect to x, but y is thought of as a function of x, the exact form of which is not required. We obtain dy/dx, and express it as a function of x and y, without knowing the function $y = f(x)$. Logically, we only need to know that the function $f(x)$ exists, and is differentiable. This can usually be guaranteed by a theorem of multivariate calculus, the implicit function theorem, which is beyond the scope of this text.

Example The equation $2y^5 - xy - x^4 = 0$ defines some kind of curve in the coordinate plane. We observe that it contains the point $(1, 1)$. To solve for y as a function of x, or for x as a function of y, is difficult (although some algebraic arguments show that there is a unique positive y for each positive x; see the nugget "Multiplicity"). Nevertheless, we can calculate the tangent to the curve at the point $(1, 1)$. Assuming that we can represent the curve around the point $(1, 1)$ as a graph $y = f(x)$ with differentiable f (a fact that can be justified using the implicit function theorem), implicit differentiation gives

$$10y^4y' - y - xy' - 4x^3 = 0 \ \Rightarrow \ y' = \frac{y + 4x^3}{10y^4 - x}$$

and therefore at the point $(1, 1)$ we have $y' = \frac{5}{9}$. Note how differentiating the middle term xy gives rise to $y + xy'$ because we are thinking of y as a function of x. The equation of the tangent line is therefore $y - 1 = \frac{5}{9}(x - 1)$, or more simply, $5x - 9y + 4 = 0$.

5.3.5 Exercises

1. (a) Determine the equation of the tangent to the parabola $y = x^2$ at the point (t, t^2).
   (b) Show that the line perpendicular to the tangent of item (a) and intersecting it on the x-axis, passes through the point $(0, \frac{1}{4})$, independently of t.
2. Show that the equation of the tangent to the ellipse

$$\frac{x^2}{a^2} + \frac{y^2}{b^2} = 1$$

at the point (x_0, y_0), assumed to be on the ellipse, is

$$\frac{x_0 x}{a^2} + \frac{y_0 y}{b^2} = 1.$$

3. Give an example of a graph $y = f(x)$ (with differentiable f) and a point $(a, f(a))$ on the graph, such that the tangent at $(a, f(a))$ crosses the graph at $(a, f(a))$.
4. A vessel has the shape of a right circular cone standing on its apex. Let h be the height of the cone and let r be the radius of its base. Mercury is poured into the vessel, not necessarily at a constant rate. Introduce variables: t for the time, v for the volume of mercury in the vessel and y for the height of the mercury in the vessel.
   (a) Find the relationship between $\dfrac{dv}{dt}$ and $\dfrac{dy}{dt}$.
   (b) Suppose that $h = 1$ m, $r = 1$ m, $y = 0.5$ m and the mercury is poured at a constant rate of 1 litre per second. Approximately, how much time is needed to raise the surface level by 1 cm?

   Note. Physics and engineering abound with problems like this one. A bunch of variables are connected by a constitutive relation. In this problem the relation between v and y is geometric. Examples from physics are pressure, volume and temperature connected by the ideal gas equation; or stress and strain connected by the law of elasticity. If the variables change with time, then the constitutive relation implies a linear connection between their derivatives with respect to time. If the variables are three or more then the problem really requires multivariate calculus, in particular partial derivatives. With two variables we can just about get by without them.

5.4 Higher Order Derivatives

If the function $f :]a, b[\rightarrow \mathbb{R}$ is differentiable, its derivative $f' :]a, b[\rightarrow \mathbb{R}$ is a new function. Now it could happen that f' is differentiable. If so, we can differentiate and produce the function $f'' :]a, b[\rightarrow \mathbb{R}$, called the second derivative of f. Continuing in this way as far as is allowable, we can define a whole sequence: second, third, fourth, fifth, ..., n^{th} derivatives of f. They are denoted by f'', f''', f'''', f'''' and so on, but around the fourth it becomes more practical to write instead $f^{(4)}$, $f^{(5)}$, ..., $f^{(n)}$... as counting those little dashes becomes tiresome and irritating. When using this notation it is often convenient to allow $n = 0$ and interpret $f^{(0)}$ to be the same as f.

The differentiation can be continued beyond $f^{(n)}$ when the latter is differentiable on the interval $]a, b[$, where f was defined. It could happen that every function produced in this way is differentiable. Then we say that f is infinitely often differentiable. If the process can be continued at least as far as $f^{(n)}$ we say that f is n-times differentiable, or that f is differentiable to order n, or that f has derivatives to order n (none of which precludes going further).

Can we give any sense to the statement that f is n-times differentiable at the point c? For a function to be differentiable at a given point it must be defined on an interval that contains that point. Therefore the meaning to be attached to this phrase is the following. There exists $\delta > 0$, such that f is $(n - 1)$-times differentiable in the interval $]c - \delta, c + \delta[$ and $f^{(n-1)}$ is differentiable at c. We sometimes say in this case that the derivatives $f^{(k)}(c)$ exist up to $k = n$; or, most briefly: f has n derivatives at c.

Leibniz's notation for the higher derivatives is

$$y = f(x), \quad \frac{dy}{dx} = f'(x), \quad \frac{d^2 y}{dx^2} = f''(x), \quad \ldots \quad \frac{d^m y}{dx^m} = f^{(m)}(x).$$

5.4.1 Exercises

1. Let f be a polynomial of degree m. Show that $f^{(k)} = 0$ for all $k > m$.
2. Let g be a function having derivatives of all orders and let a be a real number, such that $g(a) \neq 0$. Set $f(x) = (x - a)^m g(x)$, where m is a positive integer. Show that $f^{(k)}(a) = 0$ for $k = 0, 1, ..., m - 1$, but that $f^{(m)}(a) \neq 0$.
3. Show that

$$\frac{d^k}{dx^k} x^a = a(a - 1)...(a - k + 1)x^{a-k}.$$

Here you may assume that a is rational (pending the rigorous definition of irrational powers in Chap. 7). Also you may assume that $x > 0$ if a is not an integer. If a is a positive integer show that

$$\frac{d^k}{dx^k} x^a = \frac{a!}{(a-k)!} x^{a-k}$$

for $k \le a$. One can even allow $k > a$ if we interpret $1/m!$ as 0 if m is a negative integer.

4. We assume some knowledge of the exponential function e^x, namely, that its derivative is again e^x. Let $f(x) = e^{-1/x}$ for $x \ne 0$. Show that for all natural numbers n we have

$$f^{(n)}(x) = P_n\left(\frac{1}{x}\right) e^{-1/x}, \quad (x \ne 0),$$

where for each n, $P_n(t)$ is a polynomial in the variable t of degree $2n$. Find a recurrence formula for $P_n(t)$.

5. Prove Leibniz's formula for the nth derivative of a product. If u and v are functions with derivatives up to order n, then uv has derivatives to order n and

$$(uv)^{(n)} = \sum_{k=0}^{n} \binom{n}{k} u^{(k)} v^{(n-k)}.$$

6. Calculate some higher derivatives of the composite function $y = g(f(x))$, as far as your patience allows.

7. A function is defined by

$$f(x) = \begin{cases} -x^5, & \text{if } x < 0 \\ x^5, & \text{if } x \ge 0. \end{cases}$$

How many derivatives does f possess at $x = 0$?

8. A function with domain $\mathbb{R}$ is called an *even function* if it satisfies $f(-x) = f(x)$ for all x. It is called an *odd function* if it satisfies $f(-x) = -f(x)$ for all x.

   (a) Show that every function f with domain $\mathbb{R}$ has a unique decomposition $f = g + h$ where g is even and h is odd.
   (b) Suppose that f has m derivatives at $x = 0$. Show that if f is even, then all derivatives $f^{(k)}(0)$ with odd $k \le m$ are zero. Show, on the other hand, that if f is odd, then all derivatives $f^{(k)}(0)$ with even $k \le m$ are zero.

9. How many derivatives does the function $|x|^{7/2}$ possess at $x = 0$?

10. Define a function f on the domain $]-\infty, 1[$ by

$$f(x) = \begin{cases} 0, & \text{if } x < 0 \\ 1 - \sqrt{1-x^2}, & \text{if } 0 \le x < 1. \end{cases}$$

Show that f is differentiable at all points of its domain, that f' is continuous, and that f is twice differentiable at all points except at $x = 0$. At $x = 0$ the second derivative does not exist; but calculate its "jump", the quantity

$$\lim_{x \to 0+} f''(x) - \lim_{x \to 0-} f''(x).$$

Note. Where a straight section of rail track joins a curved section, it is safer if the curve is designed so that the second derivative is continuous and is 0 at the join. This is to avoid discontinuities in the acceleration normal to the track. The graph in the exercise typifies the join in a model railway, where the curves are usually arcs of circles, and that is where the model train is most likely to leave the track.

11. Variables x and y are connected by the equation $2y^5 - xy - x^4 = 0$. Calculate the second derivative d^2y/dx^2 when $x = 1$ and $y = 1$.

12. In this exercise we assume some acquaintance with determinants. Let u_1 and u_2 be differentiable functions in an interval A.

   (a) Suppose that the functions u_1 and u_2 are linearly dependent in A; by this is meant that there exist constants λ_1 and λ_2, not both 0, such that

   $$\lambda_1 u_1(x) + \lambda_2 u_2(x) = 0$$

   for all x in A. Show that, for all x in A:

   $$\begin{vmatrix} u_1(x) & u_2(x) \\ u_1'(x) & u_2'(x) \end{vmatrix} = 0.$$

   (b) The example $u_1(x) = u_2(x) = 0$ for $x < 0$ and $u_1(x) = x^2$, $u_2(x) = 2x^2$ for $x \geq 0$ shows that the converse is false.

   (c) Extend the result of item (a) to the case of m functions $u_1,..., u_m$, each $m - 1$ times differentiable. Show that a necessary condition for their linear dependence in A is that

   $$\begin{vmatrix} u_1(x) & u_2(x) & \cdots & u_m(x) \\ u_1'(x) & u_2'(x) & \cdots & u_m'(x) \\ \vdots & \vdots & \ddots & \vdots \\ u_1^{(m-1)}(x) & u_2^{(m-1)}(x) & \cdots & u_m^{(m-1)}(x) \end{vmatrix} = 0$$

   for all x in A.

5.5   Significance of the Derivative

In this section we begin to extract useful information about a function from knowledge of its derivative.

If A is an interval we call a point c in A an interior point if c is not an endpoint of the interval. In the next paragraphs a set denoted by A will always be an interval with distinct endpoints.

Proposition 5.5 *Let $f : A \to \mathbb{R}$, let c be an interior point of A and let f be differentiable at c. Then the following hold:*

(1) If $f'(c) > 0$ there exists $\delta > 0$, such that

$$f(x) < f(c) \ \text{if} \ c - \delta < x < c, \quad \text{and} \ f(x) > f(c) \ \text{if} \ c < x < c + \delta.$$

(2) If $f'(c) < 0$, then there exists $\delta > 0$, such that

$$f(x) > f(c) \ \text{if} \ c - \delta < x < c, \quad \text{and} \ f(x) < f(c) \ \text{if} \ c < x < c + \delta.$$

Proof Let $f'(c) > 0$. Now

$$f'(c) = \lim_{h \to 0} \frac{f(c + h) - f(c)}{h},$$

and taking ε to be $\frac{1}{2} f'(c)$ in the definition of limit we find that there exists $\delta > 0$, such that

$$\frac{f(c + h) - f(c)}{h} > \frac{f'(c)}{2}$$

for all h that satisfy $0 < |h| < \delta$. For such h that are negative we have

$$f(c + h) - f(c) < \frac{h f'(c)}{2} < 0$$

and for such h that are positive we have

$$f(c + h) - f(c) > \frac{h f'(c)}{2} > 0.$$

The case when $f'(c) < 0$ is treated similarly.   □

We did not assume that f was differentiable at points other than c. But even if it is, the assumption that $f'(c) > 0$ tells us little about the derivative $f'(x)$, for x near to c. We could have points x, arbitrarily near to c, at which $f'(x) < 0$, for example. Or even points at which $f'(x)$ is arbitrarily large.

5.5.1 Maxima and Minima

One of the main applications of the last paragraph is to the problem, familiar from applied mathematics, of finding maxima and minima. Problems of this nature are generally called extremal problems.

Let $f : A \to \mathbb{R}$. Recall that A denotes an interval, with or without endpoints, though the latter must be distinct. The status of a point c in A regarding the local extremal behaviour of f can be usefully, if somewhat pedantically, classified as follows:

(a) The point c is called a *local minimum point for* f if there exists $\delta > 0$, such that $f(x) \geq f(c)$ for all x in A that satisfy $|x - c| < \delta$.

(b) The point c is called a *local maximum point for* f if there exists $\delta > 0$, such that $f(x) \leq f(c)$ for all x in A that satisfy $|x - c| < \delta$.

(c) The point c is called a *strict local minimum point for* f if there exists $\delta > 0$, such that $f(x) > f(c)$ for all x in A that satisfy $0 < |x - c| < \delta$.

(d) The point c is called a *strict local maximum point for* f if there exists $\delta > 0$, such that $f(x) < f(c)$ for all x in A that satisfy $0 < |x - c| < \delta$.

Note that c could be an endpoint of the interval A in these definitions. Moreover c could belong to none of the above four classes, in which case it is of no interest as regards the extremal problem for f.

The next proposition defines precisely the notion, loosely expressed, that the derivative vanishes at a maximum or minimum.

Proposition 5.6 *Let* $f : A \to \mathbb{R}$, *let* c *be a point in* A *and assume that* c *is either a local minimum point, or a local maximum point, of* f. *If, in addition,* c *is an interior point of* A *and* f *is differentiable at* c, *then* $f'(c) = 0$.

Proof Consider the case when c is a local minimum point. If $f'(c) < 0$ then, by Proposition 5.5, there exists $\delta > 0$, such that $f(x) < f(c)$ if $c < x < c + \delta$. If $f'(c) > 0$ then there exists $\delta > 0$, such that $f(x) < f(c)$ if $c - \delta < x < c$. In neither case can c be a local minimum point, so we have a contradiction. We conclude that $f'(c) = 0$. A similar argument is used for the case when c is a local maximum point. □

That the derivative is 0, given that c is an interior point and f is differentiable at c, is only a necessary condition for c to be a local minimum or maximum point. It is not sufficient. There is a need for a term to cover the case that $f'(c) = 0$, irrespective of whether c is a local maximum or minimum point. The terms extreme point, extremal point, stationary point and critical point have been used (and there are probably others). The last two should be preferred as they do not suggest a maximum or minimum.

5.5.2 Finding Maxima and Minima in Practice

Let $f : [a, b] \to \mathbb{R}$ be a continuous function. The domain $[a, b]$ is a bounded and closed interval. We know by the extreme value theorem (Proposition 4.11) that f attains both a maximum and a minimum value in $[a, b]$. The problem of maxima and minima is to find the points where these are attained, as well as the maximum and minimum values.

Suppose that the maximum is attained at a point $x = c$ (there could be more than one such point). There are three possibilities (exclusive; each excludes the other two):

(a) The point c is either a or b.
(b) The point c is an interior point, that is, a point of the open interval $]a, b[$, f is differentiable at c and (by Proposition 5.6) $f'(c) = 0$.
(c) The point c is an interior point at which f is not differentiable.

The most usual situation is that there are only a finite number of points c in $[a, b]$ that satisfy any one of these three conditions. It may be feasible to find them, and once found, to arrange them in a list. This might begin with the endpoints a and b, continue with the points in $]a, b[$ at which f is not differentiable (if finitely many) and conclude with all the solutions of $f'(x) = 0$ in $]a, b[$ (if finitely many). Now it only remains to calculate f at each of the points in the list and find the highest and lowest of these values.

5.5.3 Exercises

1. In each of the following cases determine the maximum and minimum of the function f over the interval A:
   (a) $f(x) = x^3 - 3x^2 + x$, $A = [1, 3]$.
   (b) $f(x) = \max\left(1 - 2x - x^2, \ 2 + x - x^2, \ 1 + 3x - x^2\right)$, $A = [-1, 2]$.
   (c) $f(x) = \dfrac{5}{1 + |x - 4|} + \dfrac{4}{1 + |x - 5|}$, $A = [-6, 6]$.

   Hint. In items (b) and (c) it helps to express the functions by cases. For the numerical work in these exercises it makes sense to use a calculator and state the answers with a certain number of decimal digits, say, three.
2. Determine the minimum of the function
$$f(x) = x + \frac{1}{x^3}$$
   in the interval $]0, \infty[$.
3. Let $a_1, a_2, ..., a_n$ be a strictly increasing sequence of real numbers. Let $f(x) = \sum_{j=1}^{n} |x - a_j|$ for each real x. Determine the minimum of f over the whole real line.

4. Define the function f with domain $\mathbb{R}$ by $f(x) = x^2 \sin(1/x)$ for $x \neq 0$, and $f(0) = 0$. Show that f is everywhere differentiable, including at $x = 0$, but that f' is discontinuous at $x = 0$.

5. Find an example of a continuous function f that has a strict local minimum at $x = 0$, and for every $\delta > 0$ has also a strict local minimum in the interval $]0, \delta[$.
   Hint. To help you think about it, note that there would have to be infinitely many minima in $]0, \delta[$, each with a value higher than $f(0)$.

6. Give an example of a differentiable function f, such that $f'(0) > 0$, but there exists no $\delta > 0$ such that f is strictly increasing in the interval $]-\delta, \delta[$.
   Hint. Try to exploit the wildly oscillating function $\sin(1/x)$.

7. Give an example of a differentiable function f such that $f'(0) = 0$ and in every interval $]-\delta, \delta[$ (with $\delta > 0$) the derivative f' takes arbitrarily large positive values and arbitrarily large negative values.

5.6 The Mean Value Theorem

It has been called the most useful theorem in analysis (notably by the influential French mathematician and Bourbakiste, Jean Dieudonné, but he was probably echoing G. H. Hardy). We leave it to the reader to judge the truth or otherwise of this claim. It might seem more logical to write "mean-value theorem", as there is nothing mean about it, nor is it one of a collection of value theorems. The lack of a hyphen is sanctioned by usage, as it is in the names of other theorems with compound qualifiers (as in "small oscillation theorem").

In the following, the interval $[a, b]$ has distinct endpoints, and is manifestly bounded and closed.

Proposition 5.7 (Rolle's theorem) *Let $f : [a, b] \to \mathbb{R}$ be a continuous function that is differentiable for $a < x < b$. Assume that $f(a) = f(b)$. Then there exists c, such that $a < c < b$ and $f'(c) = 0$.*

Proof Let $m = \inf_{[a,b]} f$ and $M = \sup_{[a,b]} f$ (both m and M are attained by the extreme value theorem, so they are minimum and maximum). If $m = M$ then f is a constant and so $f'(x) = 0$ for all x in $]a, b[$ and we are done.

Assume next that $m < M$. If these values are attained at the endpoints, then, since $f(a) = f(b)$, we again have $m = M$. So at least one of them is attained at an interior point. Either there exists c in $]a, b[$ such that $f(c) = m$ or there exists c in $]a, b[$ such that $f(c) = M$. In both these cases we have $f'(c) = 0$. □

Proposition 5.8 (Mean value theorem) *Let $f : [a, b] \to \mathbb{R}$ be a continuous function that is differentiable for $a < x < b$. Then there exists c in the open interval $]a, b[$, such that*

$$f(b) - f(a) = f'(c)(b - a).$$

Proof Set

$$A = \frac{f(b) - f(a)}{b - a}.$$

Then $f(a) - Aa = f(b) - Ab$. Define the function $g(x) = f(x) - Ax$ for $a \le x \le b$. Now $g(a) = g(b)$ and we deduce that there exists c, such that $a < c < b$ and $g'(c) = 0$. Then $f'(c) - A = 0$ and we find

$$\frac{f(b) - f(a)}{b - a} = f'(c),$$

which gives $f(b) - f(a) = f'(c)(b - a)$.   □

5.6.1 First Consequences of the Mean Value Theorem

The following reformulation of the mean value theorem is often useful.

Proposition 5.9 (Mean value theorem, version 2) *Let $A =]a, b[$ and let $f : A \to \mathbb{R}$ be a differentiable function. Let x and $x + h$ both lie in A (note that h could be negative). Then there exists θ, such that $0 < \theta < 1$ and $f(x + h) = f(x) + hf'(x + \theta h)$.*

Proof Apply the mean value theorem to the interval with endpoints x and $x + h$, and write the point c in the form $c = x + \theta h$. Then $0 < \theta < 1$.   □

It is surprising that only now, with the mean value theorem in place, do we have the machinery to give a nice proof of the following "obvious" result.

Proposition 5.10 *Let A be an open interval (which could be unbounded), let f be differentiable in A and suppose that $f'(x) = 0$ for all x in A. Then f is a constant in the interval A.*

Proof Let a and b be points in A. By the mean value theorem we have $f(b) - f(a) = f'(c)(b - a)$ for some c between a and b. But then $f(a) - f(b) = 0$, that is, $f(a) = f(b)$. We deduce that f is constant in A.   □

It is important that A should be an interval. If, for example, A is the set $]0, 1[\cup]2, 3[$, then there exists a function with domain A, differentiable and satisfying $f'(x) = 0$ at every point of A, but f is not constant in A.

Another important application is to give a criterion for a function to be increasing or decreasing.

Proposition 5.11 *Let the function $f :]a, b[\to \mathbb{R}$ be differentiable and assume that $f'(x) > 0$ for all x in $]a, b[$. Then f is strictly increasing. If, on the other hand, $f'(x) < 0$ for all x in $]a, b[$, then f is strictly decreasing.*

Proof Let $f'(x) > 0$ for all x in $]a, b[$ and let $a < x_1 < x_2 < b$. Then there exists c in $]x_1, x_2[$, such that $f(x_2) - f(x_1) = f'(c)(x_2 - x_1) > 0$. The argument for decreasing is similar. □

Note that if f is strictly increasing we cannot conclude that $f'(x) > 0$ for all x. However, dropping the strictness, we obtain a correct result: a function differentiable in an open interval is increasing if and only if $f'(x) \geq 0$ for all x in its domain. See the exercises. mul

5.6.2 Exercises

1. Find a function f for which the domain is the open set $]0, 1[\cup]2, 3[$, and is such that the derivative of f is zero at each point in its domain, but f is not a constant function.
2. Show that Rolle's theorem is true in case f is defined and differentiable in the *open* interval $]a, b[$, and $\lim_{x \to a+} f(x) = \lim_{x \to b-} f(x)$. Note that a could be $-\infty$ and b could be $+\infty$. Furthermore the two limits could also be infinite.
3. Find an example of a function f, differentiable and strictly increasing in an open interval A, but for which the inequality $f'(x) > 0$ fails for at least one point x.
4. Show that a function f, differentiable in an open interval A, is increasing if and only if $f'(x) \geq 0$ for all x in A.
5. Let f be a function on the domain $]a, b[$, where b is a finite number. Suppose that f is differentiable and there exists a constant K, such that $|f'(x)| < K$ for all x in the domain. Show that $\lim_{t \to b-} f(t)$ exists and is a finite number.
6. Suppose that f is known to be continuous in an interval $]a, b[$, differentiable at all points in $]a, b[$ except possibly at a point c, and it is known that $\lim_{x \to c} f'(x) = \ell$, where ℓ is a finite number. Show that f is differentiable at c and $f'(c) = \ell$.
   Note. In cases when $f(x)$ is given by a nice formula for $x \neq c$ (such as in Sect. 5.2, Exercise 3) many a beginning student might compute the derivative $f'(c)$ correctly by taking the limit $\lim_{x \to c} f'(x)$ without appreciating that it needs justification. Should a teacher give "correct"?
7. Let the function f be defined and continuous in an open interval A. Suppose that c is a point in A and that f has derivatives up to order m on the set $A \setminus \{c\}$. Suppose further that $\lim_{x \to c} f^{(k)}(x)$ exists for $k = 1, ..., m$ and the limits are finite numbers. Show that f has derivatives up to order m in all of A. Moreover $f^{(k)}(c) = \lim_{x \to c} f^{(k)}(x)$, for $k = 1, ..., m$.
8. We have seen that a continuous function f on a domain A, where A is an open interval, has the intermediate value property: if a and b are points of A, and η lies strictly between $f(a)$ and $f(b)$, then there exists c strictly between a and b such that $f(c) = \eta$.
   There is another general class of functions that possess the intermediate value property. Suppose that f is differentiable everywhere in A. There is no reason to suppose that f' is continuous; indeed it may have discontinuities. Nevertheless

f' has the intermediate value property. Prove this.

Hint. Begin with the case when $f'(a) > 0$, $f'(b) < 0$ and show that there exists c between a and b such that $f'(c) = 0$.

Note. The intermediate value property for derivatives is often treated as an intriguing but otherwise rather abstruse fact. Actually it is often useful in conjunction with the fact embodied in Sect. 4.3, Exercise 6, when multiple applications of the mean value theorem or Taylor's theorem are used to derive remainder formulas.

9. Suppose that f is differentiable and that f' is monotonic. Prove that f' is continuous.

5.7   The Derivative as a Linear Approximation

If f is differentiable at a we can rewrite the formula

$$f'(a) = \lim_{x \to a} \frac{f(x) - f(a)}{x - a}$$

by defining $R(x, a)$ to satisfy

$$f(x) = f(a) + (x - a) f'(a) + R(x, a),$$

and viewing $R(x, a)$ as the error when $f(x)$ is approximated by the first-degree polynomial $f(a) + (x - a) f'(a)$. The approximation improves as x approaches a, not just because $\lim_{x \to a} R(x, a) = 0$, but because of the stronger conclusion (embodied in the definition of derivative as limit) that

$$\frac{R(x, a)}{x - a} = \frac{f(x) - f(a)}{x - a} - f'(a) \to 0, \quad \text{(when } x \to a\text{)}.$$

The error becomes arbitrarily small in comparison with $x - a$ as x approaches a.

5.7.1   *Higher Derivatives and Taylor Polynomials*

Now suppose that $f'(a), \ldots, f^{(m)}(a)$ all exist (for the meaning of this see Sect. 5.4). As a generalisation of the above approximation rule using the first derivative, we have the more complicated

$$f(x) = f(a) + (x - a) f'(a) + \frac{1}{2!}(x - a)^2 f''(a) + \cdots + \frac{1}{m!}(x - a)^m f^{(m)}(a)$$

$$+ R_{m+1}(x, a)$$

and the error $R_{m+1}(x, a)$ satisfies

$$\lim_{x \to a} \frac{R_{m+1}(x, a)}{(x - a)^m} = 0.$$

The polynomial

$$P_m(x, a) := f(a) + (x - a)f'(a) + \frac{1}{2!}(x - a)^2 f''(a) + \cdots + \frac{1}{m!}(x - a)^m f^{(m)}(a)$$

is called the Taylor polynomial of f with degree m centred at a. As a polynomial in x it may have degree less than m so maybe one should use the term "order" instead of "degree". Whichever term one uses it is an approximation to $f(x)$ which improves as x approaches a, in the sense that the error becomes arbitrarily small in comparison with $(x - a)^m$. The proof of this will be given shortly.

5.7.2 Comparison to Taylor's Theorem

The claim made in the last subsection is not what we nowadays call Taylor's theorem. That celebrated result, which will be proved later, consists essentially of an estimate for the error $R_m(x, a)$, that often allows us to conclude that *if x is kept constant and if f satisfies certain additional conditions, then $R_m(x, a)$ tends to 0 as $m \to \infty$.*

Sometimes the conclusion stated in the previous section is called Peano's form of Taylor's theorem, and sometimes Young's, but these terms can be confusing and it is better to maintain a clear distinction between it and Taylor's theorem. One might add another source of confusion, that most named attributions of versions of Taylor's theorem are historically questionable (including the attribution to Taylor).

Compare these conclusions. In all cases we have

$$f(x) = P_m(x, a) + R_{m+1}(x, a).$$

The conclusion stated in the previous section was

$$\lim_{x \to a} \frac{R_{m+1}(x, a)}{(x - a)^m} = 0 \quad \text{(note: } m \text{ is held constant)}.$$

Taylor's theorem can sometimes justify the conclusion, that for all x in some interval $]a - h, a + h[$, we have

$$\lim_{m \to \infty} R_m(x, a) = 0 \quad \text{(note: } x \text{ is held constant)}$$

but to obtain this conclusion a close examination of f is usually needed.

5.7.3   Cauchy's Form of the Mean Value Theorem

Proposition 5.12 *Let* $f : [a, b] \to \mathbb{R}$ *and* $g : [a, b] \to \mathbb{R}$ *be continuous functions, that are differentiable for* $a < x < b$. *Then there exists c in the open interval* $]a, b[$, *such that*

$$\big(f(b) - f(a)\big)g'(c) = \big(g(b) - g(a)\big)f'(c).$$

Proof Let $h(x) = Af(x) + Bg(x)$ for $a \leq x \leq b$. Choose the constants A and B so that they are not both zero but $h(a) = h(b)$ (possible in many ways by simple algebra). Then there exists c in $]a, b[$, such that $h'(c) = 0$, which is to say

$$Af'(c) + Bg'(c) = 0.$$

But since $h(b) - h(a) = 0$ we also have

$$A(f(b) - f(a)) + B(g(b) - g(a)) = 0.$$

Now A and B are not both 0, so by a popular rule of linear algebra we must have

$$\big(f(b) - f(a)\big)g'(c) = \big(g(b) - g(a)\big)f'(c)$$

as required.   □

The formula in Cauchy's mean value theorem can be written in the memorable form

$$\frac{f(b) - f(a)}{g(b) - g(a)} = \frac{f'(c)}{g'(c)}$$

provided neither denominator is zero.

Another way to state and prove Cauchy's mean value theorem uses determinants. We set

$$\phi(x) = \begin{vmatrix} f(b) - f(a) & f(x) \\ g(b) - g(a) & g(x) \end{vmatrix},$$

note that $\phi(a) = \phi(b)$, and deduce, by Rolle's theorem that, for some c between a and b we have

$$\begin{vmatrix} f(b) - f(a) & f'(c) \\ g(b) - g(a) & g'(c) \end{vmatrix} = 0.$$

5.7.4   Geometric Interpretation of the Mean Value Theorem

Consider the curve $y = f(x)$ in the (x, y)-plane between $x = a$ and $x = b$. The mean value theorem says that there exists a point c between a and b where the tangent line

Fig. 5.3 The mean value
theorem at a glance

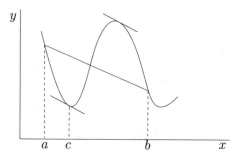

Fig. 5.4 Cauchy's mean
value theorem at a glance

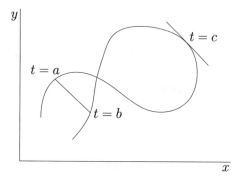

to the curve at the point $(c, f(c))$ is parallel to the chord joining the points $(a, f(a))$
and $(b, f(b))$. There may be more than one point with the property possessed by c.
This is illustrated in Fig. 5.3.

Cauchy's mean value theorem also has a geometric interpretation, but requires
the use of plane, parametrised curves. We assume that the reader is acquainted with
these and with vectors in the plane.

The equation

$$\big(f(b) - f(a)\big)g'(c) = \big(g(b) - g(a)\big)f'(c)$$

says that the vectors $\big(f(b) - f(a), g(b) - g(a)\big)$ and $\big(f'(c), g'(c)\big)$ are parallel (as
is clear from the determinantal version given in the last section). Consider now the
parametrised plane curve

$$x = f(t), \quad y = g(t).$$

Given the parameters $t = a$ and $t = b$, yielding points $(f(a), g(a))$ and $(f(b), g(b))$
on the curve, there exists a parameter $t = c$ between a and b, such that the tangent
line to the curve at parameter $t = c$ is parallel to the chord joining $(f(a), g(a))$ and
$(f(b), g(b))$. This is illustrated in Fig. 5.4.

There may be more than one tangent to the curve at the plane point $(f(c), g(c))$,
as the curve may happen to cross itself at this point. That is why we refer to the
tangent as being at parameter value c, rather than at the point $(f(c), g(c))$.

5.7.5   *Exercises*

1. The idea that $f(a) + hf'(a)$ is an approximation to $f(a + h)$ gives us a useful rule of thumb for improving an initial approximation to $f(a + h)$. Try it in the the following examples. You can test the improvement using a calculator.

   (a) $\sqrt{65}$ with initial approximation 8.
   (b) $\sqrt[3]{28}$ with initial approximation 3.
   (c) $626^{3/4}$ with initial approximation 125.

2. Let f be twice differentiable in an interval A. Show that for every distinct pair of points a and x in A there is a point ξ, strictly between a and x, such that

$$f(x) - f(a) - f'(a)(x - a) = \frac{1}{2} f''(\xi)(x - a)^2.$$

Hint. Apply Cauchy's mean value theorem twice in a row to the quotient

$$\frac{f(x) - f(a) - f'(a)(x - a)}{(x - a)^2}.$$

Note. This result is a particular case of Taylor's theorem. It enables us to estimate the error, when $f(x)$ is approximated by $f(a) + f'(a)(x - a)$, if we know some bound for the second derivative.

3. Use the result of the previous exercise to estimate the error in the approximations of Exercise 1.

4. Prove Liouville's theorem. Let α be a root of the polynomial equation $P(x) := a_n x^n + a_{n-1} x^{n-1} + \cdots + a_0 = 0$ where the coefficients $a_0, a_1, \ldots a_n$ are integers and $a_n \neq 0$. Assume also that this equation has no rational solutions. Prove that there exists a number $c > 0$, such that for all rational numbers p/q the inequality

$$\left| \alpha - \frac{p}{q} \right| > \frac{c}{q^n}$$

holds. Use this to show that the number

$$L := \sum_{n=1}^{\infty} 10^{-n!}$$

is not the root of any polynomial equation with integer coefficients.

Hint. Assuming first that $|\alpha - (p/q)| \leq 1$ apply the mean value theorem to the difference $P(\alpha) - P(p/q)$. Then remove the assumption.

Note. Real numbers can be divided into two classes. The algebraic numbers are those that are the roots of polynomial equations with integer coefficients. These include not only all the rational numbers, but also numbers like $\sqrt{2}, \sqrt{2 + \sqrt{2}}$ and $\sqrt[3]{5}$, that is, numbers expressible by radicals. They also include numbers like the positive root of $x^5 - x + 1 = 0$ which cannot be expressed by radicals. Irrational numbers that are not algebraic are called transcendental numbers. For a long time it was not known whether any transcendental numbers existed. Then finally, in 1844, Liouville exhibited an example of a transcendental number, similar to the one appearing in this exercise.

5.8 L'Hopital's Rule

In its simplest form this popular and useful rule is as follows. Let A be an open interval, and $f : A \to \mathbb{R}$, $g : A \to \mathbb{R}$ differentiable functions. Let c be a point in A, and suppose that $f(c) = g(c) = 0$, but that $g'(c) \neq 0$. Then we have

$$\lim_{x \to c} \frac{f(x)}{g(x)} = \frac{f'(c)}{g'(c)}.$$

To prove this we simply observe that for $x \neq c$ we have

$$\frac{f(x)}{g(x)} = \frac{\dfrac{f(x) - f(c)}{x - c}}{\dfrac{g(x) - g(c)}{x - c}}$$

and this tends to $f'(c)/g'(c)$ as x tends to c, by the definition of derivative and the rule for the limit of a quotient.

L'Hopital's rule can be framed in a more general form that vastly increases its usefulness. We no longer assume that the derivatives $f'(c)$ and $g'(c)$ exist. Instead we assume that $f'(x)/g'(x)$ tends to a limit as x tends to c. It is even more useful to take the limit as one-sided; after all, a two-sided limit is just a pair of one-sided limits that happen to be equal.

Proposition 5.13 (L'Hopital's rule for $0/0$) *Let* $f :]a, b[\to \mathbb{R}$, $g :]a, b[\to \mathbb{R}$ *be differentiable functions such that* $g'(x) \neq 0$ *for all* x *in the interval of definition. Suppose that*

$$\lim_{x \to a+} f(x) = 0, \quad \lim_{x \to a+} g(x) = 0, \quad \lim_{x \to a+} \frac{f'(x)}{g'(x)} = t.$$

Then

$$\lim_{x \to a+} \frac{f(x)}{g(x)} = t.$$

A similar conclusion holds for the limit $\lim_{x \to b-} f(x)/g(x)$.
The rule also holds if $a = -\infty$ *or* $b = \infty$; *or if* $t = \infty$ *or* $t = -\infty$.

Proof Consider first the right-hand limit at a in the case that a is not $-\infty$ and t is a finite number.

Let $\varepsilon > 0$. There exists $\delta > 0$, such that

$$\left| \frac{f'(x)}{g'(x)} - t \right| < \varepsilon$$

for all x that satisfy $a < x < a + \delta$. Let x and y satisfy $a < y < x < a + \delta$. By Cauchy's form of the mean value theorem there exists z between x and y, such that

$$\frac{f(x) - f(y)}{g(x) - g(y)} = \frac{f'(z)}{g'(z)}.$$

We deduce that for all such x and y we have

$$\left| \frac{f(x) - f(y)}{g(x) - g(y)} - t \right| < \varepsilon.$$

Let now $y \to a+$. We have that

$$\lim_{y \to a+} \frac{f(x) - f(y)}{g(x) - g(y)} = \frac{f(x)}{g(x)},$$

so that the inequality

$$\left| \frac{f(x)}{g(x)} - t \right| \leq \varepsilon$$

holds for all x that satisfy $a < x < a + \delta$. This proves the first assertion of L'Hopital's rule.

Next consider the case when $b = \infty$ and t is a finite number. We will determine $\lim_{x \to \infty} f(x)/g(x)$, the assumptions being that $\lim_{x \to \infty} f(x)$ and $\lim_{x \to \infty} g(x)$ are both 0, and $\lim_{x \to \infty} f'(x)/g'(x) = t$.

Let $\varepsilon > 0$. There exists K, such that

$$\left| \frac{f'(x)}{g'(x)} - t \right| < \varepsilon$$

for all $x > K$. Let $K < x < y$. There exists z between x and y, such that

$$\frac{f(x) - f(y)}{g(x) - g(y)} = \frac{f'(z)}{g'(z)},$$

and therefore, for all such x and y we have

$$\left| \frac{f(x) - f(y)}{g(x) - g(y)} - t \right| < \varepsilon.$$

Now let $y \to \infty$. We find that

$$\left| \frac{f(x)}{g(x)} - t \right| \le \varepsilon$$

for all x that satisfy $x > K$, thus proving the rule in this case.

Consider the case when $t = \infty$, and a is a finite number. We shall determine $\lim_{x \to a+} f(x)/g(x)$, the assumptions being that $\lim_{x \to a+} f(x)$ and $\lim_{x \to a+} g(x)$ are both 0, and $\lim_{x \to a+} f'(x)/g'(x) = \infty$.

Let K be a real number and choose $\delta > 0$, such that

$$\frac{f'(x)}{g'(x)} > K$$

for all x that satisfy $a < x < a + \delta$. For all x and y that satisfy $a < y < x < a + \delta$ we obtain

$$\frac{f(x) - f(y)}{g(x) - g(y)} > K.$$

Let $y \to a+$. We deduce that $f(x)/g(x) \ge K$ for all x that satisfy $a < x < a + \delta$.

The reader should write out the proofs for all the remaining cases; each is similar to one of the cases treated above. The common feature is the use of Cauchy's mean value theorem. $\qquad \square$

5.8.1   Using L'Hopital's Rule

There are two important things to bear in mind when one uses L'Hopital's rule. Firstly, $f(x)$ and $g(x)$ should both tend to 0 at the point where the limit of $f(x)/g(x)$ is sought. This is why we sometimes say that the rule resolves the indeterminate form $0/0$. Failure to observe this can lead to mistakes.

Secondly, we must observe the premise that $f'(x)/g'(x)$ has a limit. Thus it is not strictly correct, having first observed that $f(x)$ and $g(x)$ both tend to 0, to write that

$$\lim_{x \to a+} \frac{f(x)}{g(x)} = \lim_{x \to a+} \frac{f'(x)}{g'(x)},$$

before ascertaining that the limit on the right actually exists. For example there are cases when the limit on the left-hand side exists, but the limit on the right does not. Even so, we often write this, in the spirit of "let's wait and see," especially when the rule is used iteratively (more on this later) and it rarely leads to mistakes.

5.8.2   Is There an Error in the Proof?

The claim that

$$\lim_{y \to a+} \frac{f(x) - f(y)}{g(x) - g(y)} = \frac{f(x)}{g(x)}$$

appears in the proof of L'Hopital's rule. For this to be correct we must know that $g(x) \neq 0$. We are assuming that $\lim_{x \to a+} g(x) = 0$ and there appears to be a danger that $g(x)$ might be 0 for some values of x near to a, maybe even for infinitely many values.

In fact we are safe on this score. One assumption was that $g'(x) \neq 0$ for $a < x < b$. Therefore given $\varepsilon > 0$ we can (referring to the proof) find $\delta > 0$ having the properties stated in the proof, but also such that $g(x) \neq 0$ for all x that satisfy $a < x < a + \delta$. This is because the equation $g(x) = 0$ can have at most one solution in the open interval $]a, b[$, since otherwise, by Rolle's theorem, g' would have a zero in $]a, b[$.

The assumption that $g'(x) /= 0$ for all x in its domain of definition is unnecessarily strong for applying L'Hopital's rule to calculate $\lim_{x \to a+} f(x)/g(x)$. Obviously it is enough that there should exist $h > 0$, such that $g'(x) \neq 0$ for $a < x < a + h$.

5.8.3   Geometric Interpretation of L'Hopital's Rule

The two functions f and g define a parametric curve in the (x, y)-plane by letting $x = g(t)$, $y = f(t)$ for $a < t < b$. The assumptions that $\lim_{t \to a+} f(t) = \lim_{t \to a+} g(t) = 0$ have the geometrical interpretation that the initial point of the curve is the coordinate origin $O = (0, 0)$. Let $P(t)$ denote the point $\big(g(t), f(t)\big)$ in the plane, that is, the point on the curve with parameter t. Associated with the point $P(t)$ on the curve we can construct two lines. Firstly, the tangent at the point $P(t)$, corresponding to parameter t. Careful! The curve might cross itself. Secondly the chord joining $P(t)$ to the origin O.

L'Hopital's rule says the following: if the slope of the tangent has a limit as t tends to a (from the right), then the slope of the chord joining $P(t)$ to O has the same limit. The result is also valid if the limit is infinite; both chord and tangent then tend to a vertical position. The geometric interpretation of L'Hopital's rule is illustrated in Fig. 5.5.

5.8.4   Iterative Use of L'Hopital's Rule: Taylor Polynomials Again

If we wish to find the limit $\lim_{x \to a+} f(x)/g(x)$ using L'Hopital's rule we are directed to find the limit $\lim_{x \to a+} f'(x)/g'(x)$. This limit, too, may be found by L'Hopital's rule if it happens that $\lim_{x \to a+} f'(x) = \lim_{x \to a+} g'(x) = 0$; then we are directed

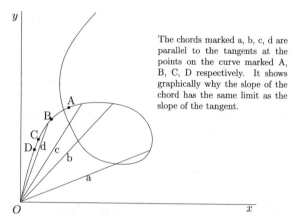

Fig. 5.5 L'Hopital's rule at a glance

The chords marked a, b, c, d are parallel to the tangents at the points on the curve marked A, B, C, D respectively. It shows graphically why the slope of the chord has the same limit as the slope of the tangent.

to the limit $\lim_{x \to a+} f''(x)/g''(x)$. Again it could happen that $\lim_{x \to a+} f''(x) = \lim_{x \to a+} g''(x) = 0$. As long as numerator and denominator have the limit 0 we may differentiate them, until a limit is found that we can easily compute.

The iterative use of L'Hopital's rule gives an easy proof of the approximation property of Taylor polynomials stated in Sect. 5.7 under the heading "Higher derivatives and Taylor polynomials".

Proposition 5.14 *Let* $f : A \to \mathbb{R}$, *where* A *is an open interval, and let* $c \in A$. *Assume that the derivatives* $f'(c), ..., f^{(m)}(c)$ *all exist and define*

$$E(h) = f(c + h) - \left(f(c) + \frac{1}{1!} f'(c)h + \frac{1}{2!} f''(c)h^2 + \cdots + \frac{1}{m!} f^{(m)}(c)h^m \right)$$

for all h *such that* $|h|$ *is sufficiently small. Then*

$$\lim_{h \to 0} \frac{E(h)}{h^m} = 0.$$

Note that h can be positive or negative, but we require that $c + h$ is in the interval A. That is why we want $|h|$ to be "sufficiently small".

The assumption that f has derivatives at c up to order m means that f is $(m - 1)$-times differentiable in some open interval containing c, and $f^{(m-1)}$ is differentiable at c.

Proof of the Proposition Differentiating $E(h)$ repeatedly with respect to h we obtain, for $j = 1, ..., m - 1$ and for all h such that $|h|$ is sufficiently small,

$$E^{(j)}(h) = f^{(j)}(c + h) - \left(f^{(j)}(c) + \frac{1}{1!} f^{(j+1)}(c)h + \cdots + \frac{1}{(m - j)!} f^{(m)}(c)h^{m-j} \right),$$

from which we see that $E^{(j)}(0) = 0$ for $j = 1, ..., m - 1$. We also have (convenient to use Leibniz's notation here)

$$\frac{d^j}{dh^j}h^m = \frac{m!}{(m-j)!}h^{m-j} \text{ for } j = 0, 1, 2, ..., m,$$

so that

$$\frac{d^j}{dh^j}h^m\bigg|_{h=0} = 0 \text{ for } j = 0, 1, 2, ..., m-1.$$

Using L'Hopital's rule iteratively (with a "wait and see" approach to the existence of the limits) now gives

$$\lim_{h\to 0}\frac{E(h)}{h^m} = \lim_{h\to 0}\frac{E^{(m-1)}(h)}{m!h} = \frac{1}{m!}\lim_{h\to 0}\frac{f^{(m-1)}(c+h) - f^{(m-1)}(c) - f^{(m)}(c)h}{h},$$

and this is 0 by definition of derivative (and it also tells us that $E^{(m)}(0) = 0$). This completes the proof.   ☐

Let us write $x - c$ for h in the formula for $E(h)$. We obtain the conclusion

$$\lim_{x\to c}\frac{f(x) - P_m(x, c)}{(x - c)^m} = 0.$$

This describes admirably how the approximation to $f(x)$ by the Taylor polynomial $P_m(x, c)$ improves sharply as x approaches c. On the other hand there is no reason to think that the approximation improves if we hold x fixed and increase m (assuming we have the derivatives). This question is partly settled by Taylor's theorem proper in a later chapter.

5.8.5   Application to Maxima and Minima

If the derivative of f at c is zero, the examination of higher derivatives at c can sometimes resolve the question as to whether c is a local maximum point or a local minimum point.

Proposition 5.15 *Let $f :]a, b[\to \mathbb{R}$ and let $a < c < b$. Assume that f is $(m-1)$-times differentiable, that $f^{(j)}(c) = 0$ for $j = 1, 2, ...m - 1$, but that $f^{(m)}(c)$ exists and is not 0. In addition to all this assume that m is an even number. The following conclusions then hold:*

(1) If $f^{(m)}(c) > 0$ then c is a strict local minimum point.
(2) If $f^{(m)}(c) < 0$ then c is a strict local maximum point.

Proof By Proposition 5.14 we have

$$f(c + h) - f(c) = \frac{1}{m!}f^{(m)}(c)h^m + E(h)$$

where the error term $E(h)$ satisfies

$$\lim_{h \to 0} \frac{E(h)}{h^m} = 0.$$

For $h \neq 0$ we can write

$$\frac{f(c+h) - f(c)}{h^m} = \frac{1}{m!} f^{(m)}(c) + \frac{E(h)}{h^m}$$

and we know that $f^{(m)}(c) \neq 0$. Since m is even we must have $h^m > 0$, both for $h > 0$ and for $h < 0$. We conclude that there exists $\delta > 0$, such that $f(c+h) - f(c) \neq 0$ and has the same sign as $f^{(m)}(c)$, for all h that satisfy $0 < |h| < \delta$. This is precisely the sought-for conclusion. ☐

5.8.6   More on L'Hopital's Rule: The ∞/∞ Version

Sometimes we consider L'Hopital's rule as resolving the indeterminate form $0/0$, an expression that is really quite meaningless. Since we are indulging in meaninglessness we might suggest some other indeterminate forms, for example

$$\frac{\infty}{\infty}, \quad 0.\infty, \quad 0^0, \quad \infty - \infty.$$

These can often be resolved by some judicious manipulations combined with L'Hopital's rule. However there is a version of the rule directly applicable to ∞/∞ and, as we shall see, it turns out to be very useful.

Proposition 5.16 *Let* $f :]a, b[\to \mathbb{R}$, $g :]a, b[\to \mathbb{R}$ *be differentiable functions such that* $g'(x) \neq 0$ *for all x in its domain of definition. Assume that*

$$\lim_{x \to a+} g(x) = \infty, \quad \lim_{x \to a+} \frac{f'(x)}{g'(x)} = t.$$

Then

$$\lim_{x \to a+} \frac{f(x)}{g(x)} = t.$$

A similar conclusion holds for $\lim_{x \to b-} f(x)/g(x)$.
   The rule also holds if $a = -\infty$ *or* $b = \infty$; *or if* $t = \infty$ *or* $t = -\infty$.

Note that we made no assumption about $\lim_{x \to a+} f(x)$. This is not a mistake.

Proof of the Proposition We shall only consider the case when t and a are finite numbers. The other cases are left to the reader to complete.
   Let $\varepsilon > 0$. Since $\lim_{x \to a+} f'(x)/g'(x) = t$, there exists $\delta_1 > 0$, such that

$$\left| \frac{f'(z)}{g'(z)} - t \right| < \varepsilon$$

for all z that satisfy $a < z < a + \delta_1$.

Let $a < x < y < a + \delta_1$. It follows by Cauchy's form of the mean value theorem, that

$$\left| \frac{f(x) - f(y)}{g(x) - g(y)} - t \right| < \varepsilon,$$

which we rewrite in the form

$$t - \varepsilon < \frac{f(x) - f(y)}{g(x) - g(y)} < t + \varepsilon.$$

Now keep y fixed (if desired we could fix y as $a + \frac{1}{2}\delta_1$, but the thing we do require is that $a < x < y < a + \delta_1$). Since $\lim_{x \to a+} g(x) = \infty$ we find that $g(x) > 0$ and $g(x) - g(y) > 0$ when x is sufficiently close to a, for example for $a < x < a + \delta_2$, and then we have, for $a < x < y < \delta_1$ and $a < x < a + \delta_2$, that

$$\big(g(x) - g(y)\big)(t - \varepsilon) < f(x) - f(y) < \big(g(x) - g(y)\big)(t + \varepsilon).$$

Dividing by $g(x)$ (which is positive) gives

$$\left(1 - \frac{g(y)}{g(x)}\right)(t - \varepsilon) < \frac{f(x) - f(y)}{g(x)} < \left(1 - \frac{g(y)}{g(x)}\right)(t + \varepsilon),$$

and therefore

$$\frac{f(y)}{g(x)} + \left(1 - \frac{g(y)}{g(x)}\right)(t - \varepsilon) < \frac{f(x)}{g(x)} < \frac{f(y)}{g(x)} + \left(1 - \frac{g(y)}{g(x)}\right)(t + \varepsilon).$$

As $x \to a+$ the left-hand member of the inequalities tends to $t - \varepsilon$ and the right-hand member to $t + \varepsilon$ (recall that we keep y constant). Hence there exists $\delta_3 > 0$, such that the left-hand member is above $t - 2\varepsilon$ and the right-hand member below $t + 2\varepsilon$ for all x that satisfy $a < x < a + \delta_3$. Let $\delta = \min(\delta_2, \delta_3)$. If $a < x < a + \delta$ we find

$$t - 2\varepsilon < \frac{f(x)}{g(x)} < t + 2\varepsilon.$$

This says that $\lim_{x \to a+} f(x)/g(x) = t$ and concludes the proof of the first claim. The proofs of the remaining claims are left to the reader.   □

The proof raises some interesting speculation about the meaning of "for each ε there exists δ". It is too simple to say that δ is supposed to be a function of ε. In the above proof we first chose δ_1, in a non-explicit fashion, from the set of all possible numbers that would work for the limit $\lim_{x \to a+} f'(x)/g'(x)$. Then y was chosen rather arbitrarily. A workable value was defined in an aside but that was not really necessary. Finally we found a δ that worked.

Where desired and possible we can try to define the quantities we use by functions (such as using the max function). But at times we have to say, as in effect we did at the beginning of the above proof, "here is a set (of usable δ's), known to be non-empty; let us choose one". Sometimes it is simply not very helpful to try to see δ as an explicitly computable function of ε.

5.8.7 Exercises

1. Calculate the following limits:

   (a) $\lim\limits_{x \to 1} \dfrac{x^4 - x^3 - x + 1}{x^4 - 3x^3 + 2x^2 + x - 1}$

   (b) $\lim\limits_{x \to 0} \dfrac{\sqrt{1+x} - 1}{\sqrt[3]{1+x} - 1}$

   (c) $\lim\limits_{x \to \infty} \sqrt{x^2 + x} - x.$

2. Calculate the following limits. Use your school knowledge of the circular functions $\sin x$ and $\cos x$ and their derivatives (or refer to Sect. 5.1, Exercise 2).

   (a) $\lim\limits_{x \to 0} \dfrac{1}{x} - \dfrac{1}{\sin x}$

   (b) $\lim\limits_{x \to 0} \dfrac{1}{x^2} - \dfrac{1}{x \sin x}$

   (c) $\lim\limits_{x \to 0} \dfrac{1}{6x} + \dfrac{1}{x^3} - \dfrac{1}{x^2 \sin x}.$

3. Exploiting only two properties of the exponential function:

$$\lim_{x \to \infty} e^x = \infty \quad \text{and} \quad \frac{d}{dx} e^x = e^x,$$

   show that for any natural number n we have $\lim_{x \to \infty} e^x / x^n = \infty$.
   Note. The conclusion demonstrates the proverbial growth of the exponential function in a graphic way; it overpowers any polynomial.

4. Let f be twice differentiable in an interval A, let a be a point in A and suppose that $f''(a) \neq 0$. Show that the tangent to the graph $y = f(x)$ at the point $(a, f(a))$ does not cross the graph at $(a, f(a))$. Show, in addition, that there exists $\delta > 0$, such that the tangent and the graph have no common point in the interval $]a - \delta, a + \delta[$, except at $x = a$.

5. Suppose the function f is differentiable in an interval A and let $a \in A$. Let λ and μ be distinct numbers. Show that

$$f'(a) = \lim_{h \to 0} \frac{f(a + \lambda h) - f(a + \mu h)}{(\lambda - \mu)h}.$$

Note. A case important for *numerical differentiation* is $\lambda = 1, \mu = -1$. More general formulas are known, approximating the first, and higher, derivatives. See the next exercises.

6. Suppose the function f is differentiable in an interval A and let $a \in A$. Show that

$$f'(a) = \lim_{h \to 0} \frac{-f(a + 2h) + 8f(a + h) - 8f(a - h) + f(a - 2h)}{12h}.$$

7. Suppose the function f is twice differentiable in an interval A and let $a \in A$. Show that

$$f''(a) = \lim_{h \to 0} \frac{f(a + h) - 2f(a) + f(a - h)}{h^2}.$$

8. (a) Show that for all positive integers n

$$\sum_{k=0}^{n} (-1)^k \binom{n}{k} k^j = \begin{cases} 0, & j = 0, 1, ..., n - 1 \\ (-1)^n n! & j = n. \end{cases}$$

   Hint. Expand $(1 - x)^n$ by the binomial rule. Repeatedly differentiate, but with a twist.

   (b) Suppose the function f is n times differentiable in an interval A and let $a \in A$. Show that

$$f^{(n)}(a) = \lim_{h \to 0} \frac{1}{h^n} \sum_{k=0}^{n} (-1)^{k+n} \binom{n}{k} f(a + kh).$$

9. We can define the left derivative of f at c, denoted by $D_l f(c)$, and the right derivative $D_r f(c)$, in the obvious way:

$$D_l f(c) = \lim_{x \to c-} \frac{f(x) - f(c)}{x - c}, \quad D_r f(c) = \lim_{x \to c+} \frac{f(x) - f(c)}{x - c}$$

   when these limits exist. Now suppose that f is differentiable in an interval $]c - \alpha, c[$, continuous in $]c - \alpha, c]$ and the limit $\lim_{x \to c-} f'(x)$ exists and is a finite number A. Show that $D_l f(c)$ exists and equals A. A similar result holds for the right derivative.
   Show that the result also holds if $A = \infty$ or $-\infty$, if we allow a derivative to be infinite (the definition should be obvious).

10. The previous exercise has an interesting consequence. Suppose that f is differentiable everywhere in an open interval A. Show that discontinuities of f', if there are any, are never jump discontinuities.

   Note. f' can be discontinuous. An example was exhibited in Sect. 5.5, Exercise 4.

11. Show that if f is differentiable everywhere in an open interval A, and if f' is increasing, then f' is continuous.

12. Find an example of differentiable functions f and g with domain $\mathbb{R}$, such that $f(0) = g(0) = 0$, $\lim_{x\to 0} f(x)/g(x)$ exists, but $\lim_{x\to 0} f'(x)/g'(x)$ does not exist.

13. Suppose that f has n derivatives at a point c (meaning that f has $n-1$ derivatives in an interval A containing c, and that $f^{(n-1)}$ is differentiable at c) and that $f(c) = 0$. One expects that the function $f(x)/(x-c)$, extended to be $f'(c)$ at c, should have (at least) $n-1$ derivatives at c. This is quite tricky with such minimalistic premises. One can prove the following proposition, in which, for simplicity we take $c = 0$:

Suppose that f has n derivatives at 0. Suppose further that $f(0) = 0$. Let $g(x) = f(x)/x$ for $x \neq 0$, and $g(0) = f'(0)$. Then g has $n-1$ derivatives at 0 and $g^{(k)}(0) = f^{(k+1)}(0)/(k+1)$ for $k = 0, 1, ..., n-1$.

You can deduce the proposition from the following two steps:

(a) Show that

$$g^{(k)}(x) = \frac{f^{(k)}(x)}{x} - \frac{kg^{(k-1)}(x)}{x},$$

and

$$\frac{d}{dx}(x^k g^{(k-1)}(x)) = x^{k-1} f^{(k)}(x),$$

for $x \neq 0$ and $k = 0, 1, ..., n-1$. The condition $f(0) = 0$ is not needed for this step.

(b) Show that

$$\lim_{x\to 0} g^{(k)}(x) = \frac{1}{k+1} f^{(k+1)}(0)$$

for $k = 0, 1, ..., n-1$.

Note. If f has continuous derivatives up to order n a much simpler proof can be given by the fundamental theorem of calculus, a key result of integration theory. This is only a slight strengthening of the premises. See Sect. 12.2, Exercise 4.

5.9 ($\Diamond$) Multiplicity

Let us assume that the function f has derivatives of all orders. Everything we are going to say can be formulated for functions with finitely many derivatives, but requires more circumlocution. Much of the material of this section will be developed through exercises.

We say that a point c is a root of $f(x) = 0$ with multiplicity m, if $f^{(k)}(c) = 0$ for $k = 0, ..., m-1$ and $f^{(m)}(c) \neq 0$. As usual $f^{(0)}$ denotes f. We also speak of c as an m-fold zero of f. A 1-fold zero is usually called a simple zero. A zero with multiplicity 2 or higher is called a multiple zero. When we talk of a zero of finite multiplicity we mean one with multiplicity m for some positive natural number m. When roots or zeros are *counted according to multiplicity*, it means that an m-fold zero is accorded the number m; it is viewed as m zeros for the purpose of counting them.

In general a point c of a set B of real numbers (not necessarily here an interval) is called an isolated point of B if it is not a limit point of B. This is equivalent to saying that there exists $h > 0$, such that c is the only point of B in the interval $]c - h, c + h[$.

We now state the following facts. For simplicity we suppose that f is defined in an open interval A.

(a) The point c is a zero with multiplicity m if and only if we can write $f(x) = (x - c)^m g(x)$, where g is a function on the same domain as f, g has derivatives of all orders and $g(c) \neq 0$.

(b) A zero of finite multiplicity is an isolated point of the set of zeros (or, in short, an isolated zero).

(c) The function f changes sign at a zero of multiplicity m if and only if m is odd.

5.9.1 Exercises

1. Prove claims (a), (b) and (c).
   Hint. Section 5.8, Exercise 13 could be useful.
2. Suppose that all zeros of f have finite multiplicity. Let a and b be points of A, such that $a < b$ and neither point is a zero. Show that f has at most finitely many zeros in $]a, b[$.
   Hint. One can use the Bolzano–Weierstrass theorem.
3. In the previous exercise, if $f(a)$ and $f(b)$ have the same sign, show that the number of zeros in $]a, b[$, counted by multiplicity, is even. If $f(a)$ and $f(b)$ have opposite signs, show that the number of zeros in $]a, b[$, counted by multiplicity, is odd.
   Note. This extension to the intermediate value theorem, valid for functions with enough derivatives and for which all zeros are known to have finite multiplicity, is very useful. It applies to all polynomials, for example. More generally it applies to so-called analytic functions, studied in complex analysis.
4. Let f be a polynomial with odd degree. Show that the number of real zeros of f, counted by multiplicity, is odd.
5. Suppose that f has m zeros and g has n zeros in A, counted by multiplicity. Show that fg has $m + n$ zeros in A, counted by multiplicity.

6. Suppose that f has m zeros in A (an open interval, recall), counted by multiplicity. Show that f' has at least $m - 1$ zeros in A, again counted by multiplicity.
7. Show that the polynomial

$$P_n(x) = \frac{d^n}{dx^n}(x^2 - 1)^n$$

has n distinct zeros in the interval $]-1, 1[$, all simple.
8. Prove Descartes' rule of signs. We consider a polynomial equation

$$a_n x^n + a_{n-1} x^{n-1} + \cdots + a_0 = 0$$

with real coefficients. Without loss of generality we can assume that $a_n > 0$. We let σ denote the number of sign changes in the sequence $(a_n, a_{n-1}, ..., a_0)$ (omitting any that are 0 for the purpose), and let r_+ be the number of strictly positive roots, counted by multiplicity. The rule of signs says: $r_+ \leq \sigma$ and $\sigma - r_+$ is even. A proof might use the following steps:

(a) Show that the difference $\sigma - r_+$ is even.
(b) Complete the proof by induction on the degree of f. The case of degree 1 is easy. Assuming that the rule of signs holds for polynomials of degree less than or equal to $n - 1$, we let r'_+ be the number of positive roots of the derivative f', and let σ' be the number of sign changes in the coefficients of f'. By the induction hypothesis, $r'_+ \leq \sigma'$. Deduce that $r_+ \leq \sigma + 1$ and use (a) to get $r_+ \leq \sigma$.

5.9.2 Sturm's Theorem

The intermediate value theorem can tell you that a continuous function has at least one root in a given interval. Consideration of multiplicity can tell you the parity of the total root-count. Sturm (1829) developed a method that can be used to compute the actual number of real roots of a polynomial equation in an interval. It exploits the Euclidean algorithm, which is used to find the highest common divisor of two polynomials of one variable. We shall summarise the Euclidean algorithm here; however, the reader unfamiliar with it should probably consult an algebra text.

Let f and g be polynomials such that $\deg(g) \leq \deg(f)$. We can divide g into f, producing a quotient q with degree $\deg(f) - \deg(g)$, and a remainder r which, if not zero, has degree strictly less than $\deg(g)$. This entails that $f = gq + r$, and either r is zero or it is the unique polynomial, with degree less than $\deg(g)$, for which this equation holds for some q. Putting it differently, r is the unique polynomial, with degree less than $\deg(g)$, such that g divides $f - r$. Let us denote the remainder, when f is divided by g, by $\text{rem}(f, g)$.

The Euclidean algorithm computes the greatest common divisor of f and g in the following way. We set $r_0 = f$ and $r_1 = g$ and define recursively $r_{k+2} = \text{rem}(r_k, r_{k+1})$, provided neither r_k nor r_{k+1} is zero. The degrees of the polynomials r_k are strictly decreasing with increasing k (except that r_0 and r_1 could have the same degree); so the process ends in a finite number of steps. This implies that there is a least integer m such that $r_{m+1} = 0$. It is then not hard to see that r_m is the highest common divisor of f and g.

Sturm's theorem counts roots *without multiplicity*. It is best described by supposing that f and f' have highest common divisor 1. If this is not already the case we can divide f by the highest common divisor of f and f'. This does not change the roots of f but converts all multiple roots into simple ones.

Given that the highest common factor of f and f' is 1, we apply the Euclidean algorithm with a slight twist. We set $p_0 = f$, $p_1 = f'$ and then, recursively, $p_{k+2} = -\text{rem}(p_k, p_{k+1})$. After a finite number of steps the sequence terminates, with $p_{m+1} = 0$, say. As it is easy to see that p_k is, up to sign, the same as r_k, and as the highest common factor of f and f' is 1, the polynomial p_m must be a non-zero constant.

Now for each k we have

$$p_{k-1} = p_k q_k - p_{k+1}$$

for a certain polynomial q_k. It follows that for each j, the polynomials p_j and p_{j+1} cannot have a common zero, for if they did, it would be a zero of every polynomial p_k, which is impossible since p_m is a non-zero constant. Moreover, if for some x, and some k in the range $0 < k < m$, we have $p_k(x) = 0$, then clearly $p_{k+1}(x)$ and $p_{k-1}(x)$ have opposite signs. These properties of the chain $p_0, p_1, ..., p_m$ are central to the proof of the following result.

Proposition 5.17 (Sturm's theorem) *Suppose that $a < b$ and neither a nor b is a root of f (where f is assumed to have only simple roots). With $p_0, p_1, ..., p_m$ defined as above, for each x we let $\sigma(x)$ be the number of sign changes in the sequence $p_0(x), p_1(x), ..., p_m(x)$ (ignoring zeros). Then the number of roots of f in the interval $]a, b[$ equals $\sigma(a) - \sigma(b)$.*

Exercise Prove Sturm's theorem. You can use the following steps:

(a) Show that at a root of f the number $\sigma(x)$ decreases by 1, (when the root is passed with increasing x).
(b) Show that if $k \geq 1$, then passing a root of p_k, that is not also a root of f, does not change $\sigma(x)$.
(c) Deduce Sturm's theorem from (a) and (b).

This result is so pretty that an illustrative example is called for.

Example Count and roughly locate the roots of the polynomial $4x^4 - 16x^3 + 11x^2 - 16x + 7$.

A calculation (done by hand) revealed the following.

$$p_0(x) = 4x^4 - 16x^3 + 11x^2 - 16x + 7$$

$$p_1(x) = 16x^3 - 24x^2 + 22x - 16$$

$$p_2(x) = \frac{13}{2}x^2 + \frac{13}{2}x - 3$$

$$p_3(x) = -\frac{1214}{13}x + \frac{592}{13}$$

$p_4(x)$ is a negative constant (indicating that all roots are simple).

Tabulating the signs at $x = 0, 1, 2, 3, 4$ and $\pm\infty$ (that is, for x sufficiently high or low to stabilise the signs) we find the following table of signs:

	$-\infty$	0	1	2	3	4	$+\infty$
p_0	+	+	−	−	−	+	+
p_1	−	−	−	−	+	+	+
p_2	+	−	+	+	+	+	+
p_3	+	+	−	−	−	−	−
p_4	−	−	−	−	−	−	−

There is therefore exactly one root between 0 and 1, exactly one root between 3 and 4, and no subsequent positive root. There is no negative root. It is fascinating to see the minuses drift upwards and vanish, like bubbles rising from a submerged wreck.

By way of comparison, Descartes' rule of signs (Exercise 8) indicates none, two or four positive roots, and no negative roots. The intermediate value theorem (with the supplement covered in Exercise 3) indicates (by the first row of the table) an odd number of roots between 0 and 1, and an odd number of roots between 3 and 4, counted with multiplicity. This does not rule out three roots between 3 and 4, for example; nor does it rule out two roots between 4 and $+\infty$.

5.9.3 Exercises (cont'd)

9. Find the number of real roots of the equation

$$x^5 - 20x + 1 = 0$$

along with their signs.

Hint. Use Descartes' rule of signs (Exercise 8) and the intermediate value theorem (Exercise 3).

10. Find the number of real roots of the equation

$$5x^4 - 10x^2 + 2x + 1 = 0$$

in the given intervals:

(a) $-\infty < x < -1$
(b) $-1 < x < 0$
(c) $0 < x < 1$
(d) $1 < x < \infty$.

11. Find the number of positive roots and the number of negative roots of the equation

$$x^4 + x^3 - 2x - 3 = 0.$$

5.9.4 Pointers to Further Study

$\rightarrow$ Theory of equations

5.10   Convex Functions

We first give a geometric definition of convex function based on the graph of the function, viewed as a curve. The line segment joining two points on a curve is called a chord, this being the standard usage in the case of a circle.

Definition Let A be an interval. A function $f : A \rightarrow \mathbb{R}$ is said to be *strictly convex*, if, for each pair of points a and b in the interval A, with $a < b$, the graph of f for $a < x < b$ lies strictly below the chord joining $(a, f(a))$ and $(b, f(b))$.

Plain convexity is a slightly, but significantly, weaker notion.

Definition Let A be an interval. A function $f : A \rightarrow \mathbb{R}$ is said to be *convex*, if, for each pair of points a and b in the interval A, with $a < b$, the graph of f between a and b does not go above the chord joining $(a, f(a))$ and $(b, f(b))$.

Our focus is entirely on strict convexity.[2] At the level of single-variable calculus it is strict convexity that has all the interesting applications. In some calculus texts a strictly convex function is called concave-up, a term that describes it admirably. Its uses explored here (some of them in the exercises) include some interesting deductions about solutions of equations, minimisation problems, the Legendre transform, inflection points and (in the next section) a sharp form of Jensen's inequality. Last

---

[2]This focus produces a tiresome need to repeat the words "strict" and "strictly". An alternative would have been to use the term "convex" instead of "strictly convex" and in the few places where convexity of the not necessarily strict kind is mentioned, to use "weakly convex". There is a precedent in some of the sources and it is consistent with the rule that the more useful version should have the simpler name. But it is not consistent with multivariate calculus where the greater usefulness of strict convexity compared to convexity is not so apparent.

but not least, an understanding about where a function is strictly convex and where strictly concave is a great aid to sketching its graph, still a useful mathematical skill.

Now we translate strict convexity into algebra. One way to write the equation of the chord is to "proceed from the point $(a, f(a))$" thus

$$y = \left(\frac{f(b) - f(a)}{b - a} \right)(x - a) + f(a).$$

Using this we can write the condition that f is strictly convex as follows. For all a, b and x in A such that $a < x < b$ we require

$$f(x) < \left(\frac{f(b) - f(a)}{b - a} \right)(x - a) + f(a),$$

or equivalently

$$\frac{f(x) - f(a)}{x - a} < \frac{f(b) - f(a)}{b - a}. \tag{5.3}$$

This inequality asserts that the slope of the chord is an increasing function of its right endpoint (just think of b as variable).

The inequality (5.3) is algebraically equivalent to each of two others; like it they each compare the slope of two chords. They are

$$\frac{f(b) - f(a)}{b - a} < \frac{f(b) - f(x)}{b - x}, \tag{5.4}$$

which asserts that the slope of the chord is an increasing function of its left endpoint, and

$$\frac{f(x) - f(a)}{x - a} < \frac{f(b) - f(x)}{b - x}. \tag{5.5}$$

It is a nice exercise for the reader to show that all three inequalities are algebraically equivalent. Any one of them implies the other two. Geometrically this is obvious, as the three quantities being compared are the slopes of three chords forming the sides of a triangle whose vertices are the points $(a, f(a))$, $(x, f(x))$ and $(b, f(b))$ on the curve $y = f(x)$. A picture makes this rather obvious.

There is even a fourth version of the same inequality, also easy to obtain, that rather obviously expresses the claim that the graph is below the chord, namely

$$f(x) < \left(\frac{b - x}{b - a} \right)f(a) + \left(\frac{x - a}{b - a} \right)f(b). \tag{5.6}$$

Exercise Prove that the inequalities (5.3)–(5.6) are algebraically equivalent.

Putting this together we can set out a rather wordy necessary and sufficient condition for strict convexity of the function f; that for every three points a, x and b in the interval of definition, such that $a < x < b$, at least one of the above four inequalities is verified (and if one is true then all are true).

If, however, f is differentiable there is a much simpler criterion.

Proposition 5.18 *A differentiable function f is strictly convex if and only if f' is strictly increasing.*

Proof Suppose that f is strictly convex and differentiable. Let $a < b$. It follows from the inequalities that the quotient $(f(x) - f(a))/(x - a)$ is a strictly increasing function of x for $x > a$, and the quotient $(f(b) - f(x))/(b - x)$ is a strictly increasing function of x for $x < b$. Hence

$$f'(a) = \lim_{x \to a+} \frac{f(x) - f(a)}{x - a} < \frac{f(b) - f(a)}{b - a} < \lim_{x \to b-} \frac{f(b) - f(x)}{b - x} = f'(b)$$

giving $f'(a) < f'(b)$.

Conversely suppose that f' is strictly increasing. Let a, x, b be in the interval of definition of f and suppose that $a < x < b$. By the mean value theorem there are points y between a and x, and z between x and b, such that

$$\frac{f(x) - f(a)}{x - a} = f'(y)$$

and

$$\frac{f(b) - f(x)}{b - x} = f'(z).$$

But $f'(y) < f'(z)$ so we find

$$\frac{f(x) - f(a)}{x - a} < \frac{f(b) - f(x)}{b - x}.$$

This is inequality (5.5) and shows that f is strictly convex. □

As an immediate consequence we have the most useful test for strict convexity; it is based on calculus rather than geometry, but requires second derivatives.

Proposition 5.19 *A sufficient condition for a twice differentiable function f to be strictly convex is that $f''(x) > 0$ for all x in the interval of definition.*

5.10.1 Tangent Lines and Convexity

Another useful conclusion, and a fifth necessary and sufficient, purely geometric condition for strict convexity, but based on the assumption that the function is differentiable, is the following.

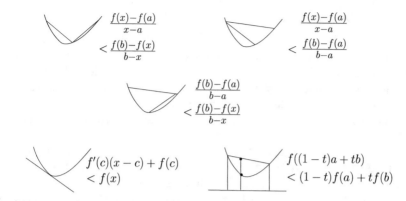

Fig. 5.6 Five views of strict convexity

Proposition 5.20 *Let f be differentiable in the open interval A. A necessary and sufficient condition for f to be strictly convex is that for every c in A, the tangent line to the curve $y = f(x)$ at the point $(c, f(c))$ lies wholly below the curve itself, except that they both contain the point $(c, f(c))$.*

Proof Suppose that f is strictly convex. We know that $(f(c) - f(x))/(c - x)$ is a strictly increasing function of x for $x < c$; and that $(f(x) - f(c))/(x - c)$ is a strictly increasing function of x for $x > c$. Hence if $x < c$ we find

$$\frac{f(c) - f(x)}{c - x} < \lim_{t \to c-} \frac{f(c) - f(t)}{c - t} = f'(c)$$

which implies

$$f(x) > f(c) + f'(c)(x - c)$$

and if $c < x$ we find

$$f'(c) = \lim_{t \to c+} \frac{f(t) - f(c)}{t - c} < \frac{f(x) - f(c)}{x - c}$$

which implies

$$f(x) > f(c) + f'(c)(x - c).$$

This shows that the condition is necessary.

The reader is invited to finish the proof by showing that the condition is sufficient for strict convexity given that f is differentiable. $\qquad \square$

The five geometrical conditions for strict convexity are illustrated in Fig. 5.6.

5.10.2   *Inflection Points*

A function f such that $-f$ is strictly convex is called strictly concave (in some calculus texts it is called concave-down). Let f be differentiable in the interval A. A point $(a, f(a))$ on the curve $y = f(x)$ is called an *inflection point* of the curve if there exists $h > 0$, such that f is strictly convex [respectively, strictly concave] in the interval $]a - h, a[$, and strictly concave [respectively, strictly convex] in the interval $]a, a + h[$.

In other words the function switches from strictly convex to strictly concave, or from strictly concave to strictly convex, at the point a. We say loosely that f has an inflection point at a. Inflection points are illustrated in Fig. 5.7.

For some reason the notion of inflection point is only applied to differentiable functions; there has to be a tangent. Properly an inflection point is a property possessed by a plane curve and not just a graph; it is a point where the curvature changes sign. The concept of curvature really belongs to the study of the differential geometry of plane curves.

A necessary condition for an inflection point at a is that f' has either a local strict maximum or a local strict minimum at a. This is not sufficient. Again if f is twice differentiable it is necessary that $f''(a) = 0$, but still not sufficient. We have to force f'' to change sign, to be strictly positive on one side of a and strictly negative on the other.

A problem left to the exercises is to find a sufficient condition that f has an inflection point at a that builds on higher derivatives of f at a alone.

We often want to sketch the graph of a given function. Nowadays there are many good software packages that do this. A good sketch prepared without the help of a computer should show roughly where the function is strictly convex and where strictly concave. This means having some idea of where f'' is positive, where negative and where the inflection points are that separate these regions.

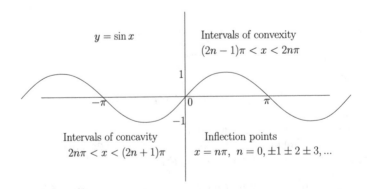

Fig. 5.7 Inflection points of $y = \sin x$

5.10.3 *Exercises*

1. Let f be a strictly convex function defined in an open interval A and let c be a point in A. Show that the limits

$$\lim_{t \to c-} \frac{f(c) - f(t)}{c - t} \quad \text{and} \quad \lim_{t \to c+} \frac{f(t) - f(c)}{t - c}$$

   both exist and that the first is less than or equal to the second. These limits are the left and right derivatives, $D_l f(c)$ and $D_r f(c)$. Give an example to show that they do not have to be equal.

2. Show that a strictly convex function, defined in an interval A, is continuous if A is open, but that continuity may fail if A is not open.
   Hint. One way is to use the previous exercise.

3. The function in Proposition 5.20 was assumed to be differentiable. Without assuming differentiability it is possible to say something similar, and obtain a sixth necessary and sufficient, purely geometric condition for strict convexity. Prove the following:

   A function f, defined in an open interval A, is strictly convex if and only if it satisfies the following condition: for every c in A there exists a straight line through the point $(c, f(c))$ that lies wholly below the graph of f, except that the line and graph both contain the point $(c, f(c))$.

4. Let f be a convex function and suppose that there exist points $a < x < b$, such that the point $(x, f(x))$ lies on the chord joining $(a, f(a))$ and $(b, f(b))$. Show that the whole of the chord lies on the graph of f. So the graph of a non-strictly convex function differs from that of a strictly convex one by including some straight line segments.
   Hint. Consider how the inequalities (5.3)–(5.6) should be modified for a function that is convex but not necessarily strictly convex.

5. Let f be a strictly convex function on the interval $[0, \infty[$ and suppose that $f(0) = 0$. Show that f satisfies

$$f(a + b) > f(a) + f(b)$$

   for all positive a and b.

6. Show that if f is a strictly convex function and a and b are constants, then the function $f(x) + ax + b$ is also strictly convex.

7. Suppose that a function f satisfies

$$f\left(\frac{a + b}{2}\right) < \frac{1}{2}f(a) + \frac{1}{2}f(b)$$

   for all a and b in its interval of definition.

(a) Show that

$$f\left(ta + (1 - t)b\right) < tf(a) + (1 - t)f(b)$$

for all a and b and for all *dyadic fractions* t in the interval $0 < t < 1$; that is, for all t of the form $t = k/2^n$ where n is a positive integer and k is an integer in the range $1 \le k \le 2^n - 1$.

(b) Show that if f is continuous then f is strictly convex.

Hint. For item (a) use induction with respect to n. To get started figure out how to handle the case $t = \frac{1}{4}$. For item (b) use the fact that every real number in the interval $[0, 1]$ can be approximated arbitrarily closely by dyadic fractions, as is shown by the binary representation of real numbers, analogous to the decimal representation but using 2 as a base instead of 10. You will need to figure out why the inequality remains strict when t is an arbitrary real number in the interval $0 < t < 1$.

8. Show that a twice differentiable function is convex (of the not necessarily strict kind) if and only if its second derivative is non-negative.

Note. This is a case where convexity is simpler than strict convexity. The counterpart of Proposition 5.19 is a necessary and sufficient condition for convexity, whereas Proposition 5.19 is only a sufficient condition for strict convexity.

9. Let f be a strictly convex function. Show that a straight line intersects the graph of f in at most two points. In other words, given constants a and b, the equation $f(x) = ax + b$ has at most two roots.

10. Let f be a strictly convex function defined in an interval A.

(a) Show that if f attains a minimum it does so at a unique point.

(b) Suppose that there exist distinct points a and b in A such that $f(a) = f(b)$. Show that f attains a minimum (which by (a) occurs at a unique point).

(c) Let c be an interior point of A (that is, c is not an endpoint). Show that f attains a minimum at c if and only if $D_l f(c) \le 0$ and $D_r f(c) \ge 0$ (see Exercise 1).

(d) Suppose that c is an interior point of A and that f is differentiable at c. Show that f attains a minimum at c if and only if $f'(c) = 0$.

11. For the purposes of this exercise we shall call a line that cuts a curve $y = f(x)$ a secant line. A secant line meets the curve and crosses it; it contains points (x_1, y_1) and (x_2, y_2), such that $y_1 < f(x_1)$ and $y_2 > f(x_2)$. Note that this is slightly different from the common usage, which requires a secant to meet the curve in two points, an assumption not made here.

Let f be strictly convex in the whole real line. Show that a secant line that is parallel to some chord of the curve $y = f(x)$ cuts the curve in two points.

12. Let f be strictly convex and defined in the whole real line. Suppose that f attains a minimum. Prove that $\lim_{x \to -\infty} f(x) = \lim_{x \to \infty} f(x) = \infty$.

13. Let f be defined in an open interval A and let c be a point in A. Show that the following is sufficient for f to have an inflection point at c: the derivatives

$f^{(j)}(c)$ exist up to $j = m$, $f^{(j)}(c) = 0$ for $j = 2, ..., m - 1$, $f^{(m)}(c) \neq 0$ and m is odd.

In the following series of exercises we study the Legendre transform. This is an important construction associated with convex functions that has many applications, both theoretical and practical.

14. Let f be strictly convex and differentiable in the open interval A. Let $c = \inf f'$ and $d = \sup f'$. For each p in the interval $B :=]c, d[$ let g_p be the function $g_p(x) = px - f(x)$.

   (a) Show that g_p attains a maximum value at a unique point x_p in A.
   (b) For each p in B we let

   $$f_*(p) = px_p - f(x_p).$$

   The function f_* is called the Legendre transform of f. Now suppose that f is twice differentiable and that $f'' > 0$. Show that

   $$(f_*)' = (f')^{-1}$$

   and deduce that f_* is strictly convex.
   (c) What if the second derivative does not exist? Can you prove the formula in item (b) from first principles, that is, by arguing from the difference quotient?
   (d) Show that $f_{**} = f$. Algebraically, the operation of passing from f to f_* is an *involution*. The same operation applied to f_* brings one back to f.

15. Prove Young's inequality. Given that f is strictly convex and differentiable, then

   $$px \leq f_*(p) + f(x)$$

   for all x in A and p in B (where A and B are the domains of f and f_* respectively).
16. Show that the power function x^a, with $a > 1$, is strictly convex in its interval of definition $0 < x < \infty$. It therefore has a Legendre transform. Obtain nice results by computing the Legendre transform of x^a/a and writing down the result of Young's inequality.
17. Try the previous exercise for the function e^x. You will need some knowledge of the exponential function and natural logarithm.
18. Let f be a strictly convex function defined in an open interval A. Let B be the set of all real numbers p, such that the graph $y = f(x)$ has a chord with slope p.

   (a) Show that B is an interval.
   (b) Show that for each p in B the function $px - f(x)$ attains a maximum at a unique point x_p in A.

This shows how the Legendre transform can be defined for strictly convex functions that are not everywhere differentiable. We set $f_*(p) = px_p - f(x_p)$ for each p in B. However, f_*, though convex, may fail to be strictly convex.

(c) Let f be defined by

$$f(x) = \begin{cases} (x-1)^2 & \text{if } x < 0 \\ (x+1)^2 & \text{if } x \geq 0. \end{cases}$$

Show that f is strictly convex and compute f_*. Show that the latter is convex but not strictly convex.

5.11   ($\diamond$) Jensen's Inequality

Jensen claimed that his inequality implied almost all known inequalities as special cases.[3] If this was only partially true it would make it a remarkable object of study. Actually Jensen's inequality is a natural enough extension of the fourth inequality characterising strictly convex functions, inequality (5.6).

Proposition 5.21 *Let f be a strictly convex function with domain A, let x_j, ($j = 1, 2, ..., n$), be points in A, and let t_j, ($j = 1, 2, ..., n$), be positive numbers such that $\sum_{j=1}^{n} t_j = 1$. Then*

$$f\left(\sum_{j=1}^{n} t_j x_j\right) \leq \sum_{j=1}^{n} t_j f(x_j).$$

Equality holds if and only if the numbers x_j are all equal.

Proof We set $c = \sum_{j=1}^{n} t_j x_j$. Because the numbers t_j are positive and sum to 1, it follows that c belongs to the interval A. By the result of Sect. 5.10, Exercise 3, there exists a line through $(c, f(c))$, that lies wholly below the graph $y = f(x)$, except that both the line and the graph contain the point $(c, f(c))$. Let the line have the equation $y = f(c) + m(x - c)$. Then for all $x \neq c$ we have

$$f(c) + m(x - c) < f(x)$$

whilst for $x = c$ we have equality. We now find

$$\sum_{j=1}^{n} t_j \big(f(c) + m(x_j - c)\big) \leq \sum_{j=1}^{n} t_j f(x_j)$$

---

[3]This is stated in the book "A Course of Analysis" by E. G. Phillips, originally published in 1930. I don't know what the author's source was; maybe he knew Jensen.

with equality if and only if all the numbers x_j are equal to c. Since $c = \sum_{j=1}^{n} t_j x_j$ and $\sum_{j=1}^{n} t_j = 1$ we obtain

$$f(c) \le \sum_{j=1}^{n} t_j f(x_j),$$

which is Jensen's inequality.   □

Jensen's inequality is also valid if f is merely convex. An analogue of Sect. 5.10, Exercise 3 holds for (not necessarily strictly) convex functions. In this case a line $y = m(x - c) + f(c)$ can be found that nowhere goes above the graph $y = f(x)$, so that Jensen's inequality results just the same. However, we no longer get the striking conclusion that equality holds if and only if all the points x_j are equal.

One of the spectacular applications of Jensen's inequality is to proving a general form of the inequality of arithmetic and geometric means. This is achieved by applying it to the exponential function a^x. We have not, so far, rigorously defined this function, but we can summarise what we need as follows:

(a) For a given positive base a, the function a^x extends to real x the power function a^x with rational x as already defined.
(b) We have the laws of exponents:

$$a^{s+t} = a^s a^t \quad \text{and} \quad (a^s)^t = a^{st}.$$

(c) The function a^x has an inverse function (a being here a fixed base), the logarithm with base a denoted by $\log_a$; that is, the equation $y = a^x$ inverts to $x = \log_a y$.
(d) The function a^x is convex.

Let x_j, $(j = 1, 2, ..., n)$, be real numbers and let t_j, $(j = 1, 2, ..., n)$, be positive numbers such that $\sum_{j=1}^{n} t_j = 1$. Applying Jensen's inequality to the exponential function 2^x we obtain

$$2^{\sum_{j=1}^{n} t_j x_j} \le \sum_{j=1}^{n} t_j 2^{x_j}$$

with equality only if all the numbers x_j are equal. By the laws of exponents this gives

$$\prod_{j=1}^{n} (2^{x_j})^{t_j} \le \sum_{j=1}^{n} t_j 2^{x_j}.$$

Now let a_j, $(j = 1, 2, ..., n)$, be positive real numbers and let $x_j = \log_2 a_j$. We obtain the generalised inequality of arithmetic and geometric means

$$\prod_{j=1}^{n} a_j^{t_j} \le \sum_{j=1}^{n} t_j a_j,$$

as well as the additional fact that equality only holds when the numbers a_j are all equal.

5.11.1   Exercises

1. Give a proof of Jensen's inequality by induction on the number n. Note that the case $n = 2$ is inequality (5.6). Some care is required to include the conclusion that asserts when equality occurs.
2. The case $n = 2$ of the generalised inequality of the arithmetic and geometric means,

$$a^s b^t \leq sa + bt$$

   is sometimes called Young's inequality. The assumptions are that a, b, s and t are all positive, and that $s + t = 1$. Equality holds if and only if $a = b$.

   (a) Use Young's inequality to prove Hölder's inequality:

$$\sum_{k=1}^{n} a_k b_k \leq \left(\sum_{k=1}^{n} a_k^p \right)^{1/p} \left(\sum_{k=1}^{n} b_k^q \right)^{1/q}$$

   where $p > 1$, $q > 1$, $(1/p) + (1/q) = 1$, and for each k we have $a_k \geq 0$, $b_k \geq 0$.
   Hint. Do it first with the assumption $\sum_{k=1}^{n} a_k^p = \sum_{k=1}^{n} b_k^q = 1$. Apply Young's inequality with $a = a_k^p$, $b = b_k^q$, $s = 1/p$, $t = 1/q$ and sum over k. Then remove the assumption.
   (b) Show that equality holds in Hölder's inequality if and only if the following is satisfied: either $b_k = 0$ for all k, or else there exists t such that $a_k = tb_k$ for all k.

5.11.2   Pointers to Further Study

→ Convexity theory
→ Inequalities

5.12   (◇) How Fast Do Iterations Converge?

Given the iteration $a_n = f(a_{n-1})$, and knowing that $\lim_{n \to \infty} a_n = t$, where t is a root of $f(x) = x$, we wish to study the speed of convergence. Can we usefully specify how fast a_n tends to t? This could be of great importance in deciding whether a method of approximating the solution to an equation is practical. For example, can we say approximately how many additional correct decimal digits are obtained in passing from a_n to a_{n+1}?

The answer depends on studying the derivatives of f. The derivative $f'(t)$ at the root being approximated plays a key role. We will assume some knowledge of the common logarithm, and once we use Taylor's theorem. More precisely we use the fact that for a twice-differentiable function f the error $f(x) - f(a) - f'(a)(x - a)$, in using the first-degree Taylor polynomial as an approximation to $f(x)$, is $\frac{1}{2} f''(\xi)(x - a)^2$, for some ξ between a and x (this special case was the content of Sect. 5.7, Exercise 2).

In the following we do not assume in advance that the iteration can be continued indefinitely. We assume throughout that f is defined in an open interval A.

Proposition 5.22 *Suppose that f is differentiable in A and that f' is continuous. Let t be a point in A, such that $f(t) = t$, and assume that $|f'(t)| < 1$. Then there exists $\delta > 0$, such that if $|a_0 - t| < \delta$ the iteration can be continued indefinitely and a_n converges to t.*

Proof Since $|f'(t)| < 1$ we can choose k so that $|f'(t)| < k < 1$. Since f' is continuous, there exists $\delta > 0$, such that the interval $]t - \delta, t + \delta[$ is included in the domain of definition of f, and such that $|f'(x)| < k$ for all x that satisfy $|x - t| < \delta$.

Suppose that $|a_0 - t| < \delta$. Then the iteration cannot quit the interval $]t - \delta, t + \delta[$, and so continues indefinitely. For suppose that we have reached a_n without quitting the interval. By the mean value theorem we have

$$a_{n+1} - t = f(a_n) - t = f(a_n) - f(t) = f'(\xi_n)(a_n - t) \tag{5.7}$$

for some number ξ_n between a_n and t. Hence

$$|a_{n+1} - t| < k|a_n - t|$$

and since $k < 1$, the number a_{n+1} is in the interval $]t - \delta, t + \delta[$. The iteration continues indefinitely and satisfies $|a_n - t| < k^n|a_0 - t|$, and so a_n converges to t. □

The estimate $|a_n - t| < k^n|a_0 - t|$ tells us something more that has practical importance but is not very precise. Each iteration step contributes at worst roughly the same number of additional correct decimal digits, the number obtained in each step being approximately $-\log_{10} k$.

Proposition 5.23 *Suppose that f is differentiable in A and that f' is continuous. Let t be a point in A, such that $f(t) = t$, and assume that $|f'(t)| > 1$. Then, however we choose a_0, provided $a_0 \neq t$ the iteration cannot converge to t.*

Proof There exists $k > 1$ and $\varepsilon > 0$, such that $|f'(x)| > k$ for all x that satisfy $t - \varepsilon < x < t + \varepsilon$. If $a_n \in]t - \varepsilon, t + \varepsilon[$ then $|a_{n+1} - t| > k|a_n - t|$ according to (5.7). After a finite number of steps a_j will quit the interval $]t - \varepsilon, t + \varepsilon[$ for some $j > n$. It is impossible that N can exist, such that for all $n \geq N$ we have $|a_n - t| < \varepsilon$.
□

We can obtain much faster convergence under some conditions. Suppose that f is twice differentiable in A and that the second derivative f'' is bounded in absolute value by a constant M. Let t be a point in A, such that $f(t) = t$, and assume that $f'(t) = 0$. We can find $\delta > 0$, such that $|f'(x)| < \frac{1}{2}$ for all x that satisfy $t - \delta < x < t + \delta$. If $|a_0 - t| < \delta$ then (by the arguments above) $|a_n - t|$ is decreasing and a_n converges to t.

The condition that $f'(t) = 0$ has a profound effect on the speed of convergence. Let us suppose that a_n converges to t (as it will do if the initial point is near enough to t as we have just seen), but we do not assume that $|a_0 - t| < \delta$. Since a_n tends to t the inequality $|a_n - t| < \delta$ will certainly hold once n is large enough. Furthermore, since $f'(t) = 0$, it follows by Taylor's theorem (see Sect. 5.7, Exercise 2), that if $|a_n - t| < \delta$ then

$$a_{n+1} - t = f(a_n) - f(t) = \frac{1}{2} f''(\xi_n)(a_n - t)^2,$$

where the number ξ_n lies between a_n and t. Then we find

$$|a_{n+1} - t| < \frac{M}{2} |a_n - t|^2, \tag{5.8}$$

which we modify to read

$$\frac{M}{2} |a_{n+1} - t| < \left(\frac{M}{2} |a_n - t| \right)^2 .$$

Since a_n tends to t, there exists n_0, such that $(M/2)|a_{n_0} - t| < 1$. Set

$$\rho = \frac{M}{2} |a_{n_0} - t|.$$

For $n \geq n_0$ we have

$$\frac{M}{2} |a_n - t| < \left(\frac{M}{2} |a_{n_0} - t| \right)^{2^{n-n_0}} = \rho^{2^{n-n_0}}.$$

To see what this means we consider the common logarithm of the error, namely $\log_{10} |a_n - t|$. We have

$$- \log_{10} |a_n - t| > \log_{10} \left(\frac{M}{2} \right) + 2^{n-n_0} (- \log_{10} \rho).$$

As before we interpret the left-hand side as the approximate number of correct decimal digits. The number on the right-hand side approximately doubles at each iteration step (if M is big we would have to add that n must be sufficiently large). With some simplification we could say that the number of correct digits approximately doubles at each step, at the very least.

The type of convergence described here, where the error at step $n + 1$ is bounded by a constant times the square of the error at step n, as indicated in the inequality (5.8), is called *quadratic convergence*. It is clearly very desirable if an efficient computation method is sought. In the next sections we shall study some examples of this.

5.12.1 The Babylonian Method

A striking example of quadratic convergence occurs with the Babylonian method for calculating the square root $\sqrt{c}$. This is the iteration

$$a_{n+1} = \frac{1}{2}\left(a_n + \frac{c}{a_n}\right).$$

Set

$$f(x) = \frac{1}{2}\left(x + \frac{c}{x}\right), \quad (x > 0).$$

Then $f(\sqrt{c}) = \sqrt{c}$ and $f'(\sqrt{c}) = 0$. The conclusions of the last section tell us that $a_n \to \sqrt{c}$ if a_0 is sufficiently near to $\sqrt{c}$, and the number of correct digits approximately doubles with each step.

Let us try this on $\sqrt{3}$. Take $a_0 = 1$. The results are, up to a_5:

$$2, \quad 1.75, \quad 1.73214285714286, \quad 1.73205081001473, \quad 1.73205080756888$$

The number of correct digits for a_2, a_3, a_4 and a_5 is successively

$$1, \quad 3, \quad 7, \quad 13$$

roughly as predicted.

For which a_0 can we assert that a_n tends to $\sqrt{c}$? This question is usually tricky. But in this case we can start at any positive a_0 whatever. The inequality of arithmetic and geometric means gives

$$\frac{1}{2}\left(x + \frac{c}{x}\right) \geq \sqrt{c}$$

(equality only if $x = \sqrt{c}$). If $a_0 \neq \sqrt{c}$ then $a_n > \sqrt{c}$ for $n \geq 1$. And then a_n is decreasing for $n \geq 1$ since

$$a_n - \frac{1}{2}\left(a_n + \frac{c}{a_n}\right) = \frac{a_n^2 - c}{2a_n} > 0.$$

The limit $\lim_{n \to \infty} a_n$ therefore exists and equals the unique positive root of $f(x) = x$, which is $\sqrt{c}$.

5.12.2 Newton's Method

The Babylonian method for calculating $\sqrt{c}$ is an instance of Newton's method of approximation, more precisely it results from using Newton's method to solve the equation $x^2 - c = 0$.

Let g be twice differentiable. Newton's method computes a solution of the equation $g(x) = 0$ by means of the iteration

$$a_{n+1} = a_n - \frac{g(a_n)}{g'(a_n)}.$$

Let

$$f(x) = x - \frac{g(x)}{g'(x)}.$$

If t is a solution of $g(x) = 0$ and $g'(t) \neq 0$, then $f(t) = t$ and $f'(t) = 0$. We therefore have quadratic convergence to t if a_1 is sufficiently near to t. How near it needs to be to ensure convergence is a sensitive and tricky question, and the reader should consult works on numerical analysis for further discussion.

Exercise Verify the claim that if $f(x) = x - (g(x)/g'(x))$ and $f(t) = t$ then $f'(t) = 0$.

Newton's method is based on the plausible notion that if $f(a_1)$ is small whilst $f'(a_1)$ is big, so that the graph is steep at $(a_1, f(a_1))$ and close to the x-axis, then the graph must cross the x-axis at a point near to a_1. Moreover, a better approximation to the crossing point is found by following the tangent at $(a_1, f(a_1))$ until it crosses the x-axis. Just what "small" and "big" mean in this context has to be made precise. Sharp turning of the graph to defeat the crossing of the x-axis is prevented by having a bound on the second derivative. This is illustrated in Fig. 5.8. Thus intuitively, the success of Newton's method beginning at a point a_1 depends on a delicate interplay between $f(a_1)$, $f'(a_1)$ and a local bound on $f''(x)$.

Nevertheless, in cases where f' and f'' do not change sign a relatively simple analysis is possible. This is presented in Exercise 4.

As an example, the equation $x^3 - 2x + 2 = 0$ has exactly one real root and it lies between -2 and -1. The reader should check that Newton's method gives rise to the iteration

$$a_{n+1} = \frac{2a_n^3 - 2}{3a_n^2 - 2}.$$

Fig. 5.8 Newton's method
for the root t of $f(x) = 0$

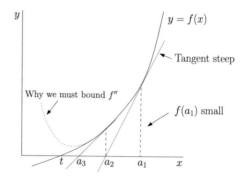

If $a_1 = 0$ then a_n jumps repeatedly back and forth between 0 and 1. If, on the other
hand, $a_1 = -2$ then a_n appears to converge fast, as $a_2 = -1.8$, $a_3 = -1.769948$,
$a_4 = -1.769292$ and it looks as if we have already reached 3 correct decimal digits.

5.12.3 Exercises

1. Let $(a_n)_{n=1}^{\infty}$ be the sequence of Fibonacci numbers, 1, 1, 2, 3, 5, 8, 13, ... and so on.
   It is known that $a_{n+1}/a_n \to \phi$, where ϕ is the Golden Ratio. The convergence is
   quite slow. Apply the heuristic analysis that followed Proposition 5.22 to estimate
   roughly how many further correct decimal digits of ϕ are obtained with each
   increment of n. Check your estimate against reality by calculating some values
   of a_{n+1}/a_n.
   Hint. Consult Sect. 3.3, Exercise 5.
2. Verify the claim that the Babylonian method results from applying Newton's
   method to the problem $x^2 - c = 0$.
3. Apply Newton's method to obtain an iteration scheme for the cube root, or, more
   generally, for the rth root of c.
4. There is a simple situation where we can always infer that Newton approximations
   converge to a solution. Suppose that f is defined in an open interval A and that
   f' and f'' are both strictly positive in A. Suppose further that there is a root t of
   $f(x) = 0$ in A.

   (a) Show that t is the only root of $f(x) = 0$ in A and that f changes sign at t.
   (b) Let a_1 be a point in A lying above t. Show that

   $$t < a_1 - \frac{f(a_1)}{f'(a_1)} < a_1.$$

   (c) Deduce that Newton approximations, beginning at a_1, form a decreasing
   sequence that converges to t.

(d) Suppose that $a_1 < t$. What can you say about Newton approximations beginning at a_1?

Note. Obviously there are variants of this conclusion in which f' and f'' are both negative, or have opposite signs. What they have in common is that neither f' nor f'' changes sign in A. It is left to the reader to explore them.

5. Calculate all the real roots of the equation $x^5 - 20x + 1 = 0$ to three correct decimal places.

6. The calculation of Gauss's arithmetic-geometric mean (Sect. 3.4, Exercise 10) provides another nice example of the fast convergence found in Newton approximations. Let a and b be distinct positive numbers and define sequences by

$$a_0 = a, \qquad\qquad b_0 = b$$
$$a_{n+1} = \tfrac{1}{2}(a_n + b_n), \qquad b_{n+1} = \sqrt{a_n b_n}$$

for $n = 0, 1, 2, \ldots$. The sequences satisfy $b_n < b_{n+1} < a_{n+1} < a_n$ for $n \geq 1$. The limits $\lim_{n \to \infty} a_n$ and $\lim_{n \to \infty} b_n$ are equal. Their common value is the arithmetic-geometric mean of a and b, which we shall denote by $M(a, b)$.

(a) Let $c_n = a_n / b_n$. Show that

$$c_{n+1} = \frac{1}{2}\left(\sqrt{c_n} + \frac{1}{\sqrt{c_n}}\right).$$

(b) Deduce from item (a) that the convergence of c_n to 1 is quadratic (that is, the error $|c_n - 1|$ satisfies an inequality like (5.8)). In fact, show that, given $\delta > 0$, the inequality

$$|c_{n+1} - 1| < \left(\frac{1}{8} + \delta\right)|c_n - 1|^2$$

holds for all sufficiently large n.

(c) Deduce that the convergence of $a_n - b_n$ to 0 is also quadratic.

Note. The fast convergence of a_n and b_n to $M(a, b)$ implied by the conclusion of item (c) has applications to the computation of so-called elliptic integrals. See the exercises in Sect. 11.2.

7. Let A be a closed interval, which may be unbounded or even all of $\mathbb{R}$. Let the function $f : A \to A$ satisfy the following condition: there exists K, such that $0 < K < 1$ and $|f(s) - f(t)| \leq K|s - t|$ for all s and t in A. Prove that there exists a unique x in A, such that $f(x) = x$.

Hint. Let a_0 be any point whatsoever in A and define the iteration $a_{n+1} = f(a_n)$, $n = 0, 1, 2, 3, \ldots$. Show that for all natural numbers n and p we have

$$|a_{n+p} - a_n| \leq |a_1 - a_0| \sum_{j=n}^{n+p-1} K^j$$

and apply Cauchy's principle to show that a_n converges. Argue that the limit x is in A, that $f(x) = x$ and that this equation can have only one solution in A. Note how crucial it is that f maps A *into itself* and observe carefully why A must be a *closed* interval.

8. Let $f : \mathbb{R} \to \mathbb{R}$. Suppose that f is differentiable and that there exists $K > 0$, such that $K < 1$ and $|f'(t)| < K$ for all t. Show that there exists a unique x, such that $f(x) = x$.

9. Find an example of a function $f : \mathbb{R} \to \mathbb{R}$, such that $|f(a) - f(b)| < |a - b|$ for all a and b, but the equation $f(x) = x$ has no solution.

5.12.4   Pointers to Further Study

→ Dynamical systems
→ Numerical analysis

Chapter 6
Integrals and Integration

*Any segment of a section of a right angled cone [i.e. a parabola]
is four-thirds of the triangle which has the same base and equal
height*

Archimedes. The method of mechanical theorems

6.1 Two Unlike Problems

Problem A. To find an antiderivative for a given function.
If $f'(x) = F(x)$, then we call the function f an antiderivative for F. Now

$$\frac{d}{dx}x^n = nx^{n-1}$$

for each integer n. This tells us that $x^{n+1}/(n+1)$ is an antiderivative for x^n in the cases
$n = 0, 1, \pm 2, \pm 3, \ldots$. But what can be an antiderivative for x^{-1}? It makes no sense
to put $n = -1$ in this formula. This question greatly exercised the mathematicians
who invented calculus in the seventeenth century.

We can ask the more general question: which functions have an antiderivative?
Our problem is to solve the simplest of all *differential equations*: given the function
F to find a function $y(x)$, such that

$$\frac{dy}{dx} = F(x).$$

Problem B. To calculate the area of a plane figure bounded by a curve.
Historically this problem was called quadrature, as assigning an area to a plane figure
meant that a square with the same area was determined. We will not give a general

Fig. 6.1 Archimedes'
parabolic segment

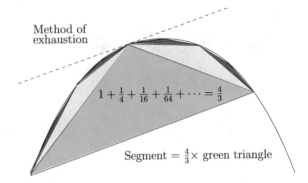

Method of
exhaustion

$$1 + \tfrac{1}{4} + \tfrac{1}{16} + \tfrac{1}{64} + \cdots = \tfrac{4}{3}$$

Segment $= \tfrac{4}{3} \times$ green triangle

definition of area; that is the subject of measure theory. Nevertheless this will not stop us from discussing it, any more than it stopped the mathematicians of antiquity.

Archimedes calculated the area of a circle and the area of a parabolic segment (the figure bounded by a parabola and one of its chords). He gave the formula πr^2 for the circle and showed that $223/71 < \pi < 22/7$. His greatest achievement in the computation of area was the parabolic segment, stating that its area was $4/3$ times the area of a certain inscribed triangle (with base the given chord and top vertex at the point on the parabola where the tangent was parallel to the chord). To reach this conclusion he had to invent a method, the method of exhaustion, that in its use of an infinite sequence of approximations from below resembles modern integration theories. He also had to compute the sum of the geometric series $\sum_{n=0}^{\infty} 1/4^n$ (Fig. 6.1).

Fast forward to the fifteenth century and we find Kepler considering the volume of a wine barrel. This is a solid of revolution and the calculation of its volume depends on calculating the area of a plane figure.

Only with the invention of calculus was a method proposed that could be used to calculate the areas of general plane figures, starting with the area under the graph of a function. In the first place we consider the area between the graph of a positive function f and the x-axis, cut off by two vertical lines $x = a$ and $x = b$ (Fig. 6.2). This leads to the definition of the Riemann integral or the Darboux integral; two different approaches that turn out to be equivalent. We shall call it the Riemann–Darboux integral, although in defining it we shall take Darboux's approach.

We therefore proceed to Problem B and only later show how it leads to a solution to Problem A.

6.2   Defining the Riemann–Darboux Integral

Let $f : [a, b] \to \mathbb{R}$ be a bounded function. Its domain is a bounded and closed interval. We do not assume that f is continuous. This is an advantage because it is necessary for practical applications to be able to integrate some discontinuous functions. But it is essential for the following considerations to make sense that f

Fig. 6.2 Problem B.
Calculate the area of a plane
figure

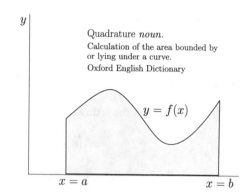

should be bounded. This, and the requirement that the domain is a bounded and closed interval, are defects of the Riemann integral that were successfully removed by the introduction of the Lebesgue integral in the early twentieth century.

We do not assume that f is positive. However, in the case that $f(x) > 0$ for all x, the integral, when successfully defined, will give an acceptable notion for the area bounded by the lines $x = a$, $y = 0$, $x = b$ and the graph $y = f(x)$.

Definition A partition P of the interval $[a, b]$ is a finite sequence $(t_j)_{j=0}^m$ (not necessarily uniformly spaced), such that

$$a = t_0 < t_1 < t_2 < \cdots < t_m = b.$$

The intervals $[t_j, t_{j+1}]$ are called the subintervals of the partition.

For a given partition $(t_0, t_1, t_2, ..., t_m)$ we set

$$m_j = \inf_{[t_j, t_{j+1}]} f, \quad M_j = \sup_{[t_j, t_{j+1}]} f, \quad j = 0, 1, ...m - 1$$

and define the lower sum $L(f, P)$ and the upper sum $U(f, P)$ by

$$L(f, P) = \sum_{j=0}^{m-1} m_j(t_{j+1} - t_j), \quad U(f, P) = \sum_{j=0}^{m-1} M_j(t_{j+1} - t_j).$$

It is clear that $L(f, P) \leq U(f, P)$, since $m_j \leq M_j$ for each j.

Definition A partition P' is said to be finer than the partition P if every point of P is also a point of P'.

In the next three propositions we assume that f is a bounded function on the interval $[a, b]$.

Proposition 6.1 *Let P and P' be partitions of $[a, b]$. If P' is finer than P then*

$$L(f, P) \leq L(f, P') \leq U(f, P') \leq U(f, P).$$

Proof Consider how $L(f, P)$ changes if an additional point r is included in the partition. Suppose that $t_j < r < t_{j+1}$. The only change in $L(f, P)$ that arises is due to the replacement of the term $m_j(t_{j+1} - t_j)$ by the sum of two terms

$$m'_j(r - t_j) + m''_j(t_{j+1} - r),$$

where $m'_j = \inf_{[t_j, r]} f$ and $m''_j = \inf_{[r, t_{j+1}]} f$. But $m'_j \geq m_j$ and $m''_j \geq m_j$ (since the new infima are taken over smaller sets), so that

$$m'_j(r - t_j) + m''_j(t_{j+1} - r) \geq m_j(t_{j+1} - t_j),$$

and therefore $L(f, P) \leq L(f, P')$. The other inequality is proved by a similar argument.   □

Proposition 6.2 *Let P_1 and P_2 be partitions of $[a, b]$. Then*

$$L(f, P_1) \leq U(f, P_2).$$

Proof Create a new partition P_3 by uniting the points in P_1 and P_2 into one sequence. Then P_3 is finer than P_1 and also finer than P_2. This implies that

$$L(f, P_1) \leq L(f, P_3) \leq U(f, P_3) \leq U(f, P_2),$$

so that $L(f, P_1) \leq U(f, P_2)$ as required.   □

Consider next all numbers $L(f, P)$, that is, all lower sums, as P ranges over all possible partitions. These form a set (we could define it by specification for example). This set is moreover bounded above; for example, if we fix a partition P_1, then $L(f, P) \leq U(f, P_1)$ for every partition P. Similarly the set of all upper sums $U(f, P)$ is bounded below. We therefore define the lower and upper integrals

$$\underline{\int} f := \sup_P L(f, P), \qquad \overline{\int} f := \inf_P U(f, P)$$

as the supremum of the lower sums and the infimum of the upper sums respectively, taken over all possible partitions.

If f is a positive function and we wish to assign an area to the region between the graph $y = f(x)$ and the x-axis, bounded by the lines $x = a$ and $x = b$, then it seems clear that whatever this area might be, it should lie between the lower and upper integrals.

Proposition 6.3

$$\underline{\int} f \;\leq\; \overline{\int} f$$

Proof Let P_1 and P_2 be partitions of $[a, b]$. Then $L(f, P_1) \leq U(f, P_2)$. Taking the supremum over all partitions P_1, we obtain

$$\underline{\int} f \leq U(f, P_2).$$

Taking next the infimum over all partitions P_2, we obtain $\underline{\int} f \leq \overline{\int} f$ as required. $\square$

Now we can define the Darboux integral. It has to be said that the process leading to this definition is remarkably short. As with the treatment of some previous concepts, such as limit or derivative, the definition singles out a class of functions, here called integrable, and for each integrable function defines a number called its integral.

Definition Let the function f be bounded on the interval $[a, b]$. If the upper and lower integrals of f are equal, we say that f is integrable (on the interval $[a, b]$). If f is integrable, the common value of its upper and lower integrals is called the integral of f (on the interval $[a, b]$). It is commonly denoted by one of the following:

$$\int f, \quad \int_{[a,b]} f, \quad \int_a^b f \quad \text{or} \quad \int_a^b f(x)\, dx.$$

6.2.1 Thoughts on the Definition

The concept of integral has a reputation for being hard to define. The definition we have just given for the Riemann–Darboux integral is actually quite short and some of its complexities may be concealed.

First of all the role of the completeness axiom comes out clearly in the repeated use of supremum and infimum. The supremum of the set of lower sums (defining the lower integral) is analogous to the supremum of a function. It is not though a function that assigns a real number to each real number in its domain, for the domain here is not a set of real numbers, but the set of partitions. The notation $L(f, P)$ reflects this and emphasises the dependence on P (whilst f remains fixed throughout the discussion).

It appears that the integral is essentially a more complex concept than the derivative. Previously the only sets we encountered were sets of real numbers, mainly intervals, or sets of natural numbers, and one could quite happily define the derivative without using more complex sets. When it comes to the integral, we have to embrace the set of all partitions of an interval. A partition is a sequence of real numbers with certain constraints; so the set of all partitions is a set of sequences of real numbers. This is a higher level of complexity than a set of real numbers. It seems that every approach to the integral involves complexity at this level.

Another approach to the integral is possible in which we approximate f from both above and below by step functions. We will then encounter *sets of step functions*.

Definition A function $g : [a, b] \to \mathbb{R}$ is called a step function if there exists a partition $(t_0, t_1, t_2, ..., t_m)$ of $[a, b]$, and numbers $(c_0, c_1, c_2, ..., c_{m-1})$, such that $g(x) = c_j$ for $t_j < x < t_{j+1}$, $j = 0, 1, 2, ..., m - 1$. In other words g is constant on each open interval $]t_j, t_{j+1}[$.

The area under the graph of a positive step function ought by rights to be $\sum_{j=0}^{m-1} c_j(t_{j+1} - t_j)$. This suggests that we first define the integral for the step function g, whether positive or not, as

$$S(g) = \sum_{j=0}^{m-1} c_j(t_{j+1} - t_j).$$

For a function f, supposed bounded on $[a, b]$, we can define the set of lower approximations as the set of all numbers $S(g)$ as g ranges through step functions such that $g \leq f$ (it is here that a set of step functions is needed). This set is not empty thanks to the boundedness of f. Similarly the set of upper approximations is the set of all numbers $S(g)$ as g ranges through step functions such that $g \geq f$.

So far neither supremum nor infimum has been used. Next, we define the lower integral as the supremum of the set of all lower approximations and the upper integral as the infimum of the set of all upper approximations. Finally, the function is called integrable when the lower and upper integrals coincide.

The idea of approximating a function from above and below by simpler functions for which the integral has an obvious definition is common to many approaches to defining integrals. In particular it recurs in the definition of the Lebesgue integral, one of the greatest achievements of analysis in the twentieth century, to which the Riemann–Darboux integral is but a halfway house, and many of its faults are thereby alleviated.

Exercise Prove that the integral defined using approximation by step functions is the same as the Riemann–Darboux integral.

6.3   First Results on Integrability

The definition of the Riemann–Darboux integral raises some questions:

(a) What functions are integrable? More precisely, what conditions can we impose on f (in addition to its being bounded) that suffice for f to be integrable?
(b) Continuous functions on the interval $[a, b]$ are necessarily bounded. Are they integrable?

(c) Are step functions integrable? If so, and if f is a step function, is $\int f = S(f)$ (as defined in the last section)?

(d) If f is integrable can we find a practical way to calculate the integral? It is clearly impractical to compute the supremum over all lower sums.

We shall devote a considerable effort and a large part of this text to answering these questions.

One step is used repeatedly in the proofs and it is useful to set it out in advance. Let $\varepsilon > 0$. Since the lower integral is the supremum of the lower sums $L(f, P)$ over all partitions P, and the upper integral is the infimum of all upper sums $U(f, P)$ over all partitions P, there exists a partition P_1, such that

$$L(f, P_1) > \underline{\int} f - \varepsilon,$$

and another partition P_2, such that

$$U(f, P_2) < \overline{\int} f + \varepsilon.$$

Now construct a partition P by uniting the points of P_1 and P_2. Then P is simultaneously finer than both P_1 and P_2. Hence in passing from P_1 and P_2 to P, the lower sum cannot decrease and the upper sum cannot increase. Therefore the above inequalities hold also for P in place of P_1 and P_2.

The convenience is that both inequalities hold for the same partition. We can even do the same for a finite set of functions. For example, for two functions f and g, and a given ε, we can find a single partition P, such that the inequalities hold for both f and g.

6.3.1 Riemann's Condition

The condition introduced here is basic for proving that given functions are integrable.

Proposition 6.4 *The function f is integrable if and only if the following condition (which we shall call Riemann's condition[1]) is satisfied: for each $\varepsilon > 0$ there exists a partition P, such that*

$$U(f, P) - L(f, P) < \varepsilon.$$

Proof Assume that f is integrable. Then $\underline{\int} f = \overline{\int} f$. Choose a partition P, such that

---

[1] The name "Riemann's condition" appears in the book "Mathematical Analysis" by T. Apostol. I do not know of any other author who names it after Riemann. It is, however, convenient to have a name for it.

$$U(f, P) < \overline{\int} f + \frac{\varepsilon}{2} \quad \text{and} \quad L(f, P) > \underline{\int} f - \frac{\varepsilon}{2}.$$

It follows that

$$U(f, P) - L(f, P) < \overline{\int} f + \frac{\varepsilon}{2} - \underline{\int} f + \frac{\varepsilon}{2} = \varepsilon.$$

Conversely, assume that Riemann's condition is satisfied. Let $\varepsilon > 0$. Choose a partition P, such that $U(f, P) - L(f, P) < \varepsilon$. Now we have

$$L(f, P) \le \underline{\int} f \le \overline{\int} f \le U(f, P)$$

so that $\overline{\int} f - \underline{\int} f < \varepsilon$. But this holds for all $\varepsilon > 0$. We conclude that $\overline{\int} f = \underline{\int} f$. $\square$

The great strength of Riemann's condition is that we only have to find a single partition that satisfies $U(f, P) - L(f, P) < \varepsilon$. At this point it is useful to note that

$$U(f, P) - L(f, P) = \sum_{j=0}^{m-1} \Omega_j(f)(t_{j+1} - t_j)$$

where $\Omega_j(f)$ denotes the oscillation of f on the interval $[t_j, t_{j+1}]$, that is, the difference between the supremum and the infimum (see Sect. 4.5). We recall (Sect. 4.5 Exercise 4) that the oscillation of f on the interval $[c_1, c_2]$ is the same as the quantity

$$\sup_{c_1 \le x, y \le c_2} |f(x) - f(y)|.$$

The supremum here is taken over all pairs of points, x and y, in the interval $[c_1, c_2]$. This formula is very useful for comparing the oscillation of two functions, especially when it is required to deduce the integrability of one of them from the known integrability of the other, as we shall see.

6.3.2 Integrability of Continuous Functions and Monotonic Functions

We begin to answer the question as to which functions are integrable. We shall show that, loosely paraphrased, continuous functions and monotonic functions are integrable.

Proposition 6.5 *Let $f : [a, b] \to \mathbb{R}$ be continuous. Then f is integrable.*

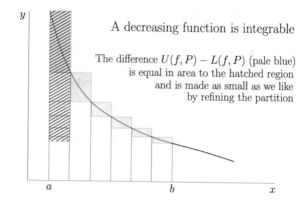

Fig. 6.3 Picture of the proof, adapted from Newton's Principia

A decreasing function is integrable

The difference $U(f, P) - L(f, P)$ (pale blue) is equal in area to the hatched region and is made as small as we like by refining the partition

Proof Let $\varepsilon > 0$. We use the small oscillation theorem (Proposition 4.14; now is the time to read it). There exists a partition P, such that $M_j - m_j < \varepsilon$ for each subinterval of the partition. But then

$$U(f, P) - L(f, P) = \sum_{j=0}^{m-1}(M_j - m_j)(t_{j+1} - t_j) < \sum_{j=0}^{m-1}\varepsilon(t_{j+1} - t_j) = \varepsilon(b - a)$$

and Riemann's condition is satisfied. $\qquad\square$

Newton's pictorial proof of the integrability of monotonic functions is illustrated in Fig. 6.3.

Proposition 6.6 *Let $f : [a, b] \to \mathbb{R}$ be monotonic. Then f is integrable.*

Proof Assume for example that f is increasing (though not necessarily strictly). If $f(a) = f(b)$ then f is constant and obviously integrable; see the next section. So we may suppose that $f(a) < f(b)$.

Let $\varepsilon > 0$. Construct a partition $P = (t_0, t_1, ..., t_m)$, such that

$$t_{j+1} - t_j < \frac{\varepsilon}{f(b) - f(a)}$$

for $j = 0, 1, 2, ..., m$. Since f is increasing we have $m_j = f(t_j)$ and $M_j = f(t_{j+1})$, and we verify Riemann's condition by the calculation

$$
\begin{aligned}
U(f, P) - L(f, P) &= \sum_{j=0}^{m-1}(M_j - m_j)(t_{j+1} - t_j) \\
&= \sum_{j=0}^{m-1}(f(t_{j+1}) - f(t_j))(t_{j+1} - t_j) \\
&\leq \frac{\varepsilon}{f(b) - f(a)} \sum_{j=0}^{m-1}(f(t_{j+1}) - f(t_j)) \\
&\leq \frac{\varepsilon}{f(b) - f(a)}(f(b) - f(a)) = \varepsilon.
\end{aligned}
$$
$\qquad\square$

6.3.3   Two Simple Integrals Computed

In this short section we shall compute our first integrals. The two results are not very impressive, and the treatment of the first function may seem tortuous, but a wait and see attitude is required. They will be used to find the integrals of step functions in Sect. 6.4.

Function A. Let $f : [a, b] \to \mathbb{R}$ where $f(x) = 0$ for $a < x < b$ but $f(a)$ and $f(b)$ are not necessarily 0. Then f is integrable and $\int f = 0$.

For each $\varepsilon > 0$ we consider the partition $P_\varepsilon = (a, a + \varepsilon, b - \varepsilon, b)$. If $f(a)$ and $f(b)$ are positive, then, for all ε, we have

$$U(f, P_\varepsilon) = \varepsilon(f(a) + f(b)).$$

If $f(a) > 0 \geq f(b)$, then, for all ε, we have

$$U(f, P_\varepsilon) = \varepsilon f(a).$$

If $f(b) > 0 \geq f(a)$, then, for all ε, we have

$$U(f, P_\varepsilon) = \varepsilon f(b).$$

Finally, if neither $f(a)$ nor $f(b)$ is positive, then, for all ε, we have

$$U(f, P_\varepsilon) = 0.$$

From these facts it is clear that

$$\overline{\int} f = \inf_P U(f, P) \leq \inf_{\varepsilon > 0} U(f, P_\varepsilon) = 0.$$

That is, $\overline{\int} f \leq 0$. Similar considerations apply to $L(f, P)$ and show that $\underline{\int} f \geq 0$. Hence $\underline{\int} f = \overline{\int} f = 0$ and so we have $\int f = 0$. The argument is illustrated in Fig. 6.4.

Fig. 6.4 An upper sum for function A

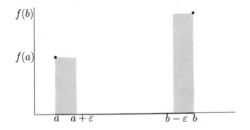

Function B. Let $g : [a, b] \to \mathbb{R}$ be the constant C. Then $\int g = C(b - a)$.
Now $U(g, P) = L(g, P) = C(b - a)$ for every partition and therefore g is integrable with $\int g = C(b - a)$.

6.4 Basic Integration Rules

The rules proved in this section enable us to build new integrable functions from the old ones. Loosely described, the sum and product of integrable functions are integrable. Moreover integration is a linear operation in the space of functions integrable on a given interval.

In the preamble to rules and propositions, we shall often write that the functions are bounded before assuming that they are integrable. Though logically unnecessary, it could be useful to emphasise that Riemann–Darboux integration applies only to bounded functions.

Proposition 6.7 (Sum of functions) *Let $f : [a, b] \to \mathbb{R}$ and $g : [a, b] \to \mathbb{R}$ be bounded functions and assume that they are both integrable. Then $f + g$ is integrable and*

$$\int (f + g) = \int f + \int g.$$

Proof Let $P = (t_0, t_1, ..., t_m)$ be a partition of $[a, b]$. Set

$$m_j = \inf_{[t_j, t_{j+1}]} (f + g), \quad m'_j = \inf_{[t_j, t_{j+1}]} f, \quad m''_j = \inf_{[t_j, t_{j+1}]} g,$$

with similar definitions for M_j, M'_j, M''_j using suprema instead of infima.

For x in $[t_j, t_{j+1}]$ we have $f(x) + g(x) \le M'_j + M''_j$, so that we find $M_j \le M'_j + M''_j$. Similarly $m_j \ge m'_j + m''_j$. These give the inequalities

$$U(f + g, P) \le U(f, P) + U(g, P), \quad L(f + g, P) \ge L(f, P) + L(g, P).$$

Let $\varepsilon > 0$. There exists a partition P (see the discussion in Sect. 6.3 on this point), such that

$$U(f, P) < \int f + \varepsilon, \quad U(g, P) < \int g + \varepsilon$$

$$L(f, P) > \int f - \varepsilon, \quad L(g, P) > \int g - \varepsilon.$$

We obtain

$$U(f + g, P) - L(f + g, P) \le U(f, P) - L(f, P) + U(g, P) - L(g, P) < 4\varepsilon.$$

This shows that Riemann's condition holds for $f + g$. In addition we have

$$\int f + \int g - 2\varepsilon < L(f, P) + L(g, P) \le L(f + g, P)$$

$$\le \int (f + g) \le U(f + g, P) \le U(f, P) + U(g, P) < \int f + \int g + 2\varepsilon$$

so that the inequality

$$\int f + \int g - 2\varepsilon < \int (f + g) < \int f + \int g + 2\varepsilon$$

holds for all $\varepsilon > 0$. We conclude that $\int (f + g) = \int f + \int g$.   □

Proposition 6.8 (Multiplication by scalars) *Let* $f : [a, b] \to \mathbb{R}$ *be bounded and integrable. Let* α *be a real number. Then the function* αf *is integrable on* $[a, b]$ *and*

$$\int \alpha f = \alpha \int f.$$

Proof For an arbitrary set B we have the equalities

$$\sup_{B}(\alpha f) = \alpha \sup_{B} f, \quad \inf_{B}(\alpha f) = \alpha \inf_{B} f \quad (\alpha > 0) \tag{6.1}$$

and

$$\sup_{B}(\alpha f) = \alpha \inf_{B} f, \quad \inf_{B}(\alpha f) = \alpha \sup_{B} f \quad (\alpha < 0). \tag{6.2}$$

Hence

$$U(\alpha f, P) = \alpha U(f, P), \quad L(\alpha f, P) = \alpha L(f, P) \quad (\alpha > 0)$$

and

$$U(\alpha f, P) = \alpha L(f, P), \quad L(\alpha f, P) = \alpha U(f, P) \quad (\alpha < 0).$$

In the case $\alpha > 0$ we therefore have

$$\sup_{P} L(\alpha f, P) = \alpha \sup_{P} L(f, P) = \alpha \int f = \alpha \inf_{P} U(f, P) = \inf_{P} U(\alpha f, P).$$

The extreme terms are therefore equal. Hence each is the same as $\int \alpha f$ and at the same time $\alpha \int f$.

In the case $\alpha < 0$ we have

$$\sup_{P} L(\alpha f, P) = \alpha \inf_{P} U(f, P) = \alpha \int f = \alpha \sup_{P} L(f, P) = \inf_{P} U(\alpha f, P)$$

with the same conclusion.   □

Exercise Prove the formulas (6.1) and (6.2) in the proof of Proposition 6.8.

Proposition 6.9 (Join of intervals) *Let $f : [a, b] \to \mathbb{R}$ be bounded and let $a < c < b$. If f is integrable on $[a, c]$ and also on $[c, b]$ then f is integrable on $[a, b]$ and*

$$\int_{[a,b]} f = \int_{[a,c]} f + \int_{[c,b]} f.$$

Conversely if f is integrable on $[a, b]$, then f is also integrable on $[a, c]$ and on $[c, b]$ and the same equation holds.

Proof Consider the first assertion. Let f be integrable both on $[a, c]$ and on $[c, b]$. Denote by f_1 the restriction of f to $[a, c]$ and by f_2 the restriction of f to $[c, b]$.

Let $\varepsilon > 0$. Choose partitions P_1 on $[a, c]$ and P_2 on $[c, b]$, such that

$$U(f_1, P_1) - \varepsilon < \int_{[a,c]} f < L(f_1, P_1) + \varepsilon$$

and

$$U(f_2, P_2) - \varepsilon < \int_{[c,b]} f < L(f_2, P_2) + \varepsilon.$$

Next construct a partition P on $[a, b]$ by uniting P_1 and P_2. It is clear that

$$L(f, P) = L(f_1, P_1) + L(f_2, P_2)$$

and

$$U(f, P) = U(f_1, P_1) + U(f_2, P_2).$$

But then we get

$$U(f, P) - 2\varepsilon < \int_{[a,c]} f + \int_{[c,b]} f < L(f, P) + 2\varepsilon.$$

This gives $U(f, P) - L(f, P) < 4\varepsilon$ and Riemann's condition is satisfied for f on $[a, b]$. This allows us to expand the last inequalities to

$$\int_{[a,c]} f + \int_{[c,b]} f - 2\varepsilon < L(f, P) \leq \int_{[a,b]} f \leq U(f, P) < \int_{[a,c]} f + \int_{[c,b]} f + 2\varepsilon$$

which are valid for all $\varepsilon > 0$. The first claim of the proposition now follows.

For the second assertion we must show that f_1 and f_2 are integrable given that f is integrable. Let $\varepsilon > 0$. We consider a partition P of $[a, b]$, which contains the point c and satisfies $U(f, P) - L(f, P) < \varepsilon$. From P we make in an obvious way partitions P_1 of $[a, c]$ and P_2 of $[c, b]$ which satisfy $U(f_1, P_1) - L(f_1, P_1) < \varepsilon$ and $U(f_2, P_2) - L(f_2, P_2) < \varepsilon$. □

6.4.1   Integration of Step Functions

Let $f : [a, b] \to \mathbb{R}$ be a step function. There is a partition $(t_0, t_1, t_2, ..., t_m)$ of $[a, b]$, and numbers $(c_0, c_1, c_2, ..., c_{m-1})$, such that $f(x) = c_j$ for $t_j < x < t_{j+1}$, $j = 0, 1, 2, ..., m - 1$.

Consider the restriction of f to the interval $[t_j, t_{j+1}]$. This is a constant c_j, plus a function that is 0 in the open interval $]t_j, t_{j+1}[$, though not necessarily 0 at its endpoints.

We conclude, by Proposition 6.9 and the two simple integrals calculated in Sect. 6.3, that f is integrable on each subinterval of the partition, and hence also on $[a, b]$, and moreover

$$\int_a^b f = \sum_{j=0}^{m-1} \int_{t_j}^{t_{j+1}} f = \sum_{j=0}^{m-1} c_j (t_{j+1} - t_j).$$

The Riemann–Darboux integral gives the "right answer" for the integral of a step function. Note that the values taken by f at the points of the partition do not influence the outcome.

6.4.2   The Integral from a to b

Up to now the integral has been defined over the set $[a, b]$. A new twist introduces integrals over directed intervals; the integral from a to b, or the integral with *lower limit* a and *upper limit* b. The terminology is not supposed to imply that $a < b$; indeed, we could have $b < a$ or $a = b$. The use of the term "limit" is customary here.

Definition Let A be a closed and bounded interval and $f : A \to \mathbb{R}$ a bounded function that is integrable on A. Let a and b be points in A. We define

$$\int_a^b f = \int_{[a,b]} f \quad \text{if} \ \ a < b,$$

$$\int_a^b f = -\int_{[b,a]} f \quad \text{if} \ \ a > b,$$

and

$$\int_a^b f = 0 \quad \text{if} \ \ a = b.$$

Proposition 6.10 *Let a, b and c be points of A in any order. Then*

$$\int_a^c f = \int_a^b f + \int_b^c f.$$

Proof One can consider all six possibilities for the ordering of a, b and c and use Proposition 6.9 on the join of intervals. It all works out and the reader is invited to check it. □

6.4.3 Leibniz's Notation for Integrals

Leibniz denoted the integral $\int_a^b f$ by $\int_a^b f(x)\,dx$. We often use this notation; it has many advantages comparable to the advantages of using Leibniz's notation for derivatives.

As an example we can write

$$\int_0^\pi \sqrt{\sin x}\,dx$$

to denote $\int_0^\pi f$ where f is the function $f(x) = \sqrt{\sin x}$. This could then be read as

The integral of the square root of the sine of x with respect to x from 0 to π.

But I am sure most English-speaking mathematicians read it according to the following phonetics:

The integral of the square root of sine ex dee ex from nought [zero in US] to pie.

In differential geometry expressions like $f(x)\,dx$ can be precisely defined and are called differential forms. It is differential forms that are integrated, rather than functions. But that is a whole new topic beyond fundamental analysis. In this text the expression "dx" has no independent meaning, other than indicating how the integral should be understood. Consider for example the two integrals

$$\int_0^1 \sqrt{x^2 + a^4}\,dx, \qquad \int_0^1 \sqrt{x^2 + a^4}\,da.$$

Here, two unlike functions are to be integrated. In the first place $f(x) = \sqrt{x^2 + a^4}$ where a is a constant; in the second place $g(a) = \sqrt{x^2 + a^4}$ where x is a constant.

6.4.4 Useful Estimates

The reader who has studied vector spaces may recognise that the two integration rules, Propositions 6.7 and 6.8, assert that the set of all functions integrable on the interval $[a, b]$ constitute a vector space over the field $\mathbb{R}$ and that the integral is a linear functional.

Now functions also possess an ordering, where $f \leq g$ means that $f(x) \leq g(x)$ for all x in the common domain. We shall see that integration is not only linear, respecting addition and scalar multiplication, but it is a positive linear functional, by which we mean that it respects the ordering.

Proposition 6.11 *Let $f : [a, b] \to \mathbb{R}$ be bounded and integrable, and assume that $f(x) \geq 0$ for all $x \in [a, b]$. Then*

$$\int_a^b f \geq 0.$$

Proof It is obvious that $L(f, P) \geq 0$ for every partition P. Hence also $\int_a^b f \geq 0$.

$\square$

Proposition 6.12 *Let $f : [a, b] \to \mathbb{R}$, $g : [a, b] \to \mathbb{R}$ be bounded and integrable, and suppose that $f(x) \leq g(x)$ for all $x \in [a, b]$. Then*

$$\int_a^b f \leq \int_a^b g.$$

Proof Because $\int_a^b (g - f) \geq 0$ and it equals $\int_a^b g - \int_a^b f$. $\square$

Proposition 6.13 *Let $f : [a, b] \to \mathbb{R}$ be bounded and integrable. Suppose that*

$$m \leq f(x) \leq M$$

for all x in $[a, b]$. Then

$$m(b - a) \leq \int_a^b f \leq M(b - a).$$

Proof Integrating the inequalities $m \leq f(x) \leq M$ we find

$$m(b - a) \leq \int_a^b f \leq M(b - a).$$

$\square$

The inequalities appearing in the next two propositions are immensely important.

Proposition 6.14 *Let $f : [a, b] \to \mathbb{R}$ be bounded and integrable. Then the function $|f|$ is also integrable and*

$$\left| \int_a^b f \right| \leq \int_a^b |f|.$$

Proof The proof that $|f|$ is integrable is an exercise (see below). Now $f \leq |f|$ and $-f \leq |f|$ on $[a, b]$, so that

$$\int_a^b f \leq \int_a^b |f| \quad \text{and} \quad -\int_a^b f \leq \int_a^b |f|.$$

One of the left-hand sides is equal to $\left| \int_a^b f \right|$.   □

Proposition 6.15 (Mean value theorem for integrals) *Let* $f : [a, b] \to \mathbb{R}$ *be continuous and let* $g : [a, b] \to \mathbb{R}$ *be bounded, integrable and non-negative. Then there exists* ξ *in* $[a, b]$, *such that*

$$\int_a^b fg = f(\xi) \int_a^b g.$$

Proof Let $M = \max_{[a,b]} f$ and $m = \min_{[a,b]} f$. Then, since $g(x) \geq 0$, we have

$$mg(x) \leq f(x)g(x) \leq Mg(x)$$

for all x, and hence

$$m \int_a^b g = \int_a^b mg \leq \int_a^b fg \leq \int_a^b Mg = M \int_a^b g.$$

The number $\int_a^b fg$ lies between the minimum and the maximum of $f(x) \int_a^b g$ on $[a, b]$. Since f is continuous, we can apply the intermediate value theorem and conclude that there exists ξ in the interval $[a, b]$, such that

$$\int_a^b fg = f(\xi) \int_a^b g.$$ □

6.4.5 Exercises

1. Show that if f and g are bounded functions, both integrable on the interval $[a, b]$, then so is fg.
   Hint. First show that the square of an integrable function is integrable. Then use the identity $4fg = (f + g)^2 - (f - g)^2$. To show that f^2 is integrable, given that f is bounded and integrable, you may want to compare the oscillation $\Omega_{[c_1,c_2]}(f^2)$ with $\Omega_{[c_1,c_2]}(f)$.
2. In the mean value theorem for integrals (Proposition 6.15) it was assumed that the function g was non-negative. Obviously the same result holds if it is assumed instead that g is non-positive. Show, however, by means of an example, that the conclusion may not hold if g takes both positive and negative values.

3. Let f be bounded on $[a, b]$. Show that for any subinterval $[c_1, c_2]$ of $[a, b]$ we have
$$\Omega_{[c_1, c_2]}(|f|) \leq \Omega_{[c_1, c_2]}(f).$$

   Deduce that if f is integrable then so is $|f|$.
4. Suppose that f is integrable on the interval $[a, b]$ and that there exists $\alpha > 0$ such that $f(x) > \alpha$ for all x in $[a, b]$. Show that $\sqrt{f}$ is integrable on $[a, b]$.
   Hint. This is similar to the cases $|f|$ and f^2. One can compare the oscillation of $\sqrt{f}$ on an interval with that of f. Try using the identity
$$\sqrt{x} - \sqrt{y} = \frac{x - y}{\sqrt{x} + \sqrt{y}}.$$

   See also Exercise 12 below.
5. Suppose that f is bounded and integrable on the interval $[a, b]$. Suppose that g is obtained from f by changing the values of f at a finite number of points. Show that g is integrable and that $\int g = \int f$.
6. Suppose that f is bounded on $[a, b]$, and continuous except at a finite number of points. Show that f is integrable.
   Hint. By using the join of intervals one may assume that f is continuous except at a.
7. Suppose that f is bounded on $[a, b]$ and for each $h > 0$ it is integrable on $[a + h, b]$. Show that f is integrable on $[a, b]$.
8. Let f be defined on all of $\mathbb{R}$. For each real number c, we define the translated function f_c by $f_c(x) = f(x - c)$. Suppose that f is integrable on all bounded intervals. Show that
$$\int_a^b f = \int_{a+c}^{b+c} f_c$$

   for all a and b.
9. Let f be integrable on the interval $[-L, L]$ and define
$$F(x) = \int_0^x f, \quad (-L < x < L).$$

   Show that if f is an even function then F is an odd function, whereas if f is an odd function then F is an even function. This results in the often useful observation that
$$\int_{-L}^{L} f = 0$$

   if f is an odd function.
10. ($\Diamond$) Prove Hölder's inequality for integrals. Let f and g be positive functions, integrable on the interval $[a, b]$. Let p and q be positive numbers that satisfy $(1/p) + (1/q) = 1$. Then

$$\int_a^b fg \le \left(\int_a^b f^p\right)^{1/p} \left(\int_a^b g^q\right)^{1/q}.$$

Hint. See the hint for Hölder's inequality for series, Sect. 5.11 Exercise 2.

11. A function can be integrable though it is discontinuous at infinitely many points. Consider the function f with domain $[0, 1]$ defined as follows. Firstly we set $f(0) = f(1) = 0$ (these values do not of course matter). Secondly, for $0 < x < 1$ and x irrational we set $f(x) = 0$. Thirdly for $0 < x < 1$ and x rational we write $x = a/b$, where a and b are positive integers with highest common factor 1, and set $f(x) = 1/b$. Show that f is integrable.

     Hint. Obviously $L(f, P) = 0$ for any partition P. The thing is to show that $U(f, P)$ can be made as small as we like by choosing an appropriate partition.

12. The proofs that f^2 and $|f|$ were integrable (Exercises 1 and 3) depended on the Lipschitz continuity (see Sect. 4.5 Exercise 1 for the definition) of the functions x^2 and $|x|$, for the former function on a bounded interval $[-M, M]$. The same was true of $\sqrt{f}$ given that $f > \alpha > 0$ for some constant α (Exercise 4). This procedure will not work to prove that $\sqrt{f}$ is integrable given only that f is integrable on $[a, b]$ and $f \ge 0$. The problem is that $\sqrt{x}$ is not Lipschitz continuous on the interval $[0, M]$ (where $M = \sup f$). To prove integrability of $\sqrt{f}$, we need a more powerful approach. We frame a general proposition and invite the reader to prove it. It includes probably all cases of Riemann integrability that are met with in practice.

     Let f be integrable on $[a, b]$, let $M = \sup |f|$ and let g be continuous on $[-M, M]$. Then $g \circ f$ is integrable on $[a, b]$.

     The following steps are suggested:

     (a) Show that for every $\varepsilon > 0$ there exists $\delta > 0$, such that for every interval $[c_1, c_2] \subset [a, b]$, if $\Omega_{[c_1, c_2]}(f) < \delta$ then $\Omega_{[c_1, c_2]}(g \circ f) < \varepsilon$.
         Hint. Use uniform continuity of g on $[-M, M]$ (Proposition 4.15).

     (b) Let $K = \sup |g \circ f|$. Let $\varepsilon > 0$. Let δ correspond to ε as in part (a). Choose a partition P, such that $U(f, P) - L(f, P) < \varepsilon\delta$. Show that

$$U(g \circ f, P) - L(g \circ f, P) < (2K + b - a)\varepsilon$$

     and deduce that $g \circ f$ is integrable.

13. Use the result of the previous exercise to show that if f is integrable and non-negative, then $\sqrt{f}$ is integrable.

14. After all the preceding exercises, it is useful to have an example of a bounded function that is not integrable. An oft quoted one is the function f on the interval $[0, 1]$ defined by

$$f(x) = \begin{cases} 0, & \text{if } x \text{ is irrational} \\ 1, & \text{if } x \text{ is rational.} \end{cases}$$

Show that $\overline{\int} f = 1$ and $\underline{\int} f = 0$.

Note. This is a standard example of a function that is not Riemann integrable, but *is Lebesgue integrable.*

15. Let f and g be integrable on the interval $[a, b]$. Show that the functions $\min(f, g)$ and $\max(f, g)$ are integrable. In particular $f_+ := \max(f, 0)$ and $f_- := \max(-f, 0)$, called the positive and negative parts of f, are integrable.

    Note. These functions are defined pointwise: $\min(f, g)(x) = \min(f(x), g(x))$.

16. Find an example of a function f such that $|f|$ is integrable but f is not.

6.5 The Connection Between Integration and Differentiation

In this section we show that in some sense integration and differentiation are operations inverse to each other. This finds its expression in the fundamental theorem of calculus. It means that integration can be used to solve the problem of antiderivatives (Problem A) and conversely antiderivatives can be used to compute integrals. More generally integrals can be used to solve differential equations.

An important role is played by the inequality

$$\left| \int_a^b f \right| \leq \int_a^b |f|,$$

which holds if $a < b$. One has to be careful; if $a > b$ the correct inequality is

$$\left| \int_a^b f \right| \leq \int_b^a |f|.$$

Proposition 6.16 *Let $f : [a, b] \to \mathbb{R}$ be bounded and integrable. Let*

$$F(x) = \int_a^x f$$

for all x in $[a, b]$. Then the following hold:

(1) $F : [a, b] \to \mathbb{R}$ is continuous.

(2) If $a < x_0 < b$ and f is continuous at x_0, then F is differentiable at x_0 and $F'(x_0) = f(x_0)$.

Proof (1) Let $K = \sup |f|$. Suppose first that $a \leq x < b$ and consider $\lim_{h \to 0+} F(x + h)$. For $h > 0$ we have

$$F(x + h) - F(x) = \int_a^{x+h} f - \int_a^x f = \int_x^{x+h} f$$

$$|F(x+h) - F(x)| = \left| \int_x^{x+h} f \right| \le \int_x^{x+h} |f| \le Kh.$$

We conclude that $\lim_{h \to 0+} F(x+h) = F(x)$. Similarly $\lim_{h \to 0-} F(x+h) = F(x)$ for $a < x \le b$. This shows that F is continuous.

(2) Let f be continuous at x_0. We will show that both left and right limits of the difference quotient of F at x_0 are equal to $f(x_0)$.

Let $\varepsilon > 0$. There exists $\delta > 0$, such that $|f(x) - f(x_0)| < \varepsilon$ whenever $|x - x_0| < \delta$. For the right limit of the difference quotient, we let $0 < h < \delta$ and consider

$$\begin{aligned}
\frac{F(x_0 + h) - F(x_0)}{h} - f(x_0) &= \frac{1}{h} \int_{x_0}^{x_0+h} f(t)\, dt - f(x_0) \\
&= \frac{1}{h} \int_{x_0}^{x_0+h} f(t)\, dt - \frac{1}{h} \int_{x_0}^{x_0+h} f(x_0)\, dt \\
&= \frac{1}{h} \int_{x_0}^{x_0+h} (f(t) - f(x_0))\, dt.
\end{aligned}$$

We know that $|f(t) - f(x_0)| < \varepsilon$ if $x_0 < t < x_0 + h$. Hence for $0 < h < \delta$ we have

$$\left| \frac{F(x_0 + h) - F(x_0)}{h} - f(x_0) \right| \le \frac{1}{h} \int_{x_0}^{x_0+h} |f(t) - f(x_0)|\, dt \le \frac{1}{h} \varepsilon h = \varepsilon$$

and conclude that

$$\lim_{h \to 0+} \frac{F(x_0 + h) - F(x_0)}{h} = f(x_0).$$

Next we consider the left limit. Whatever the sign of h, we always have

$$\frac{F(x_0 + h) - F(x_0)}{h} - f(x_0) = \frac{1}{h} \int_{x_0}^{x_0+h} (f(t) - f(x_0))\, dt,$$

but for $h < 0$ the estimate of the integral is trickier since $x_0 + h < x_0$. For $-\delta < h < 0$ we have

$$\left| \frac{1}{h} \int_{x_0}^{x_0+h} (f(t) - f(x_0))\, dt \right| = \left| -\frac{1}{h} \int_{x_0+h}^{x_0} (f(t) - f(x_0))\, dt \right| \le \frac{1}{|h|} \varepsilon |h| = \varepsilon$$

and find that

$$\lim_{h \to 0-} \frac{F(x_0 + h) - F(x_0)}{h} = f(x_0).$$

Putting together the left and right limits we conclude that $F'(x_0) = f(x_0)$. $\square$

Here are some further, often used, rules. They are simple consequences of Proposition 6.16 and the definition of integral with limits. The reader should supply the proofs.

(i) If $G(x) = \int_x^b f$ for all x in $[a, b]$, x_0 is in the open interval $]a, b[$ and f is continuous at x_0, then $G'(x_0) = -f(x_0)$.
(ii) If $a < c < b$, $G(x) = \int_c^x f$ for all x in $[a, b]$, x_0 is in the open interval $]a, b[$ and f is continuous at x_0, then $G'(x_0) = f(x_0)$.

Proposition 6.17 (The fundamental theorem of calculus) *Let $f : [a, b] \to \mathbb{R}$ be bounded in $[a, b]$ and continuous in $]a, b[$. Suppose that there exists a function $g : [a, b] \to \mathbb{R}$, continuous in $[a, b]$ and differentiable in $]a, b[$, such that $f = g'$ in $]a, b[$. Then*

$$\int_a^b f = g(b) - g(a).$$

Proof Set

$$F(x) = \int_a^x f, \quad (a \le x \le b).$$

Then F is differentiable with $F'(x) = f(x)$ for $a < x < b$, continuous for $a \le x \le b$, and $F(a) = 0$. But $f(x) = g'(x)$ for $a < x < b$. Hence $(F - g)' = 0$ and $F - g$ is therefore constant in $]a, b[$. Now F and g are continuous in $[a, b]$ so that we can pass to the endpoints and deduce

$$F(b) - g(b) = F(a) - g(a),$$

that is, $F(b) = g(b) - g(a)$.   □

When the fundamental theorem is used to calculate an integral, it is usually applied in the following way. The given function f (the integrand) is continuous in an open interval A, possibly unbounded. An antiderivative g is known or found; that is g is defined in A and $g' = f$. Then for all a and b in A (their order does not matter) we have, in the conventional notation:

$$\int_a^b f = g(x)\Big|_a^b = g(b) - g(a).$$

Rephrasing this slightly, let g be differentiable in the open interval A and let g' be continuous in A. Let a be a point of A. Then for all x in A we have

$$\int_a^x g' = g(x) - g(a).$$

This brings out strongly the extent to which integration and differentiation are inverse operations. But some emphasis falls on the requirement that g' should be continuous. If we drop this requirement it may happen that g' is not even integrable on the interval $[a, x]$. This may be regarded as a defect of the Riemann–Darboux integral, though opinions are divided on this point.

The fundamental theorem shows how an antiderivative, when known, can help us calculate an integral. The converse of this is that integrals show that antiderivatives exist. We can solve the simplest of all differential equations, $dy/dx = f(x)$, using an integral. The proof of the following proposition should now be obvious.

Proposition 6.18 *Let f be continuous in the open interval A. We have the following conclusions:*

(1) The function f has an antiderivative, that is, there exists a function g with domain A such that $g' = f$.
(2) If g is one antiderivative, then the most general antiderivative is $g + C$ where C is a constant.
(3) Let $a \in A$. The function $F(x) = \int_a^x f$, $(x \in A)$, is an antiderivative for f.

The solution of the differential equation $dy/dx = f(x)$ is often written as

$$y(x) = \int f(x)\,dx + C.$$

The formal integral here, *written in Leibniz's notation but without the limits*, is called an *indefinite integral*. It represents *any antiderivative of f*. Finding an antiderivative is often called solving the integral $\int f(x)\,dx$, especially when it is achieved by applying a set of techniques described later (in Chap. 8). In contrast to this, an integral with limits is sometimes called a *definite integral*.

6.5.1   Thoughts About the Fundamental Theorem

The fundamental theorem asserts that the formula $\int_a^b f = F(b) - F(a)$ holds, when f is continuous and F is an antiderivative of f. This is the form of the fundamental theorem as usually presented in analysis texts. However, the requirement that f is continuous is restrictive since many discontinuous functions are integrable.

In applications (for example, Fourier series, technology and engineering) discontinuous integrands arise frequently and it is therefore useful to extend the fundamental theorem to a larger class of integrands. Ideally we would like to extend it to all integrable functions.

It is perhaps unhelpful to overemphasise the role of antiderivative as a central concept in integration theory. It could be preferable to introduce instead a notion of *primitive function*, distinct from that of antiderivative. A small warning: "primitive function" is often used as a synonym for "antiderivative", but in this text a different usage is proposed.

Definition Given an integrable function f on $[a, b]$, by a *primitive* of f is meant any function that differs from the function

$$F(x) := \int_a^x f$$

by a constant.

We immediately extend this definition to the case of a function defined on an interval A, *which could be unbounded*. Useful cases, to be considered later, are $A = [0, \infty[$ and $A =]-\infty, \infty[$. Suppose that f is integrable on every bounded subinterval $[a, b]$ of A. We shall say that a function g with domain A is a primitive of f, if there exists a in A, such that the function

$$g(x) - \int_a^x f$$

is a constant.

Exercise We can fix the point a in the definition of primitive in advance. Let $c \in A$. Show that g is a primitive of f if and only if the function $g(x) - \int_c^x f$ is a constant.

The fundamental theorem states that if f is continuous in an interval A, then any function g, continuous in A, and differentiable and satisfying $g'(x) = f(x)$ at all the interior points of A, is a primitive of f. This opens up the question of how to find primitives for more general types of integrable function. An ideal version of the fundamental theorem would enable us to identify the primitives of a given integrable function quite generally.

We shall describe a class of functions that often arise in practical applications and identify their primitive functions, thereby extending the fundamental theorem.

Definition A function $f : [a, b] \to \mathbb{R}$ is said to be *piece-wise continuous* if there exists a partition $a = t_0 < t_1 < \cdots < t_m = b$, such that for each k, the restriction of f to the open interval $]t_k, t_{k+1}[$ is continuous, and extends to a continuous function in the closed interval $[t_k, t_{k+1}]$.

Putting it differently a piece-wise continuous function f has only a finite number of discontinuities and at each one, the left and right limits exist (though only the right limit at a and only the left limit b). The values of f at the points t_j are unimportant.

We extend the definition of piece-wise continuous function to unbounded domains. A function f defined in an interval A (possibly unbounded) is said to be piece-wise continuous if its restriction to each bounded interval $[a, b]$, included in A, is piece-wise continuous in the sense of the previous paragraph. A piece-wise continuous function in an unbounded interval can have infinitely many points of discontinuity, but there are only finitely many in each bounded closed interval. An example of such a function of some importance in technology is the infinite square wave.

We next extend the fundamental theorem and thereby identify the primitives of all piece-wise continuous functions.

Fig. 6.5 An infinite square wave and one of its primitives

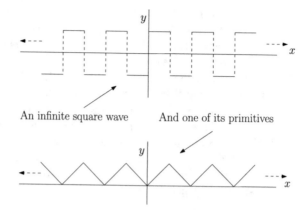

An infinite square wave   And one of its primitives

Proposition 6.19 *Let* $f : [a, b] \to \mathbb{R}$ *be piece-wise continuous. Let* $F : [a, b] \to \mathbb{R}$ *be continuous in* $[a, b]$, *differentiable except at a finite set of points (possibly empty), and satisfy* $F'(x) = f(x)$ *for each* x *at which* F *is differentiable. Then*

$$\int_a^b f = F(b) - F(a).$$

In other words the function F *is a primitive of* f.

Proof The points $t_0 < t_1 < \cdots < t_m$, at which F is not differentiable, form a partition of $[a, b]$ and we may obviously include the endpoints, so that $t_0 = a$ and $t_m = b$. Moreover these points include all discontinuities of f (by, for example, Sect. 5.8 Exercise 10). Now we have, by the fundamental theorem:

$$\int_a^b f = \sum_{k=0}^{m-1} \int_{t_k}^{t_{k+1}} f = \sum_{k=0}^{m-1} \left(F(t_{k+1}) - F(t_k) \right) = F(b) - F(a).$$

$\square$

Several rules of integral calculus can be extended by using primitives instead of antiderivatives as we shall see.

6.5.2 Exercises

1. Let f be a continuous function on an open interval A. We have seen that for every choice of a in A, the function $F(x) = \int_a^x f$ is an antiderivative for f. It can happen that f has antiderivatives that cannot be expressed in this form. Find an example of this.

2. Suppose that f is bounded on $[a, b]$ and for each $\varepsilon > 0$, it is integrable on
   $[a + \varepsilon, b]$. Show that f is integrable on $[a, b]$ and $\lim_{\varepsilon \to 0+} \int_{a+\varepsilon}^b f = \int_a^b f$.
3. Give an example to show that Proposition 6.19 does not hold if the requirement
   that F is continuous is omitted.
4. Let $f(x) = x^2 \sin(x^{-2})$ for $x \neq 0$ and $f(0) = 0$. Show that f is everywhere
   differentiable, but that f' is unbounded in any interval containing 0. Thus the
   integral $\int_{-1}^1 f'$ does not exist (at least not as a Riemann–Darboux integral).
5. In this set of exercises we study periodic functions and their primitives. Recall
   (Sect. 4.2 Exercise 20) that a function f is periodic if there exists $T \neq 0$ (a period)
   such that $f(x + T) = f(x)$ for all x. Let f be a periodic function with fundamental period T (that is, T is the lowest positive period, supposing that one such
   exists). Suppose that f is bounded and integrable on the interval $[0, T]$.

   (a) Show that f is integrable on any bounded interval.
   (b) Show that for any interval $[a, b]$, such that $b - a = T$ we have

   $$\int_a^b f = \int_0^T f.$$

   (c) Show that

   $$\lim_{|y| \to \infty} \frac{1}{y} \int_0^y f = \frac{1}{T} \int_0^T f.$$

   (d) The quantity $(1/T) \int_0^T f$ appearing in item (c) is called the mean of the
   periodic function f. Show that there exists a unique constant C such that
   $f + C$ has mean 0.
   (e) Let $F(x) = \int_0^x f$. Show that F is periodic if and only if f has mean zero.
   (f) Suppose that f has mean zero and set $F_0 = f$. Show that one may define
   uniquely a sequence of functions $(F_n)_{n=0}^\infty$, such that each function is periodic
   with period T, each has mean zero, F_1 is a primitive of F_0, and $F_n' = F_{n-1}$
   for $n = 2, 3, \ldots$ (so, in fact, F_n is a primitive of F_{n-1} for $n \geq 1$).

6.6   (◊) Riemann Sums

We have defined the integral of a bounded function f on an interval $[a, b]$ as the
supremum over all lower sums or the infimum over all upper sums, provided these
two happen to be the same. This was not Riemann's definition; actually it is due to
Darboux. However, it turns out that Riemann's integral and Darboux's are the same;
the same functions are integrable and the integral has the same value when it exists.
This is reflected in our choice of name: the Riemann–Darboux integral. In this nugget
we are going to consider how Riemann defined his integral, and show that the result
is equivalent to that of Darboux.

Riemann approached the integral through what we call Riemann sums. Given a bounded function f on a closed interval $[a, b]$ the Riemann sums are defined in the following way. Let P be a partition of the interval $[a, b]$, let us say it is the sequence $P = (t_0, t_1, ..., t_m)$. For each j we choose quite arbitrarily a point r_j in $[t_j, t_{j+1}]$, for each $j = 0, 1, ...m - 1$. Now we form the sum

$$S = \sum_{j=0}^{m-1} f(r_j)(t_{j+1} - t_j).$$

A sum formed in this way is a Riemann sum for the function f, corresponding to the partition P and the choice of points r_j. In an obvious way it is a candidate for an approximation to the area under the graph $y = f(x)$ (in the case when f is positive) that is just as plausible as an upper or lower sum, if not more so.

The quantity $\max_{0 \leq j \leq m-1}(t_{j+1} - t_j)$ is called the mesh size of the partition P (the analogy is with a fishing net that lets fish below a certain size escape; how appropriate this is remains moot). Riemann defined the integral as a kind of limit, when it exists. More precisely it is a number A that has the following property: for each $\varepsilon > 0$ there exists $\delta > 0$, such that

$$\left| A - \sum_{j=0}^{m-1} f(r_j)(t_{j+1} - t_j) \right| < \varepsilon$$

for all partitions $P = (t_0, t_1, ..., t_m)$ with mesh size less than δ, and for all possible choices of the points r_j in the intervals $[t_j, t_{j+1}]$.

Proposition 6.20 *The Riemann integral of a bounded function on an interval $[a, b]$ exists if and only if its Darboux integral exists. When the integrals exist, they are equal.*

Proof The proof of this is long and extends to the end of this subsection. We take the shorter part first.

We assume that the integral of f on $[a, b]$ exists according to Riemann's definition, and that the integral has the value A. We wish to show that the Darboux integral exists and also has the value A.

Let $\varepsilon > 0$. Choose $\delta > 0$, such that

$$\left| A - \sum_{j=0}^{m-1} f(r_j)(t_{j+1} - t_j) \right| < \varepsilon$$

for all partitions $P = (t_0, t_1, ..., t_m)$ with mesh size less than δ, and for all possible choices of the points r_j in $[t_j, t_{j+1}]$.

Consider one such partition P and let

$$m_j = \inf_{[t_j, t_{j+1}]} f \quad \text{and} \quad M_j = \sup_{[t_j, t_{j+1}]} f.$$

For each subinterval we can choose a point r_j, such that $f(r_j) < m_j + \varepsilon$. Then we find

$$\sum_{j=0}^{m-1} f(r_j)(t_{j+1} - t_j) < \sum_{j=0}^{m-1} (m_j + \varepsilon)(t_{j+1} - t_j) = L(f, P) + \varepsilon(b - a)$$

so that

$$A < L(f, P) + \varepsilon(b - a + 1).$$

In a similar way we can obtain

$$A > U(f, P) - \varepsilon(b - a + 1).$$

But from this it follows that

$$A - \varepsilon(b - a + 1) < L(f, P) \le U(f, P) < A + \varepsilon(b - a + 1).$$

This holds for all $\varepsilon > 0$, so that we conclude that f is integrable in the sense of Darboux and the integral equals A. This concludes the first, and shorter, part of the proof.

The proof that if f is Darboux-integrable, it is also Riemann-integrable, is more complicated. We begin with some general considerations before turning to the actual proof.

Let K be a constant, such that $|f(x)| < K$ for all x in $[a, b]$ (recall that we are assuming that f is bounded). Suppose $\delta > 0$ and consider a partition with mesh size less than δ. We wish to give an upper estimate for how much the lower sum $L(f, P)$ and the upper sum $U(f, P)$ change, if P is replaced by a new partition P', which is formed by adding p new points to P.

The new points land in at most p subintervals of P. First we will estimate the contribution of these intervals to the lower and upper sums, *before the inclusion of the new points*.

The absolute value of the contribution is less than $pK\delta$, for there are at most p intervals in question, each has length at most δ, and m_j and M_j have absolute value less than K.

Next we estimate the contribution to the lower and upper sums of these same intervals *after the insertion of the new points*. The intervals in question get replaced by at most $2p$ new intervals (how many they are will depend on how the new points fall), and their contribution to $L(f, P')$ and $U(f, P')$ has absolute value at most $2pK\delta$.

The conclusion therefore is that

$$|L(f, P) - L(f, P')| \leq 3pK\delta, \quad |U(f, P) - U(f, P')| \leq 3pK\delta.$$

Now we can prove that a Darboux-integrable function is Riemann-integrable. Suppose that f is Darboux-integrable and that $|f| < K$. Let $\varepsilon > 0$. Choose a partition P_1, such that

$$U(f, P_1) - \frac{\varepsilon}{2} < \int f < L(f, P_1) + \frac{\varepsilon}{2}.$$

Suppose that P_1 has p points. Let $\delta = \varepsilon/6pK$. Observe that δ depends only on ε and f (the somewhat arbitrary choice of p depended only on ε and f).

Let P be a partition with mesh size less than δ. Let P' be the partition that is formed by uniting P and P_1. Then P' is finer than P_1 so that

$$U(f, P') - \frac{\varepsilon}{2} < \int f < L(f, P') + \frac{\varepsilon}{2}.$$

But P' is obtained by adding p points to P, and the latter has mesh size less than δ, so that we find

$$|L(f, P) - L(f, P')| \leq 3pK\delta, \quad |U(f, P) - U(f, P')| \leq 3pK\delta,$$

and combining this with the previous inequalities we obtain

$$U(f, P) - 3pK\delta - \frac{\varepsilon}{2} < \int f < L(f, P) + 3pK\delta + \frac{\varepsilon}{2},$$

or, recalling the definition of δ:

$$U(f, P) - \varepsilon < \int f < L(f, P) + \varepsilon.$$

To summarise, this holds on the sole premise that the mesh size of P is less than δ. Furthermore a Riemann sum $\sum_{j=0}^{m-1} f(r_j)(t_{j+1} - t_j)$ for the partition P, formed by taking points r_j in the intervals $[t_j, t_{j+1}]$, lies between $L(f, P)$ and $U(f, P)$. We conclude that

$$\left| \sum_{j=0}^{m-1} f(r_j)(t_{j+1} - t_j) - \int f \right| < \varepsilon$$

whenever the mesh size of P is less than δ; that is, the Darboux-integral $\int f$ is also the Riemann-integral. $\qquad\square$

6.6.1 Things You Can Do with Riemann Sums

Riemann sums are highly versatile tools with both practical and theoretical uses. We give one example of each.

Approximate an Integral

A Riemann sum is a simple approximation to the corresponding integral. Better approximation methods have been developed but they are largely refinements of the Riemann sum.

We can estimate the error if we know a little more about the function f. Suppose that f is differentiable and that $|f'(x)| < M$ for all x in $[a, b]$. Then, by the mean value theorem, we have

$$|f(x) - f(r_j)| < M|x - r_j| \le M(t_{j+1} - t_j)$$

for all x in the interval $[t_j, t_{j+1}]$. Hence

$$\left| \int f - \sum_{j=0}^{m-1} f(r_j)(t_{j+1} - t_j) \right| = \left| \sum_{j=0}^{m-1} \int_{t_j}^{t_{j+1}} (f - f(r_j)) \right| < \sum_{j=0}^{m-1} M(t_{j+1} - t_j)^2.$$

To see what this means, let us suppose the partition P divides the interval $[a, b]$ into m equal intervals. Then the above error bound is

$$\frac{M(b-a)^2}{m}.$$

A natural way to implement such an approximation is to double the number of partition points in each step. After n steps the error is bounded by

$$\frac{M(b-a)^2}{2^n}.$$

This means that, at worst, each doubling contributes roughly 0.3 (approximately $\log_{10} 2$) further correct decimal places. Plainly room for improvement!

Prove a Refinement of the Fundamental Theorem

Proposition 6.21 *Let $f : [a, b] \to \mathbb{R}$ be continuous in $[a, b]$, differentiable in $]a, b[$, and suppose that f' is bounded in $]a, b[$ and integrable on $[a, b]$. Then*

$$\int f' = f(b) - f(a).$$

Proof For each partition $a = t_0 < t_1 < \cdots < t_m = b$ we have, by the mean value theorem,

$$f(b) - f(a) = \sum_{k=0}^{m-1} \left(f(t_{k+1}) - f(t_k) \right) = \sum_{k=0}^{m-1} f'(s_k)(t_{k+1} - t_k)$$

for certain numbers s_k in $[t_k, t_{k+1}], k = 0, 1, \dots m - 1$. On the right we have a Riemann sum for $\int f'$, and it tends to $\int f'$ as the mesh size tends to 0. We conclude that the integral is equal to $f(b) - f(a)$. □

The proposition says that given an integrable function g, a function f, continuous in $[a, b]$ and satisfying $f'(x) = g(x)$ at every point of $]a, b[$, is a primitive (in the sense of Sect. 6.5) of g. A nice result, but not nearly as practical as it might appear, as the requirement that $f'(x) = g(x)$ at every point of the open interval means that g, if discontinuous, cannot have jump discontinuities. In this connection see Sect. 5.8 Exercise 10.

6.6.2 Exercises

1. Compute the following limits by interpreting them as Riemann sums:

   (a) $\displaystyle \lim_{n \to \infty} \frac{1}{n^2} \sum_{k=1}^{n} k$

   (b) $\displaystyle \lim_{n \to \infty} \frac{1}{n^3} \sum_{k=1}^{n} k^2$

   (c) $\displaystyle \lim_{n \to \infty} \frac{1}{n^{p+1}} \sum_{k=1}^{n} k^p$ where $p > 0$.

   (d) $\displaystyle \lim_{n \to \infty} \sum_{k=1}^{n} \frac{n}{n^2 + k^2}$

   (e) $\displaystyle \lim_{n \to \infty} \sum_{k=1}^{n} \frac{1}{\sqrt{n^2 + k^2}}$.

6.6.3 Pointers to Further Study

→ Lebesgue integral
→ Numerical integration

6.7 (◊) The Arc Length, Volume and Surface of Revolution Integrals

Much of the motivation for the definition of the integral was derived from computing the area of a particular kind of plane figure, and a fairly simple one at that, bounded on three sides by segments of the straight lines, $y = 0$, $x = a$ and $x = b$, the fourth side being a part of the graph $y = f(x)$. One can easily extend this to the area between two graphs $y = f(x)$ and $y = g(x)$ between $x = a$ and $x = b$, assuming that $f(x) < g(x)$ in the interval $]a, b[$. The area is then $\int_a^b (g - f)$. To go beyond this in the calculation of area, we need to give a general definition of area of a plane figure. This is not so simple and requires a study of the topological properties of the plane.

The notion of arc length is by nature simpler than that of area, being essentially one-dimensional. We can give a treatment of the length of a reasonably well-behaved curve using approximations analogous to Riemann sums.

First let us look at a graph. Given a function f with domain $[a, b]$ we obtain a "curve", the graph $y = f(x)$. We can approximate what should turn out to be its length by taking a partition $P = (t_0, t_1, ..., t_m)$ of $[a, b]$ and writing down the sum

$$S(P) = \sum_{k=0}^{m-1} \sqrt{(t_{k+1} - t_k)^2 + (f(t_{k+1}) - f(t_k))^2}.$$

Geometrically $S(P)$ is the length of a polygonal curve inscribed in the curve $y = f(x)$. Now we can define the arc length in imitation of Riemann's definition of the integral, as the number L that has the following property, if such a number should exist: for all $\varepsilon > 0$ there exists $\delta > 0$, such that for all partitions of $[a, b]$ with mesh-size less than δ we have

$$|L - S(P)| < \varepsilon.$$

This construction is illustrated in Fig. 6.6.

The most important case for applications is when the curve has a continuously varying tangent. Then we obtain the arc length integral.

Fig. 6.6 Approximating arc length

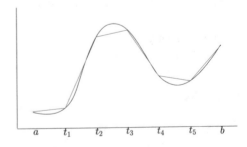

Proposition 6.22 *Let f, with domain $[a, b]$, be continuous in $[a, b]$ and differen- tiable in $]a, b[$, and assume that f' extends to a continuous function in $[a, b]$. Then the length of the curve $y = f(x)$ is given by the arc length integral*

$$L = \int_a^b \sqrt{1 + f'(x)^2}\, dx.$$

Proof For a given partition P, we can use the mean value theorem to find numbers r_k in $[t_k, t_{k+1}]$ (for each k) so that we can write

$$S(P) = \sum_{k=0}^{m-1} \sqrt{(t_{k+1} - t_k)^2 + (f(t_{k+1}) - f(t_k))^2}$$

$$= \sum_{k=0}^{m-1} \sqrt{1 + f'(r_k)^2}\, (t_{k+1} - t_k).$$

We have here a Riemann sum for the arc length integral so the result follows once we show that the integrand is integrable (see the nugget on Riemann sums). The integrability was covered in Sect. 6.4 Exercise 4.   □

The formula extends easily to the case of a curve that may have corners; more precisely to the case when f is continuous but is only piece-wise continuously differentiable. By the latter we mean that there is a partition $(s_0, s_1, ..., s_m)$ of $[a, b]$, such that the derivative exists except at the partition points s_k, and for each open interval $]s_k, s_{k+1}[$ the derivative extends continuously to the closed interval $[s_k, s_{k+1}]$. This includes the case of polygons and most curves that arise in practical applications.

6.7.1 Length of Parametric Curves

A plane parametric curve, expressed by $x = f(t)$, $y = g(t)$, where $a \le t \le b$, can cross itself, double back along itself, or worse. As a geometric object it is tricky to define its length. However we can compute the distance travelled as t goes from a to b. This has obvious practical applications.

For each partition $P = (t_0, t_1, ..., t_m)$ of $[a, b]$, we can approximate the distance travelled by

$$d(P) = \sum_{k=0}^{m-1} \sqrt{(f(t_{k+1}) - f(t_k))^2 + (g(t_{k+1}) - g(t_k))^2}.$$

Then we can say that the distance D is a number that has the following property, if such a number exists: for all $\varepsilon > 0$ there exists $\delta > 0$, such that for all partitions P with mesh-size less than δ we have

Fig. 6.7 Approximating the
length of a parametric curve

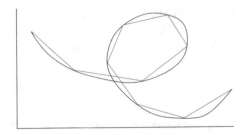

$$|D - d(P)| < \varepsilon.$$

How this leads to an integral is explored in the next exercises. The construction
is illustrated in Fig. 6.7.

6.7.2   Exercises

1. The upper unit semicircle is the curve $y = \sqrt{1 - x^2}$ for $-1 < x < 1$. Write the
   integral for the arc length from $x = 0$ to $x = c$, where $0 < c < 1$.
   Note. The arc length is of course arcsin c. The integral obtained offers a geometrically appealing
   way to define the circular functions rigorously and will be used for this purpose in the next
   chapter.
2. The upper arc of the ellipse $x^2/a^2 + y^2/b^2 = 1$, with semi-major axis a and
   semi-minor axis b is the curve

$$y = \frac{b}{a}\sqrt{a^2 - x^2}, \quad -a < x < a.$$

It is often convenient to express properties of the ellipse in terms of a, and the
eccentricity e, defined as

$$e = \sqrt{1 - \frac{b^2}{a^2}}.$$

Given c such that $0 < c < a$, express the integral for the arc length of the ellipse,
from $x = 0$ to $x = c$, in terms of a, e and c.
   Note. The apparent difficulty of computing this integral (given that $a \neq b$) culminated in the
   theory of elliptic functions early in the nineteenth century.
3. Show that the length of the curve $y = f(x)$ between $x = a$ and $x = b$, when it
   exists, is actually the supremum of $S(P)$ taken over all partitions P. Show the
   same for the distance travelled along a parametric curve, that D is the supremum
   of $d(P)$ taken over all partitions.
   Note. The arc length is often defined as the supremum of the lengths of inscribed polygons.
   One might suppose that one could define the area of a surface as the supremum of the areas of
   inscribed polyhedra. However, this does not work, as it leads to infinities.

4. Suppose that the parametric curve $x = f(t)$, $y = g(t)$, $a \leq t \leq b$ has the properties that f and g have continuous derivatives in $]a, b[$ that extend to continuous functions in $[a, b]$. Show that

$$D = \int_a^b \sqrt{f'(t)^2 + g'(t)^2}\, dt.$$

This exceedingly natural formula shows that the distance travelled is the integral of the speed, as we always knew it was.

Hint. One can write

$$d(P) = \sum_{k=0}^{m-1} \sqrt{f'(\alpha_k)^2 + g'(\beta_k)^2}\,(t_{k+1} - t_k),$$

where, for each k, the numbers α_k and β_k are in the interval $]t_k, t_{k+1}[$. As it stands this is not a Riemann sum. One has to move the point β_k to α_k, thus obtaining a Riemann sum, and estimate the total change. One can use the uniform continuity of the function $(g')^2$ on the interval $[a, b]$, and that of the function $\sqrt{x}$ on the interval $[0, M]$, is an upper bound for $f'^2 + g'^2$ on the interval $[a, b]$.

6.7.3 Volumes and Surfaces of Revolution

Calculus books intended for users of mathematics introduce methods for calculating the volume of a body of revolution and its surface area. Although these concepts belong properly to multivariable calculus, they reduce to integrals of functions of one variable. They can be motivated by the same type of approximation as we used to motivate the integral $\int f$ as the area under the graph $y = f(x)$. The full details will not be given here but the reader who has tackled Exercise 4 should be able to supply them.

Volume of Revolution Integral

Let f be a positive function with the domain $[a, b]$. The plane region bounded by the graph $y = f(x)$, the x-axis, and the lines $x = a$ and $x = b$ is rotated in three dimensions about the x-axis to form a solid. One can introduce a third coordinate axis, the z-axis, at right angles to the (x, y)-plane to effect the rotation analytically. Let $P = (t_0, t_1, ..., t_m)$ be a partition of $[a, b]$. We can approximate the volume from below by the total volume of a collection of cylinders, thus,

$$V_{\text{lower}}(P) = \sum_{k=0}^{m-1} \pi m_k^2 (t_{k+1} - t_k)$$

and from above thus,

$$V_{\text{upper}}(P) = \sum_{k=0}^{m-1} \pi M_k^2 (t_{k+1} - t_k)$$

where, we recall, $m_k = \inf_{[t_k, t_{k+1}]} f$ and $M_k = \sup_{[t_k, t_{k+1}]} f$. We are proceeding intuitively here; a definition of π building on analysis alone will be given in the next chapter.

Refinement of the partition leads to the volume of revolution integral

$$V = \pi \int_a^b f(x)^2 \, dx.$$

Surface of Revolution Integral

A similar process leads to an integral for the surface of revolution. Using a partition P we inscribe a polygon in the curve $y = f(x)$, the same as we used to obtain arc length. Rotate this polygon about the x-axis. The bit between $x = t_k$ and $x = t_{k+1}$ turns into the frustum of a cone with surface area

$$\pi \left(f(t_k) + f(t_{k+1}) \right) \sqrt{(t_{k+1} - t_k)^2 + (f(t_{k+1}) - f(t_k))^2}.$$

The sum of these leads to the surface of revolution integral

$$A = 2\pi \int_a^b f(x) \sqrt{1 + f'(x)^2} \, dx.$$

6.7.4 Exercises (cont'd)

5. Prove a theorem of Archimedes: the area of a parabolic segment cut off by a chord is 4/3 times the area of the triangle whose base is the chord and whose height equals the height of the segment measured from the chord.

   Hint. It can help to set up coordinates in a convenient way. The following is only a suggestion. Take the origin at the midpoint of the chord and the y-axis parallel to the axis of the parabola. Let the equation of the chord be $y = mx$ and its endpoints (l, ml), $(-l, -ml)$. Show that the parabola is one of a one-parameter family

$$y = c(l^2 - x^2) + mx$$

   where we may assume that the parameter $c > 0$ (this just means that the parabola opens downwards). Now compute the area A of the segment by integration, and the maximum area of a triangle with vertices at (l, ml), $(-l, -ml)$ and the third vertex on the arc joining (l, ml) and $(-l, -ml)$.

6. Prove a theorem of Archimedes: when a sphere is inscribed in a cylinder (which then has the same radius as the sphere) and both are cut by two parallel planes at

right angles to the cylinder, the two planes cut equal areas off the sphere and the cylinder.

6.7.5 Pointers to Further Study

→ Multivariable calculus
→ Differential geometry

6.8 (◊) Approximation by Step Functions

The small oscillation theorem (Proposition 4.14) admits another interpretation. It says that every continuous function on the interval $[a, b]$ can be approximated by a step function in a rather precise sense. Let f be continuous on $[a, b]$. Then for all $\varepsilon > 0$ there exists a step function h with the same domain, such that for all x in $[a, b]$ we have

$$|f(x) - h(x)| < \varepsilon.$$

All we need to do to construct h is form a partition $t_0 < t_1 < \cdots < t_m$ of $[a, b]$, such that the oscillation of f on each subinterval is less than ε. Then we define h to be constant on each subinterval, with value equal to f at its midpoint, for example.

The graphs of the two functions, f and h, remain close, with error less than ε throughout the whole interval $[a, b]$. This is called a *uniform approximation* of f by a step function.

Certain functions other than continuous ones can be uniformly approximated by step functions to arbitrary accuracy, monotonic functions for example. Suppose that f is increasing on $[a, b]$. Let $\varepsilon > 0$. Partition the interval $[f(a), f(b)]$ into subintervals of length less than ε. For example we can let $y_0 = f(a)$, $y_m = f(b)$, choosing m so that $(f(b) - f(a))/m < \varepsilon$. Then we construct the partition $y_0 < y_1 < y_2 < \cdots < y_m$ with subintervals of equal length. Next we let A_0 be the set of points in $[a, b]$, such that $y_0 \leq f(x) \leq y_1$ and then, for $k = 1, 2, ..., m - 1$, we set A_k equal to the set of points in $[a, b]$ such that $y_k < f(x) \leq y_{k+1}$.

Exercise Show that each set A_k is an interval. It may be empty; if it is not empty, it may contain neither of its endpoints, one of its endpoints, or both of them. Draw some pictures illustrating each of these possibilities. Show also that if $j < k$, then for all s in A_j and t in A_k we have $s < t$. Show finally that the union of the sets A_k is all of $[a, b]$.

To construct a step function h that approximates f uniformly with error less than ε, we define $h(x)$, for x in A_k, to be equal to $\frac{1}{2}(y_k + y_{k+1})$, except at endpoints of A_k (should either of them be in A_k). At every endpoint we let h be equal to f.

This construction leads to a step function h, such that for all x in $[a, b]$ we have $|f(x) - h(x)| < \varepsilon$. We even have $h(a) = f(a)$ and $h(b) = f(b)$, and h is also increasing, which all turns out to be quite useful as we shall see. We can use it to prove the *second mean value theorem for integrals*, a result that one can sometimes turn to when all else seems to fail.

Proposition 6.23 *Let g be integrable on the interval $[a, b]$ and let f be monotonic on the same interval. Then there exists ξ in $[a, b]$, such that*

$$\int_a^b fg = f(a) \int_a^\xi g + f(b) \int_\xi^b g.$$

Proof The proof is lengthy, so be prepared. We may suppose that f is increasing. The proof is in two main steps.

Step 1. We prove the result in the case that f is a step function. Let $a = t_0 < t_1 < \cdots < t_m = b$ be a partition of $[a, b]$ and suppose that

$$f(x) = c_k, \quad t_k < x < t_{k+1}, \quad k = 0, 1, \ldots m - 1.$$

The values of f at the endpoints can be quite arbitrary, but we require that f is increasing, which implies that the sequence c_k is increasing.

Let $G(x) = \int_a^x g$ for each x in $[a, b]$. Then G is continuous and satisfies $G(a) = 0$. Now we have

$$\int_a^b fg = \sum_{k=0}^{m-1} c_k \int_{t_k}^{t_{k+1}} g = \sum_{k=0}^{m-1} c_k (G(t_{k+1}) - G(t_k))$$

$$= \sum_{k=0}^{m-1} \left(c_{k+1} G(t_{k+1}) - c_k G(t_k) \right) + \sum_{k=0}^{m-1} (c_k - c_{k+1}) G(t_{k+1})$$

$$= c_m G(t_m) + \sum_{k=0}^{m-1} (c_k - c_{k+1}) G(t_{k+1}).$$

Recalling that $t_m = b$ and $G(a) = 0$ we obtain

$$f(b)G(b) - \int_a^b fg = \sum_{k=0}^{m-1} (c_{k+1} - c_k) G(t_{k+1}) + (f(b) - c_m) G(b),$$

so that

$$\frac{f(b)G(b) - \int_a^b fg}{f(b) - f(a)}$$

$$= \frac{(c_0 - f(a))G(a) + \sum_{k=0}^{m-1} (c_{k+1} - c_k) G(t_{k+1}) + (f(b) - c_m) G(b)}{f(b) - f(a)}.$$

On the right-hand side the coefficients in the numerator that multiply values of G, that is, the factors $c_{k+1} - c_k$, together with $c_0 - f(a)$ and $f(b) - c_m$, are all non-negative, and they add up to $f(b) - f(a)$. So the right-hand side is a weighted average of the values $G(a)$, $G(t_1)$, $G(t_2)$,...,$G(b)$, and it must therefore lie between the maximum and minimum of $G(x)$ on the interval $[a, b]$. Hence, by the intermediate value theorem (and recall here that G is continuous), there exists ξ in $[a, b]$, such that $G(\xi)$ is equal to this weighted average (compare Sect. 4.3 Exercise 6). It follows that

$$\frac{f(b)G(b) - \int_a^b fg}{f(b) - f(a)} = G(\xi)$$

or, equivalently

$$\int_a^b fg = f(b)G(b) - \big(f(b) - f(a)\big)G(\xi) = f(a)\int_a^\xi g + f(b)\int_\xi^b g.$$

This completes the proof in the case that f is a step function.

Step 2. To tackle the general case, we let f be an increasing function and approximate it uniformly by a step function. We also introduce a constant K such that $|g(x)| \le K$ for all x in $[a, b]$.

Now let $\varepsilon > 0$. There exists an increasing step function h, such that for all x in $[a, b]$ we have

$$|f(x) - h(x)| < \varepsilon,$$

and, moreover, $h(a) = f(a)$, $h(b) = f(b)$, a pair of equalities that should be borne in mind while elucidating the remainder of the proof.

By the case of a step function (we refer to the penultimate equation in step 1), there exists ξ in $[a, b]$, such that

$$\frac{h(b)G(b) - \int_a^b hg}{h(b) - h(a)} = G(\xi),$$

and since

$$\left| \int_a^b hg - \int_a^b fg \right| \le \int_a^b |f - h||g| < K(b - a)\varepsilon$$

we have that

$$\left| \frac{f(b)G(b) - \int_a^b fg}{f(b) - f(a)} - G(\xi) \right| = \left| \frac{f(b)G(b) - \int_a^b fg}{f(b) - f(a)} - \frac{h(b)G(b) - \int_a^b hg}{h(b) - h(a)} \right|$$

$$= \left| \frac{\int_a^b (f - h)g}{f(b) - f(a)} \right|$$

$$\le \frac{K(b - a)\varepsilon}{f(b) - f(a)},$$

and rearranging this we obtain

$$G(\xi) - \frac{K(b-a)\varepsilon}{f(b)-f(a)} < \frac{f(b)G(b) - \int_a^b fg}{f(b)-f(a)} < G(\xi) + \frac{K(b-a)\varepsilon}{f(b)-f(a)}.$$

Therefore

$$\min_{[a,b]} G - \frac{K(b-a)\varepsilon}{f(b)-f(a)} < \frac{f(b)G(b) - \int_a^b fg}{f(b)-f(a)} < \max_{[a,b]} G + \frac{K(b-a)\varepsilon}{f(b)-f(a)}.$$

These inequalities hold for all ε. So we conclude that

$$\min_{[a,b]} G \leq \frac{f(b)G(b) - \int_a^b fg}{f(b)-f(a)} \leq \max_{[a,b]} G.$$

Hence, by the intermediate value theorem, there exists η in $[a, b]$, such that

$$\frac{f(b)G(b) - \int_a^b fg}{f(b)-f(a)} = G(\eta)$$

which leads to the required conclusion.   □

A much simpler proof of the second mean value theorem for integrals can be given using integration by parts, but using the stronger assumptions that g is continuous, f differentiable, f' is continuous and positive (see Sect. 8.2). However, the theorem is often useful when g has discontinuities, as we shall see in the chapter on improper integrals.

A corollary, the proof of which is left as an exercise, is called Bonnet's theorem.

Proposition 6.24 *Let g be integrable on the interval* $[a, b]$ *and let f be increasing and positive on the same interval. Then there exists* ξ *in* $[a, b]$ *such that*

$$\int_a^b fg = f(b) \int_\xi^b g.$$

We now have two classes of functions that can be uniformly approximated by step functions: the continuous functions and the monotonic functions. We can reasonably ask what these classes have in common. Functions of both classes are integrable. However there are integrable functions that cannot be uniformly approximated by step functions. This is because there is a very simple necessary and sufficient condition for uniform approximation by step functions to be possible: that at every point the one-sided limits $f(x-)$ and $f(x+)$ exist (though only $f(a+)$ at a and $f(b-)$ at b). This is, of course, satisfied by continuous functions and by monotonic functions. The proof of this is not hard but uses the Heine–Borel theorem, which lies just outside the scope of this work.

Exercise Find an integrable function for which one of the one-sided limits fails to exist for at least one point.

Even though we cannot always approximate an integrable function uniformly by step functions, all is not lost. A different type of approximation is available, also by step functions, known as approximation in the mean.

Let f be integrable on $[a, b]$ and let $\varepsilon > 0$. Then there exists a step function h such that

$$\int_a^b |f - h| < \varepsilon.$$

This means that, on average, or to use the correct term, in the mean, h is close to f. At the same time there may be points where $f - h$ is big, even arbitrarily big, but this can only occur on "small" sets of points. All we have to do to produce h is to find a partition P for which $U(f, P) - L(f, P) < \varepsilon$ and set h equal to m_k (or M_k) on the subinterval $]t_k, t_{k+1}[$.

In fact, approximation in the mean could have been used instead of uniform approximation to prove the second mean value theorem, as the reader should be able to check by looking over step 2 of the proof. However, nothing is gained in generality as monotonicity of the function f seems to be essential.

The ideas explored in this nugget are very valuable and capable of much variation; in order to prove something about a whole class of functions we may first be able to prove it for a class of simpler functions (in this case step functions) and then use an approximation technique to obtain the conclusion in general.

6.8.1 Exercises

1. Prove Bonnet's theorem (Proposition 6.24).
2. A function f possessing one-sided limits at each point of its domain has been called a regulated function (notably by Bourbaki). Prove that the following condition is necessary and sufficient for a function f with domain $[a, b]$ to be regulated: for each $\varepsilon > 0$ and x in $[a, b]$ there exists $\delta > 0$, such that for all s and t in $[a, b]$ that satisfy either $x - \delta < s < t < x$ or $x < s < t < x + \delta$ we have $|f(s) - f(t)| < \varepsilon$.
3. Let f be integrable on $[0, 1]$. Prove the following limits:

   (a) $\displaystyle\lim_{n \to \infty} \int_0^1 f(x) x^n \, dx = 0.$

   (b) $\displaystyle\lim_{n \to \infty} \int_0^1 f(x) \sin n\pi x \, dx = 0.$

   Hint. Let $0 \le a < b \le 1$ and do them for a function f equal to 1 for $a \le x \le b$ and equal to 0 otherwise. Extend the conclusions to the case when f is a step function and finally use approximation by step functions in the mean. Use your school knowledge of the circular functions $\sin x$ and $\cos x$ and their derivatives.

Note. The second limit is a key result in the theory of Fourier series.

6.8.2   Pointers to Further Study

→ Functional analysis
→ Heine–Borel theorem

Chapter 7
The Elementary Transcendental Functions

Whatever sines, tangents and secants
Present you with after lengthy and heavy labour;
Fair Reader, this little table of logarithms will give you,
Without serious toil immediately.

J. Napier

It is a part of all analysis courses to give definitions of the elementary transcendental functions, namely trigonometric functions, logarithms and exponentials, that build on ideas of analysis only and do not rely on geometry. In spite of the last denial, the definition of the trigonometric functions in the analysis literature may often owe something to geometric intuition, whilst not needing it logically. This is the case here.

Before that though a word of explanation on the terminology is due. The term "transcendental" does not reflect any particular wow-factor nor is it related to mysticism. It refers to functions that cannot be built up starting with the function x, using constants and algebraic processes. For example polynomials and rational functions, together with roots of polynomial equations such as $\sqrt{x}$, are not transcendental; they are algebraic. Even the function $y = f(x)$, defined as the unique real root of the equation $y^5 + y + x = 0$, is algebraic, although we have no closed expression for it.

A precise definition of transcendental function is that it is a differentiable function f, that does not satisfy a polynomial equation; that is, there is no polynomial $P(x, y)$ of two variables, such that $P(x, f(x)) = 0$ for all x. One really needs more generality here by allowing the coefficients of P, as well as the variables x and y to be complex numbers. See Chap. 9 for a brief introduction to complex numbers.

The term "elementary" refers here, rather arbitrarily it must be said, to a set of functions that were available to mathematicians before the advent of calculus, and were needed in geometry and arithmetic. They are of course just those functions that are met with in high-school algebra, the trigonometric functions, exponential functions and logarithms. As new transcendental functions were introduced, spurred largely by the work of Euler in the eighteenth century, who defined for example the Gamma function, the epithet "elementary" began to be attached to the previously known ones, whilst the new functions began to be called special functions, a term which should probably be disparaged as having no useful meaning at all, unless it is "non-elementary transcendental function that has received a name".

Much of the material of this chapter, though not the way it is developed, will be familiar from school mathematics. Where possible we move the text along quite briskly, with short paragraphs producing the well known classical formulas, whilst dwelling longer on the less familiar parts.

7.1 Trigonometric Functions

In the simplest geometric manifestation that goes beyond their use to solve triangles, the functions sine and cosine, known as the circular functions, provide a parametrisation of the unit circle in the Euclidean plane, that starts at the point $(1, 0)$, travels anticlockwise and has speed 1. This is the reason for the name "circular functions".

To be more precise the circle $x^2 + y^2 = 1$ is parametrised by setting $x = \cos t$ and $y = \sin t$; the parametrisation satisfies $\sqrt{x'(t)^2 + y'(t)^2} = 1$ (that is, the speed is 1), the starting point $(x(0), y(0))$ is $(1, 0)$, and the direction of increasing t is anticlockwise. In this context, anticlockwise simply means passing the points $(1, 0)$, $(0, 1), (-1, 0), (0, -1)$ in that order; this choice of direction probably seems arbitrary to all but mathematicians.

We are going to define these functions using analysis alone, but underlying our approach is the idea of moving with speed 1 along the unit circle. Speed is a familiar everyday concept; a car's speedometer measures it for example. Underlying it is arc length; it is measured by the car's milometer and is also an everyday concept. Arc length is a much simpler concept than area, which many authors have used to motivate a rigorous definition of sine and cosine. One suspects a shift from thinking that area is simpler than arc length to the opposite, that reflects a society in ever-increasing motion.

7.1.1 First Steps Towards Defining Sine and Cosine

First we define arcsine. For all x in the interval $]-1, 1[$ we define

$$\arcsin x = \int_0^x \frac{1}{\sqrt{1 - t^2}}\, dt.$$

Underlying this is the idea that for $x > 0$ the integral is the length of the arc of the unit circle $x^2 + y^2 = 1$ from the point $(0, 1)$ to the point $(x, \sqrt{1 - x^2})$. The angle

Fig. 7.1 Defining arcsine by
arc length

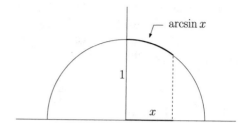

that this arc spans is measured in radians by the length of the arc (this is just the
definition of radian), and its sine by trigonometry is x (Fig. 7.1).

The function arcsin x as just defined as strictly increasing on the interval $]-1, 1[$
is an odd function, (that is $\arcsin(-x) = -\arcsin x$, compare Sect. 6.4 Exercise 9),
and by the fundamental theorem satisfies

$$\frac{d}{dx} \arcsin x = \frac{1}{\sqrt{1-x^2}}, \quad (-1 < x < 1).$$

These facts are immediate consequences of the definition that the reader is invited to
check.

Since for $0 < t < 1$ we plainly have

$$\frac{1}{\sqrt{1-t^2}} < \frac{1}{\sqrt{1-t}}$$

it follows, for $0 < x < 1$, that

$$\arcsin x < \int_0^x \frac{1}{\sqrt{1-t}}\, dt = -2\sqrt{1-t}\,\Big|_0^x = 2 - 2\sqrt{1-x} < 2.$$

The limit $\lim_{x\to 1-} \arcsin x$ therefore exists (since the function is bounded above and
increasing), and is less than or equal to 2. We define the number π by setting

$$\frac{\pi}{2} := \lim_{x\to 1-} \arcsin x.$$

Since arcsin x is an odd function we also have

$$\lim_{x\to -1+} \arcsin x = -\frac{\pi}{2}.$$

This definition of π virtually establishes it as half the perimeter of the unit circle;
about as classical a definition as one could wish for.

We define sin x (the sine of x) for $-\pi/2 < x < \pi/2$ as the inverse function to
arcsine. The function sin x is then strictly increasing odd, and carries the interval

$]-\pi/2, \pi/2[$ on to the interval $]-1, 1[$. The reader can convince themselves that this is a sensible definition by trying the geometry exercise illustrated in Fig. 7.2.

Consider the relation $y = \sin x$, still only defined for $-\pi/2 < x < \pi/2$. We are going to compute the derivative dy/dx. Inverting, we find that $x = \arcsin y$ and hence

$$\frac{dx}{dy} = \frac{1}{\sqrt{1-y^2}},$$

so that

$$\frac{dy}{dx} = \sqrt{1-y^2} = (1-y^2)^{\frac{1}{2}}.$$

Differentiate again, as we obviously may, using the chain rule. We find

$$\frac{d^2y}{dx^2} = \frac{1}{2}(1-y^2)^{-\frac{1}{2}}(-2y)\frac{dy}{dx} = \frac{1}{2}(1-y^2)^{-\frac{1}{2}}(-2y)(1-y^2)^{\frac{1}{2}} = -y.$$

The function $\sin x$ therefore satisfies the *differential equation*

$$\frac{d^2y}{dx^2} + y = 0$$

on the interval $-\pi/2 < x < \pi/2$. The equation is usually written as $y'' + y = 0$.

We observe that $\sin x$ satisfies the conditions (using the notation explained in Sect. 5.3)

$$y(0) = 0, \quad \left.\frac{dy}{dx}\right|_{x=0} = 1.$$

It is usual to write the second condition as $y'(0) = 1$. Another thing to note is that

$$\lim_{x \to \frac{\pi}{2}^-} \frac{d}{dx}\sin x = \lim_{x \to -\frac{\pi}{2}^+} \frac{d}{dx}\sin x = 0.$$

Fig. 7.2 A geometry exercise

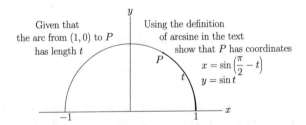

Given that the arc from $(1,0)$ to P has length t

Using the definition of arcsine in the text show that P has coordinates
$$x = \sin\left(\frac{\pi}{2} - t\right)$$
$$y = \sin t$$

7.1.2 The Differential Equation $y'' + y = 0$

The function $f(x) = \sin x$, defined at present in the interval $-\pi/2 < x < \pi/2$, is only one of infinitely many functions that satisfy the differential equation $y'' + y = 0$.

In general, by a *solution* to the differential equation $y'' + y = 0$ in an interval $]a, b[$, we shall mean a twice differentiable function $f :]a, b[\to \mathbb{R}$, that satisfies $f''(x) + f(x) = 0$ for all x in $]a, b[$. There are infinitely many such solutions. From one we can make others by the operations of *translation, reflection* and *differentiation*. From two we can make others by taking *linear combinations*. These claims are summarised in the next proposition, and their proofs are left to the reader.

Proposition 7.1

(1) *If $f(x)$ is a solution of the differential equation $y'' + y = 0$ in the interval $]a, b[$, then the function $g(x) := f(x + c)$ is a solution in the interval $]a - c, b - c[$ and the function $h(x) := f(-x)$ is a solution in the interval $]-b, -a[$.*
(2) *All solutions to $y'' + y = 0$ have derivatives of all orders and they are themselves solutions.*
(3) *If f and g are solutions in the interval $]a, b[$, and A and B are constants, then $Af + Bg$ is also a solution in $]a, b[$.*

7.1.3 Extending $\sin x$

We extend the function $\sin x$ beyond the interval $]-\pi/2, \pi/2[$ on to the whole number line $\mathbb{R}$, in such a way that the extended function satisfies the differential equation $y'' + y = 0$.

Begin with the function $\sin x$ on the interval $]-\pi/2, \pi/2[$. We are going to describe operations on the graph $y = \sin x$. Each operation is one of the transformations

Fig. 7.3 Extending $\sin x$

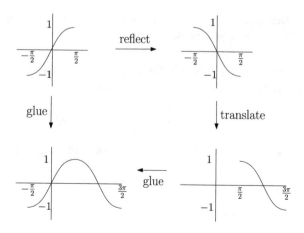

described in Proposition 7.1. First reflect the graph about the line $x = 0$. Next translate the reflection rightwards to give a new graph on the interval $]\pi/2, 3\pi/2[$. Glue this graph to the first graph. This produces an extension to the interval $]-\pi/2, 3\pi/2[$. The extension process is illustrated in Fig. 7.3.

At the join $x = \pi/2$ the two graphs (the first graph and the reflected and translated graph) have the same y-value, namely 1, and the same y'-value 0. Both graphs also satisfy $y'' + y = 0$ so that they have the same second derivative also. The extension therefore satisfies $y'' + y = 0$ in the interval $]-\pi/2, 3\pi/2[$.

The extension just defined has the same values of y, y' and y'' at the endpoints $-\pi/2$ and $3\pi/2$. The length of this interval is 2π. We can therefore extend it by repeated translation to the whole number line $\mathbb{R}$ in such a way that we obtain a function with period 2π. This function will also satisfy the differential equation $y'' + y = 0$.

Exercise Explain why the result of glueing together the pieces of the graph in the above construction produces a function that has derivatives of all orders everywhere, unlike the function constructed in an apparently similar way in Sect. 5.4 Exercise 10.

7.1.4 Defining Cosine

We define cosine by setting

$$\cos x = \sin\left(\frac{\pi}{2} - x\right), \quad x \in \mathbb{R}.$$

Then by Proposition 7.1 the function $\cos x$ is a solution to $y'' + y = 0$. We note that it satisfies the conditions $y(0) = 1$, $y'(0) = 0$.

Now we conclude that the function $A\cos x + B\sin x$ is a solution to the differential equation $y'' + y = 0$ that satisfies the initial conditions $y(0) = A$, $y'(0) = B$. Much more is true.

Proposition 7.2   *The function* $A\cos x + B\sin x$ *is the* unique *solution to* $y'' + y = 0$ *that satisfies* $y(0) = A$, $y'(0) = B$.

Proof Let f be a solution of $y'' + y = 0$ that satisfies the same initial conditions as $A\cos x + B\sin x$. Set

$$g(x) = f(x) - (A\cos x + B\sin x).$$

Then g is a solution to $y'' + y = 0$ that satisfies $g(0) = g'(0) = 0$. Let

$$\phi(x) = g(x)^2 + g'(x)^2.$$

We then have

$$\phi'(x) = 2g(x)g'(x) + 2g'(x)g''(x) = 2g'(x)(g''(x) + g(x)) = 0.$$

We conclude that ϕ is a constant. But this constant is 0 because $\phi(0) = 0$. Since we now have $g(x)^2 + g'(x)^2 = 0$ for all x we conclude that $g(x) = 0$ for all x. $\qquad\square$

We shall rely greatly on this proposition to derive properties of the circular functions. The method serves as a model for how to obtain properties of transcendental functions from the differential equations that they satisfy.

7.1.5 Differentiating cos x and sin x

Proposition 7.3

$$\frac{d}{dx}\sin x = \cos x, \qquad \frac{d}{dx}\cos x = -\sin x.$$

Proof Since the function $y = \sin x$ satisfies $y' = \sqrt{1-y^2}$ and $y'' = -y$, we have

$$\frac{d}{dx}\sin x = \sqrt{1-(\sin x)^2}, \qquad \frac{d^2}{dx^2}\sin x = -\sin x.$$

From this we get

$$\frac{d}{dx}\sin x\Big|_{x=0} = 1, \qquad \frac{d^2}{dx^2}\sin x\Big|_{x=0} = 0.$$

By Proposition 7.1 (item 2) the function $(d/dx)\sin x$ is a solution to $y'' + y = 0$, and we have just seen that it satisfies the same initial conditions at $x = 0$ as does $\cos x$. It therefore equals $\cos x$ by Proposition 7.2.

Finally we obtain

$$\frac{d}{dx}\cos x = \frac{d}{dx}\sin\left(\frac{\pi}{2} - x\right) = -\cos\left(\frac{\pi}{2} - x\right) = -\sin x.$$

$\qquad\square$

7.1.6 Addition Rules for Sine and Cosine

As usual we shall use the notation $\sin^2 x$ to mean $(\sin x)^2$ (and not $\sin(\sin(x))$, and similarly for all positive integral powers. Negative powers are written differently; $\sin^{-1}x$, if used at all, denotes the inverse function $\arcsin x$ and not $1/\sin x$.

Proposition 7.4 *For all x and y we have the addition formulas*

$$\sin(x + y) = \sin x \cos y + \cos x \sin y$$

$$\cos(x + y) = \cos x \cos y - \sin x \sin y.$$

Proof Hold y fixed. Set

$$F(x) := \sin x \cos y + \cos x \sin y$$
$$G(x) := \sin(x + y).$$

The function F satisfies $F'' + F = 0$, $F(0) = \sin y$, $F'(0) = \cos y$ as the reader can easily check. The function G satisfies the same differential equation and the same initial conditions. We conclude that F and G are the same function. This gives the first rule.

For the second rule write

$$\cos(x + y) = \sin\left(\frac{\pi}{2} - x - y\right) = \sin\left(\left(\frac{\pi}{2} - x\right) + (-y)\right)$$

and apply the first rule. ☐

Proposition 7.5

$$\cos{}^2x + \sin{}^2x = 1, \quad (x \in \mathbb{R}).$$

Proof We have $\cos{}^2x + \sin{}^2x = \cos(x - x) = \cos 0 = 1$. ☐

7.1.7 Parametrising the Circle

Proposition 7.6 *Cosine and sine provide a parametrisation of the unit circle. If $x = \cos t$ and $y = \sin t$ then the point (x, y) travels once around the circle $x^2 + y^2 = 1$ as t goes from 0 to 2π; more precisely each point on the circle is passed once as t ranges over the interval $[0, 2\pi[$.*

Proof Let (a, b) satisfy $a^2 + b^2 = 1$. We have to show that there exists a unique t in $[0, 2\pi[$, such that $a = \cos t$ and $b = \sin t$. Assume first that $-1 < a < 1$. There exist a unique $t_1 \in \,]0, \pi[$ and a unique $t_2 \in \,]\pi, 2\pi[$, such that $\cos t_1 = a$ and $\cos t_2 = a$, since cosine is strictly monotonic on each of these intervals and maps each of them on to $]-1, 1[$. Then we must have either $\sin t_1 = b$ or $\sin t_2 = b$, but not both because $\sin t_1 = -\sin t_2$. Finally if $a = -1$ we must have $t = \pi$, and if $a = 1$ we must have $t = 0$. ☐

7.1.8 The Trigonometric Functions $\tan x$, $\cot x$, $\sec x$, $\csc x$

We define tangent, cotangent, secant and cosecant as

$$\tan x = \frac{\sin x}{\cos x}, \quad \cot x = \frac{\cos x}{\sin x}, \quad \sec x = \frac{1}{\cos x}, \quad \csc x = \frac{1}{\sin x}.$$

Each of these functions is undefined at some points, namely at the points where the denominator is zero. Together with the circular functions $\sin x$ and $\cos x$ and their inverses, these functions and their inverses make up the collection of trigonometric functions. The name "circular function" is widely applied to all these functions. Their derivatives are (the reader may check them)

$$\frac{d}{dx}\tan x = \sec^2 x, \quad \frac{d}{dx}\sec x = \sec x \tan x$$

$$\frac{d}{dx}\cot x = -\csc^2 x, \quad \frac{d}{dx}\csc x = -\csc x \cot x.$$

The tangent function $\tan x$ has period π and its graph has vertical asymptotes at odd multiples of $\pi/2$. Its restriction to the interval $]-\pi/2, \pi/2[$ is strictly increasing and maps that interval on to $\mathbb{R}$. Its inverse function, arctangent, is important. It maps $\mathbb{R}$ on to $]-\pi/2, \pi/2[$. Let us find its derivative. Begin with $y = \arctan x$. Then $x = \tan y$ and $dx/dy = \sec^2 y$. Thus we find

$$\frac{dy}{dx} = \frac{1}{\sec^2 y} = \cos^2 y = \frac{\cos^2 y}{\cos^2 y + \sin^2 y} = \frac{1}{1 + \tan^2 y} = \frac{1}{1 + x^2}.$$

7.1.9 The Derivatives of arcsin x, arccos x and arctan x

All of these derivatives are important for integration because they provide us with some new and useful antiderivatives. They are

$$\frac{d}{dx}\arcsin x = \frac{1}{\sqrt{1 - x^2}}, \quad \frac{d}{dx}\arccos x = -\frac{1}{\sqrt{1 - x^2}},$$

$$\frac{d}{dx}\arctan x = \frac{1}{1 + x^2}.$$

Firstly, these formulas provide the antiderivative

$$\int \frac{1}{\sqrt{1 - x^2}}\,dx = \arcsin x + C \quad \text{or} \quad -\arccos x + C.$$

These two versions indicate why mathematics teachers lay so much emphasis on including the constant C. On the interval $]-1, 1[$ (where the integrand is defined) we have $\arcsin x = \frac{\pi}{2} - \arccos x$.

Secondly, and it is one of the most useful antiderivatives, we have

$$\int \frac{1}{1+x^2}\,dx = \arctan x + C.$$

The interval of definition here is all of $\mathbb{R}$.

7.1.10 Exercises

1. Check the formulas for the derivatives of $\tan x$, $\sec x$, $\csc x$ and $\cot x$.
2. Obtain the addition formula for $\tan x$ (familiar from school mathematics)

$$\tan(u+v) = \frac{\tan u + \tan v}{1 - \tan u \tan v}.$$

3. Let $t = \tan(x/2)$. Express $\sin x$, $\cos x$ and $\tan x$ as rational functions of t.
4. There is an addition formula for arctangent. It is often written simply as

$$\arctan x + \arctan y = \arctan\left(\frac{x+y}{1-xy}\right).$$

Care is required because the range of arctangent is the interval $]-\pi/2, \pi/2[$.

   (a) Prove the formula in the case that

$$-\frac{\pi}{2} < \arctan x + \arctan y < \frac{\pi}{2}.$$

   (b) How would you modify the formula in the cases that $\arctan x + \arctan y$ falls outside the interval $]-\pi/2, \pi/2[$?

5. Prove the following formulas, no doubt familiar to you from school mathematics. They are important tools for integration.

   (a) $1 + \tan^2 x = \sec^2 x$
   (b) $\sin 2x = 2 \sin x \cos x$
   (c) $\cos 2x = \cos^2 x - \sin^2 x$
   (d) $\cos 2x = 2 \cos^2 x - 1$
   (e) $\cos 2x = 1 - 2 \sin^2 x$
   (f) $\cos 3x = 4 \cos^3 x - 3 \cos x.$
   (g) $\sin 3x = 3 \sin x - 4 \sin^3 x.$

6. Let a and b be distinct real numbers.

   (a) Show that the function $a \cos x + b \sin x$ is periodic with period 2π, and oscillates between the values $\sqrt{a^2+b^2}$ and $-\sqrt{a^2+b^2}$.
   (b) Show that the function $a \cos^2 x + b \sin^2 x$ is periodic with period π, and oscillates between the values a and b.

(c) Perform a similar analysis on the function $a \cos^3 x + b \sin^3 x$. In addition to finding its maximum and minimum values determine all other local maxima and minima.

7. We cannot assign an angle t to each point P of the unit circle so that t is a continuous function of P. We have to omit one point of the circle. Most commonly it is the point $(-1, 0)$ (for some reason this is thought to be the point one is least likely to visit; but that varies of course). Omitting this point we assign an angle t to the rest of the circle such that t is in the open interval $-\pi < t < \pi$. With this definition for t, derive the following formulas. It is assumed that (x, y) lies on the unit circle.

(a) $t = \arctan \dfrac{y}{x}$ if $x > 0$;

(b) $t = \arctan \dfrac{y}{x} + \pi$ if $x < 0$ and $y > 0$;

(c) $t = \arctan \dfrac{y}{x} - \pi$ if $x < 0$ and $y < 0$;

(d) $t = 2 \arctan \dfrac{y}{1+x}$ if $x \neq -1$.

8. Prove that $\sin x$ is not an algebraic function.
   Hint. Suppose that a formula $f(x, \sin x) = 0$ is valid where $f(x, y) = \sum_{k=0}^{m} p_k(x) y^k$ and the coefficients $p_k(x)$ are polynomials. We can assume that $p_0(x)$ is not the zero polynomial (if it was we could lower m). From your knowledge of $\sin x$ derive an impossible property of $p_0(x)$.

7.2 Logarithms and Exponentials

As in the case of the circular functions we use an integral to give the primary definition, which means that the logarithm is defined first and the exponential function then appears as its inverse. Another important differential equation is introduced.

The exponential function with base a, denoted by a^x, generalises the rational power $a^{m/n}$. Most commonly when we talk about the exponential function, without mentioning the base, we have in mind a particular base e, which has the property that e^x is a solution of the differential equation $dy/dx = y$. This makes e^x the most important function in analysis.

7.2.1 *Defining the Natural Logarithm and the Exponential Function*

We begin by defining the natural logarithm by

$$\ln x = \int_1^x \frac{1}{t}\, dt, \quad (x > 0).$$

Then $\ln 1 = 0$, $\ln x > 0$ for $x > 1$, $\ln x < 0$ for $0 < x < 1$ and

$$\frac{d}{dx} \ln x = \frac{1}{x}, \quad (x > 0).$$

This makes $\ln x$ an antiderivative for $1/x$ on the interval $]0, \infty[$, thus filling in an important gap in the list of antiderivatives (by cheating as one might say), and solving the differential equation $y' = 1/x$.

A point of notation. Some denote the natural logarithm by $\log x$. Others use $\log x$ for the common logarithm, that is, the logarithm to base 10. We prefer the formation "ln", in which the "n" could refer to "natural" or, and this seems more likely, to Napier, the originator of logarithms. In fact natural logarithms were often called Napierian logarithms. The situation is made all the more confusing by the fact that Napier did not use what we now call the natural logarithm but another quantity closely related to it.

The laws of logarithms follow from the differential equation that $\ln x$ satisfies.

Proposition 7.7 (First law of logarithms) *For all $x > 0$ and $y > 0$ we have*

$$\ln(xy) = \ln x + \ln y.$$

Proof Fix y and set $f(x) = \ln(xy)$. Then

$$f'(x) = \frac{y}{xy} = \frac{1}{x} = \frac{d}{dx} \ln x.$$

We conclude that $f(x) - \ln x$ is a constant C. By putting $x = 1$ we see that $C = \ln y$. □

Proposition 7.8 *The function $\ln x$ is strictly increasing on the interval $]0, \infty[$, and we have the limits*

$$\lim_{x \to \infty} \ln x = \infty \quad and \quad \lim_{x \to 0+} \ln x = -\infty.$$

Proof We have

$$\frac{d}{dx} \ln x = \frac{1}{x} > 0$$

and therefore $\ln x$ is strictly increasing. Now $\ln(2^n) = n \ln 2$ for each natural number n by the first law of logarithms, and since $\ln 1 = 0$ and $\ln x$ is increasing we have $\ln 2 > 0$. We conclude, since $\ln x$ is increasing, that $\lim_{x \to \infty} \ln x = \infty$. Set $\frac{1}{2}$ in place of 2 and conclude, since $\ln \frac{1}{2} < 0$, that $\lim_{x \to 0+} \ln x = -\infty$. □

We define the exponential function $\exp : \mathbb{R} \to \,]0, \infty[$ as the inverse function to $\ln x$. Let us differentiate $\exp x$. Set $y = \exp x$. Then $x = \ln y$ and we find

$$\frac{dx}{dy} = \frac{1}{y}$$

giving

$$\frac{dy}{dx} = y = \exp x.$$

We have proved the following.

Proposition 7.9 *The function $y = \exp x$ is a solution to the differential equation*

$$\frac{dy}{dx} = y.$$

The first law of logarithms turns into the first law of exponentials.

Proposition 7.10 *For all real numbers x and y we have*

$$\exp(x + y) = \exp x . \exp y.$$

Proof Set $x = \ln s$, $y = \ln t$ where $s > 0$ and $t > 0$. Now

$$\exp(x + y) = \exp(\ln s + \ln t) = \exp(\ln(st)) = st = \exp x . \exp y. \qquad \square$$

We define the number $e := \exp 1$. We will soon see that

$$e = 2.7 \, \overbrace{1828} \, \overbrace{1828} \, \overbrace{459045} \ldots.$$

The brackets are intended as an aid to memorising the digits.

The numbers π and e are irrational (we will prove these claims later), and are the most important irrational numbers, though no doubt opinions may differ on this, and on whether $\sqrt{2}$ should be included for historical reasons.

We now have $\exp n = e^n$ for all natural numbers n and it is easy to see that $\exp(m/n) = e^{m/n}$ for all rationals m/n. The function $\exp x$ therefore extends the power $e^{m/n}$ to all the reals. We therefore define $e^x := \exp x$ for all real numbers x.

7.2.2 Exponentials and Logarithms with Base a

Let $a > 0$. We define $a^x := \exp(x \ln a)$ for all real numbers x. This is the exponential function with base a. The general power a^x is not defined in real analysis for negative a, at least not as a function. Even so, certain values exist; for example -1 has a real cube root.

The case $a = 1$ is uninteresting, since then a^x is the constant 1. If $a > 1$ then $\ln a > 0$; a^x is then strictly increasing and maps $\mathbb{R}$ on to $]0, \infty[$. If $0 < a < 1$ then $\ln a < 0$; a^x is strictly decreasing and maps $\mathbb{R}$ on to $]0, \infty[$.

The inverse function to a^x in the case $a \neq 1$ is the logarithm with base a. It is denoted by the function symbol $\log_a$. In other words the equation $y = a^x$ inverts to $x = \log_a y$. In practice we only use logarithms with base higher than 1, most frequently e (the natural logarithm) or 10 (the common logarithm). We continue to denote $\log_e x$ by $\ln x$.

If $a > 1$ then the function a^x is strictly increasing and maps $\mathbb{R}$ on to $]0, \infty[$. It follows that its inverse function $\log_a$ is also strictly increasing and maps $]0, \infty[$ on to $\mathbb{R}$.

7.2.3 The Laws of Logarithms and Exponents

Proposition 7.11 *For all $a > 0$ and all x, y in $\mathbb{R}$, we have*

(1) $a^{x+y} = a^x \cdot a^y$, *(first law of exponents).*
(2) $(a^x)^y = a^{xy}$, *(second law of exponents).*

For all $a > 0$, $b > 0$ and all real x, we have

(3) $(ab)^x = a^x b^x$, *(third law of exponents).*

For all $a > 0$ except $a \neq 1$, and all $x > 0$ and $y > 0$, we have

(4) $\log_a(xy) = \log_a x + \log_a y$, *(first law of logarithms).*

For all $x > 0$ and all real y we have

(5) $\log_a(x^y) = y \log_a x$, *(second law of logarithms).*

Proofs (1) $a^{x+y} = e^{(x+y)\ln a} = e^{x \ln a} e^{y \ln a} = a^x a^y$.
(2) $(a^x)^y = e^{y \ln(a^x)} = e^{y \ln(e^{x \ln a})} = e^{yx \ln a} = a^{xy}$.
(3) $(ab)^x = e^{x \ln(ab)} = e^{x \ln a + x \ln b} = e^{x \ln a} e^{x \ln b} = a^x b^x$.
(4)   Let $\log_a x = s$ and $\log_a y = t$. Then

$$xy = a^s a^t = a^{s+t},$$

so that

$$s + t = \log_a(xy).$$

(5)   Let $\log_a x = t$. Then

$$x^y = (a^t)^y = a^{ty},$$

so that

$$yt = \log_a(x^y).$$

$\square$

7.2.4 Differentiating a^x and x^a

Let $a > 0$ be a constant. Then

$$\frac{d}{dx}a^x = \frac{d}{dx}e^{x\ln a} = (\ln a)e^{x\ln a} = (\ln a)a^x, \quad (-\infty < x < \infty).$$

On the other hand (and here we conclude the story of differentiating x^a begun when a was an integer):

$$\frac{d}{dx}x^a = \frac{d}{dx}e^{a\ln x} = e^{a\ln x} \cdot \frac{a}{x} = x^a \cdot \frac{a}{x} = ax^{a-1}, \quad (0 < x < \infty).$$

The restriction to positive x is only made because a is unspecified. For certain values, for example when a is an integer or a rational with odd denominator, we can allow negative x.

7.2.5 Exponential Growth

The functions e^x and e^{-x} overpower all power functions x^a as $x \to \infty$. Conversely the power functions overpower $\ln x$.

Proposition 7.12 *Let $a > 0$. Then we have the limits:*

(1) $\quad \lim\limits_{x\to\infty} \dfrac{e^x}{x^a} = \infty$

(2) $\quad \lim\limits_{x\to\infty} x^a e^{-x} = 0$

(3) $\quad \lim\limits_{x\to 0+} x^a \ln x = 0$

(4) $\quad \lim\limits_{x\to\infty} \dfrac{\ln x}{x^a} = 0.$

Proofs (1) The quickest way to obtain this, and indeed the other limits, is to use L'Hopital's rule in the ∞/∞ version (Proposition 5.16). Repeated differentiation of numerator and denominator of e^x/x^a leads eventually to e^x/cx^b with $b \le 0$ and $c > 0$. This tends to ∞.

Another, and very natural, proof of this limit builds on the power series $e^x = \sum_{n=0}^{\infty} x^n/n!$. However we do not yet have this at our disposal.

(2) This is the reciprocal of the first limit.

(3) We have $x^a \ln x = \ln x/x^{-a}$ and the denominator tends to ∞ as $x \to 0+$. Differentiating numerator and denominator gives $-x^a/a$ with the limit 0 as $x \to 0+$.

(4) Differentiating numerator and denominator leads to $1/ax^a$ with limit 0 as $x \to \infty$. $\qquad \square$

7.2.6 Hyperbolic Functions

We define the hyperbolic sine, hyperbolic cosine and hyperbolic tangent. These are defined for all real x by

$$\sinh x = \frac{e^x - e^{-x}}{2}, \quad \cosh x = \frac{e^x + e^{-x}}{2}, \quad \tanh x = \frac{e^x - e^{-x}}{e^x + e^{-x}}.$$

Occasionally one sees also the functions

$$\operatorname{sech} x = \frac{1}{\cosh x}, \quad \operatorname{csch} x = \frac{1}{\sinh x}, \quad \coth x = \frac{1}{\tanh x}.$$

We state the most important properties of the hyperbolic functions for ready reference, leaving the proofs to the exercises.

For all x we have the important formula

$$\cosh^2 x - \sinh^2 x = 1.$$

This means that the parametrised curve $x = \cosh t$, $y = \sinh t$ is the hyperbola $x^2 - y^2 = 1$, and it explains the epithet hyperbolic.

The derivatives of the three important hyperbolic functions are

$$\frac{d}{dx}\sinh x = \cosh x, \quad \frac{d}{dx}\cosh x = \sinh x, \quad \frac{d}{dx}\tanh x = \frac{1}{\cosh^2 x}.$$

The inverses of the hyperbolic functions are used in integration, as they furnish new antiderivatives. The function sinh is strictly increasing, odd and maps $\mathbb{R}$ on to $\mathbb{R}$. The function cosh is even, strictly increasing on $]0, \infty[$, and maps this interval on to $]1, \infty[$. The function tanh is strictly increasing, odd and maps $\mathbb{R}$ on to $]-1, 1[$. We therefore have three inverse functions, each defined with a different domain, and bijective with the indicated codomains:

$$\sinh^{-1} : \mathbb{R} \to \mathbb{R}, \quad \cosh^{-1} :]1, \infty[\to]0, \infty[, \quad \tanh^{-1} :]-1, 1[\to \mathbb{R}.$$

Their derivatives are important as they provide valuable antiderivatives:

$$\frac{d}{dx}\sinh^{-1}x = \frac{1}{\sqrt{x^2 + 1}},$$

$$\frac{d}{dx}\cosh^{-1}x = \frac{1}{\sqrt{x^2 - 1}}, \quad (x > 1),$$

$$\frac{d}{dx}\tanh^{-1}x = \frac{1}{1 - x^2}, \quad (-1 < x < 1).$$

The following rules are frequently used to provide alternative versions of the antiderivatives of the derivatives in the previous display:

$$\sinh^{-1}x = \ln(x + \sqrt{x^2 + 1})$$
$$\cosh^{-1}x = \ln(x + \sqrt{x^2 - 1}), \quad (x > 1),$$
$$\tanh^{-1}x = \frac{1}{2}\ln\left(\frac{1+x}{1-x}\right), \quad (-1 < x < 1).$$

7.2.7 The Differential Equation $y' = ky$

We know that $(d/dx)e^{kx} = ke^{kx}$. The function $y = e^{kx}$ is therefore a solution of the differential equation $dy/dx = ky$. The function $y = Ce^{kx}$ (where C is a constant) is also a solution, and it satisfies the condition $y(0) = C$. It is an important fact that it is the unique solution that satisfies this condition. We write the differential equation in the short form $y' = ky$.

Proposition 7.13 Let $k \neq 0$. The differential equation $y' = ky$ has a unique solution, defined for all $\mathbb{R}$, that satisfies the condition $y(0) = C$. This solution is $y = Ce^{kx}$.

Proof Let $y = \phi(x)$ be a solution that satisfies $\phi(0) = C$. Then

$$\frac{d}{dx}(e^{-kx}\phi(x)) = e^{-kx}\phi'(x) - ke^{-kx}\phi(x) = e^{-kx}(\phi'(x) - k\phi(x)) = 0.$$

We conclude that $e^{-kx}\phi(x)$ is a constant, and by considering $x = 0$ we see that the constant is C. That is, $\phi(x) = Ce^{kx}$. □

7.2.8 The Antiderivative $\int (1/x)\,dx$

We have seen that $\ln x$ is an antiderivative for the function $1/x$ on the interval $]0, \infty[$. What then is an antiderivative for $1/x$ on the interval $]-\infty, 0[$? For $x < 0$ we have

$$\frac{d}{dx}\ln(-x) = -\frac{1}{(-x)} = \frac{1}{x}.$$

We have therefore the two antiderivatives, each for the appropriate interval:

$$\int \frac{1}{x}\,dx = \ln x + C \quad (x > 0), \qquad \int \frac{1}{x}\,dx = \ln(-x) + C \quad (x < 0).$$

Some recommend writing this as one formula $\int (1/x)\, dx = \ln |x| + C$, considered as valid for all $x \neq 0$. However this is not correct and should be discouraged.

Here's why. The domain of $1/x$ is the union $]-\infty, 0[\cup]0, \infty[$. We could construct an antiderivative on this domain, not of the form $\ln |x| + C$, by letting $g(x) = \ln x$ for $x > 0$ and $g(x) = \ln(-x) + 1$ for $x < 0$. This is possible because the domain is disconnected.

It is possible, by using a more general interpretation of differentiation than the usual one, to make the equation $(d/dx) \ln |x| = 1/x$ correct over the whole of $\mathbb{R}$, not excluding 0. This is accomplished by the theory of distributions (or generalised functions). It is tricky; even $1/x$ has to be reinterpreted, but not as an ordinary function assigning to each number its reciprocal, but as a distribution.

7.2.9   Exercises

1. Show that for all $x > 0$ we have

$$\log_{10} x = \frac{\ln x}{\ln 10}.$$

Note. This has the practical significance that once logarithms to base e have been tabulated the logarithms to base 10 can be found by a straightforward multiplicative conversion, requiring the single number $\ln 10$. A good approximation to $\ln 10$ is 2.3026 and a rough value 2.3, the usefulness of knowing which is recounted in an anecdote in "You must be joking Mr Feynman".

2. Find a formula for the derivative $(d/dx) \log_a x$.

3. The natural logarithm enables us to give a more accurate estimate of the divergence of the harmonic series than could be obtained in Chap. 3.

   (a) Show that

$$\ln(n+1) < 1 + \frac{1}{2} + \frac{1}{3} + \cdots + \frac{1}{n} < 1 + \ln n.$$

   Hint. Find lower and upper sums for the integral $\int_1^n (1/x)\, dx$ with a suitable partition.

   (b) Give an estimate of how many terms of the harmonic series are needed to exceed 100.

4. Let a, b and c be real numbers and suppose that $b < c$. Show that there exists K, such that $x^a e^{bx} < e^{cx}$ for all $x > K$.

5. Prove that

$$\lim_{x \to \infty} \left(1 + \frac{1}{x}\right)^x = e.$$

Hint. Take the logarithm. This is often a good idea in limits involving powers.

6. Continuation of the previous exercise. Refer to Sect. 3.8 Exercise 2 and deduce that

$$e = \sum_{n=0}^{\infty} \frac{1}{n!}.$$

Use this to compute e to 12 places of decimals using a simple calculator, more precisely, using only the numerical buttons together with $\boxed{M+}$ $\boxed{\div}$ $\boxed{=}$ and $\boxed{MR}$. Keep off the factorial button!

Note. This is a beautiful example of how a highly impractical formula can be transformed into highly practical one. To compute e to any accuracy from the limit formula of the previous exercise one would have to calculate for example $1.00000001^{100000000}$, and that only gives 7 places of decimals and can hardly be computed without using the exponential function in some form. From the series about 15 terms are enough to get 12 decimal places, and can be computed rapidly.

7. Calculate the following limits (including proofs that they exist):

   (a) $\displaystyle \lim_{x \to 0+} x^x$

   (b) $\displaystyle \lim_{x \to \frac{\pi}{2}} \frac{\cos^2 x}{(x - \frac{\pi}{2})^2}$

   (c) $\displaystyle \lim_{x \to \pi} \frac{\sin mx}{\sin x}$, where m is an integer.

   (d) $\displaystyle \lim_{x \to \frac{\pi}{4}} (\tan x)^{\tan 2x}$

   (e) $\lim_{x \to \infty} \ln |P(x)| / \ln |Q(x)|$, where P and Q are polynomials, such that P has degree m and Q has degree n.

   (f) $\displaystyle \lim_{x \to \infty} \frac{(x + 1)^s - x^s}{x^{s+1}}$, where s is a real constant.

8. ($\lozenge$) Determine the limit

$$\lim_{n \to \infty} \sum_{k=1}^{n} \frac{1}{n + k}$$

   by viewing it as a Riemann sum (Sect. 6.6).

9. (a) Show that for every power function x^m we have

$$\lim_{x \to \infty} x^m e^{-x^2} = \lim_{x \to -\infty} x^m e^{-x^2} = 0.$$

   (b) Show that for all positive integers n we can write

$$\frac{d^n}{dx^n} e^{-x^2} = P_n(x) e^{-x^2}$$

   where $P_n(x)$ is a polynomial with degree n.

(c) Show that, for all positive integers m and n

$$\lim_{x \to \infty} x^m \frac{d^n}{dx^n} e^{-x^2} = \lim_{x \to -\infty} x^m \frac{d^n}{dx^n} e^{-x^2} = 0.$$

Note. The property of e^{-x^2}, that all its derivatives overpower all powers of x at $\pm\infty$, puts it into a class of functions that have been called, rather prosaically, *smooth functions of rapid decrease.* They are important in the theory of distributions, briefly mentioned in the text.

10. The equation $x^y = y^x$ defines a curve (of a sort) in the quadrant of the (x, y)-plane for which $x > 0$ and $y > 0$. Sketch this curve without the help of a calculating device.

11. Derive the law of exponents $e^{s+t} = e^s e^t$ directly from the differential equation $y' = y$ that is satisfied by e^x. Note the role played by the uniqueness part of Proposition 7.13.

12. Let g be a continuous function on the domain $\mathbb{R}$ and suppose that for each c the differential equation $y' = g(y)$ has a unique solution, that satisfies the initial condition $y(0) = c$, and is defined on the domain $\mathbb{R}$. We can define the function $\phi(x, c)$ of the two variables x and c, so that, as a function of x, it is the solution that satisfies the condition $y(0) = c$.

   (a) Show that the translate of a solution of $y' = g(y)$ is again a solution; that is, if $y = f(x)$ is a solution and α a number then $f(x - \alpha)$ is also a solution.
   (b) Show that there is a unique solution that satisfies $y'(\alpha) = c$, and that it is $\phi(x - \alpha, c)$.
   (c) Prove that for all c, s and t we have the formula $\phi(s + t, c) = \phi(s, \phi(t, c))$.
   Note. In case $g(y) = y$ the formula in (c) is the law of exponents $e^{s+t} = e^s e^t$. As the existence and uniqueness of solutions are widely applicable properties of differential equations, this indicates that the exponential function is capable of great generalisation.

13. Prove the following (doubtlessly familiar) formulas. They are important tools for integration.

   (a) $\cosh^2 x - \sinh^2 x = 1$
   (b) $1 + \tan^2 x = \sec^2 x.$

14. Test the truth of the formula $\cosh^2 x - \sinh^2 x = 1$ on a calculator that has buttons for computing the hyperbolic functions. You will probably find that the output changes from 1 to 0 somewhere between $x = 10$ and $x = 20$. Why is this? Should we conclude that the formula is false in the "real world"?

15. Check the formulas for the derivatives of $\sinh x$, $\cosh x$ and $\tanh x$.

16. Check the formulas for the derivatives of $\sinh^{-1}$, $\cosh^{-1}$ and $\tanh^{-1}$.

17. Prove the formulas already given in the text:

   (a) $\sinh^{-1} x = \ln(x + \sqrt{x^2 + 1})$
   (b) $\cosh^{-1} x = \ln(x + \sqrt{x^2 - 1}), \quad (1 < x < \infty)$

(c) $\tanh^{-1}x = \dfrac{1}{2}\ln\dfrac{1+x}{1-x}$, $\quad(-1 < x < 1)$.

18. We saw that the function $\cosh^{-1}x$ is an antiderivative of $(x^2 - 1)^{-1/2}$ on the interval $1 < x < \infty$. Write down an antiderivative for the function $(x^2 - 1)^{-1/2}$ on the interval $-\infty < x < -1$, utilising the function $\cosh^{-1}$ as it was defined in the text with domain $]1, \infty[$.

19. Show that the addition formulas for the hyperbolic functions are as follows:

(a) $\cosh(x + y) = \cosh x \cosh y + \sinh x \sinh y$
(b) $\sinh(x + y) = \sinh x \cosh y + \cosh x \sinh y$
(c) $\tanh(x + y) = \dfrac{\tanh x + \tanh y}{1 + \tanh x \tanh y}$

and the duplication formulas (useful for integration):

(d) $\cosh 2x = \cosh^2 x + \sinh^2 x = 2\cosh^2 x - 1 = 1 + 2\sinh^2 x$
(e) $\sinh 2x = 2\sinh x \cosh x$
(f) $\tanh 2x = \dfrac{2\tanh x}{1 + \tanh^2 x}$

20. Show that the formulas $x = \cosh t$, $y = \sinh t$, $(-\infty < t < \infty)$, provide a parametrisation of the right-hand branch of the hyperbola $x^2 - y^2 = 1$. You will need to show that every point on the curve corresponds to a unique value of t.

21. Prove that e^x is not an algebraic function.
Hint. Suppose that $f(x, e^x) = 0$ where $f(x, y) = \sum_{k=0}^{m} p_k(x)y^k$ and the coefficients $p_k(x)$ are polynomials. One may assume that neither $p_0(x)$ nor $p_m(x)$ is the zero polynomial, and that m is the lowest number for which such a formula is valid. What happens if one differentiates the formula $f(x, e^x) = 0$?

7.3 ($\Diamond$) Defining Transcendental Functions

An important role was played by differential equations in the way we defined the elementary transcendental functions and obtained their basic properties. The question arises as to whether we could have gone further and actually defined cosine and sine as the solutions of the differential equation $y'' + y = 0$ that satisfy the initial conditions, respectively, $y(0) = 1$, $y'(0) = 0$ and $y(0) = 0$, $y'(0) = 1$. Similarly whether we could have defined $\exp x$ as the solution of $y' = y$ that satisfies $y(0) = 1$.

The answer is that this is a perfectly feasible procedure, but to carry it through we need to know that these differential equations have unique solutions satisfying given initial conditions. This is a proposition that typically comes later in courses of analysis, relying, as it does, on the notion of uniform convergence. One does not want to delay the definition of the elementary functions longer than is absolutely necessary.

When the existence and uniqueness theorem for differential equations is in place it becomes an invaluable source for defining new transcendental functions. We give here a brief preview, without proofs, of what is needed.

A linear homogeneous differential equation of order n has the form

$$p_n(x)y^{(n)} + p_{n-1}(x)y^{(n-1)} + \cdots + p_1(x)y' + p_0(x)y = 0. \qquad (7.1)$$

The coefficient functions $p_k(x)$, $k = 0, 1, ..., n$ are supposed to be continuous functions on a common open interval A. A solution of the equation on A is an n-times differentiable function $\phi(x)$ that satisfies

$$p_n(x)\phi^{(n)}(x) + p_{n-1}(x)\phi^{(n-1)}(x) + \cdots + p_0(x)\phi(x) = 0, \quad (x \text{ in } A). \qquad (7.2)$$

Special cases are the first-order equation $y' - y = 0$ and the second-order equation $y'' + y = 0$ that we have already studied and solved. In both these cases $A = \mathbb{R}$.

The existence theorem for the problem (7.1) applies to an interval A on which the leading coefficient function $p_n(x)$ has no zeros. If $p_n(x)$ has zeros one has to restrict to an interval that excludes them before applying the existence theorem.

Proposition 7.14 *If $p_n(x)$ has no zeros in A then the set of all solutions of (7.1) on A is an n-dimensional vector space of functions over the real field $\mathbb{R}$.*

The proposition implies that if we can find n solutions on A that are linearly independent over the reals, then they form a basis for the space of all solutions. Every other solution is a linear combination of these n solutions in precisely one way. For the first-order equation $y' - y = 0$ the solution space is one-dimensional and, for example, the function $\exp x$, taken alone, forms a basis. Every solution is of the form $C \exp x$ for some C. For the second-order equation $y'' + y = 0$, the two solutions $\cos x$ and $\sin x$ form a basis; every solution has the form $C \cos x + D \sin x$, for some constants C and D.

The second result concerns uniqueness.

Proposition 7.15 *If $p_n(x)$ has no zeros in A, if x_0 is a point in A and $c_0, c_1,...,c_{n-1}$ given numbers, then there exists a unique solution of (7.1) that satisfies the initial conditions*

$$y(x_0) = c_0, \quad y'(x_0) = c_1, \quad ..., \quad y^{(n-1)}(x_0) = c_{n-1}.$$

Examples of transcendental functions that are defined by differential equations of this kind include the following:

(a) Bessel functions. These are solutions on the interval $]0, \infty[$ of Bessel's equation:

$$x^2 y'' + xy' + (x^2 - \alpha^2)y = 0.$$

The constant α is called the order of the Bessel function.

(b) Legendre functions. These are solutions, usually studied on the interval $]-1, 1[$, of Legendre's equation:

$$(1 - x^2)y'' - 2xy' + \ell(\ell + 1)y = 0$$

where ℓ is a constant.

(c) Hypergeometric functions. These are solutions of the hypergeometric equation:

$$x(1 - x)y'' + (c - (a + b + 1)x)y' - aby = 0$$

where a, b and c are constants. They were extensively studied by Gauss and include as special cases a vast range of transcendental functions.

Non-linear differential equations are also a major source of new transcendental functions. Indeed we recall that in the interval $]-\pi/2, \pi/2[$ the function $\sin x$ is a solution of the non-linear differential equation

$$y' = \sqrt{1 - y^2}.$$

Problems in classical mechanics give rise to problems of the form

$$y' = \sqrt{P(y)}$$

where $P(y)$ is a third-degree or fourth-degree polynomial. The solutions can be expressed by a new class of periodic functions, the elliptic functions. The methods used by Abel and Jacobi, to introduce these functions early in the nineteenth century, showed that analysis was an inexhaustible source of new functions.

7.3.1 Exercises

1. We defined the circular functions by studying the integral

$$\int_0^x \frac{1}{\sqrt{1 - t^2}} \, dt$$

and inverting the function defined by it. In a similar way we can study the more general integral

$$F(x) := \int_0^x \frac{1}{\sqrt{(1 - t^2)(1 - k^2 t^2)}} \, dt, \quad -1 < x < 1$$

where k is a constant in the range $0 \le k < 1$. This so-called elliptic integral was a major puzzle to mathematicians until Abel and Jacobi, independently so it seems, pointed out that one should think of F as an inverse function to an *elliptic function*. In the following sequence of exercises the reader is invited to construct an elliptic function using the same steps as were used in the text to construct $\sin x$.

(a) Show that $F(x)$ is an odd function and is strictly increasing on the interval $]-1, 1[$.
(b) Show that the limit

$$L := \lim_{x \to 1-} F(x)$$

exists and that $L < 2/\sqrt{1 - k^2}$.
(c) The inverse function to F maps the interval $]-L, L[$ on to the interval $]-1, 1[$. We have here the function $\mathrm{sn}_k(x)$, one of a group of functions called Jacobi elliptic functions with parameter k. Show that $\mathrm{sn}_k(x)$ satisfies the differential equation

$$y'' + (1 + k^2)y - 2k^2 y^3 = 0 \qquad (7.3)$$

in the interval $]-L, L[$.
(d) The differential equation (7.3) shares many properties with the equation $y'' + y = 0$. Show that if $y = f(x)$ is a solution in an interval $]a, b[$, then the translation $y = f(x - c)$ is a solution in $]a + c, b + c[$, and the reflection $y = f(-x)$ a solution in $]-b, -a[$.
(e) Show that the function $\mathrm{sn}_k(x)$, initially defined in the interval $]-L, L[$, extends to a function, also denoted by $\mathrm{sn}_k(x)$, on all of $\mathbb{R}$, that satisfies (7.3), and has period $4L$.

2. For each natural number n we define the function

$$f_n(x) = e^{x^2} \frac{d^n}{dx^n} e^{-x^2}.$$

(a) Show that f_n is a polynomial of degree n, and is moreover an even function when n is even and an odd function when n is odd (for the definitions of even and odd functions see Sect. 5.4 Exercise 8).
(b) Show that $f_{n+1}(x) = f_n'(x) - 2x f_n(x)$ for $n = 0, 1, 2, ...$
(c) Show that f_n has n distinct real roots.
(d) Show that f_n satisfies the differential equation

$$y'' - 2xy' + 2ny = 0.$$

Note. The functions f_n are (up to a normalisation constant) the Hermite polynomials. The differential equation $y'' - 2xy' + 2\lambda y = 0$, where λ is a real parameter, is known as Hermite's equation. Non-polynomial solutions to Hermite's equation exist; they are transcendental functions and some are non-elementary.

3. For each natural number n define

$$\phi_n(x) = \frac{d^n}{dx^n}(x^2 - 1)^n.$$

Obviously ϕ_n is a polynomial of degree n. Up to a normalisation constant it is the nth Legendre polynomial. Its zeros are all real and simple, and they lie in the interval $]-1, 1[$ (see Sect. 5.9 Exercise 7).

(a) Derive the formulas

$$\phi'_{n+1}(x) = 2(n + 1)x\phi'_n(x) + 2(n + 1)^2\phi_n(x)$$
$$\phi'_{n+1}(x) = \big((x^2 - 1)\phi'_n(x)\big)' + 2(n + 1)x\phi'_n(x) + (n + 1)(n + 2)\phi_n(x).$$

Hint. Attack the expressions

$$\frac{d^{n+2}}{dx^{n+2}}(x^2 - 1)^{n+1} \quad \text{and} \quad \frac{d^{n+1}}{dx^{n+1}}(x^2 - 1)^{n+1}$$

with Leibniz's formula (Sect. 5.4 Exercise 5).

(b) Deduce that $\phi_n(x)$ satisfies Legendre's equation with $\ell = n$.

Note. Legendre's equation also has transcendental solutions, some of which are non-elementary. The standard form of the Legendre polynomial is $P_n(x) := \phi_n(x)/2^n n!$.

4. Some Bessel functions are elementary.

(a) Show that the functions $x^{-1/2}\cos x$ and $x^{-1/2}\sin x$ satisfy Bessel's equation with $\alpha = 1/2$ on the interval $]0, \infty[$.

To conjure up more examples you can use the following steps:

(b) We make a change of variables in Bessel's equation. More precisely we introduce a new variable u (really a function of x) related to the variable y by $u = x^{-\alpha}y$. Show that $y(x)$ satisfies Bessel's equation with order α (as always we mean in the interval $]0, \infty[$) if and only if $u(x)$ satisfies the equation

$$xu'' + (2\alpha + 1)u' + xu = 0. \tag{7.4}$$

(c) Show that if $u(x)$ is a solution of (7.4) for a given α then the function

$$v(x) = \frac{1}{x}u'(x)$$

is a solution of (7.4) in which $\alpha + 1$ replaces α.

(d) Deduce that if $y(x)$ satisfies Bessel's equation with order α then the function

$$x^{\alpha+1}\left(\frac{1}{x}\frac{d}{dx}\right)\big(x^{-\alpha}y(x)\big)$$

satisfies Bessel's equation with order $\alpha + 1$.

(e) Deduce that if $y(x)$ satisfies Bessel's equation with order α then the function

$$x^{\alpha+n} \left(\frac{1}{x} \frac{d}{dx} \right)^n \left(x^{-\alpha} y(x) \right)$$

satisfies Bessel's equation with order $\alpha + n$.

(f) Deduce that the functions

$$x^{n+\frac{1}{2}} \left(\frac{1}{x} \frac{d}{dx} \right)^n \left(\frac{\cos x}{x} \right) \quad \text{and} \quad x^{n+\frac{1}{2}} \left(\frac{1}{x} \frac{d}{dx} \right)^n \left(\frac{\sin x}{x} \right)$$

satisfy Bessel's equation with $\alpha = n + \frac{1}{2}$.

In these formulas the expression $\frac{1}{x} \frac{d}{dx}$ is a *differential operator* that converts the following function u to u'/x. More precisely

$$\left(\frac{1}{x} \frac{d}{dx} \right) u(x) = \frac{1}{x} u'(x).$$

Its nth power is the differential operator that applies $\frac{1}{x} \frac{d}{dx}$ to the succeeding function n times in a row. More precisely, we have the inductive definition

$$\left(\frac{1}{x} \frac{d}{dx} \right)^{n+1} u(x) := \left(\frac{1}{x} \frac{d}{dx} \right) \left(\left(\frac{1}{x} \frac{d}{dx} \right)^n u(x) \right).$$

7.3.2   Pointers to Further Study

→ Differential equations.
→ Special functions.

Chapter 8
The Techniques of Integration

> *Design is not making beauty, beauty emerges from selection, affinities, integration, love.*
>
> *Louis Kahn, architect*

This chapter covers a classical set of techniques which essentially can be used to find antiderivatives of common functions. The time was when a mathematical education placed great emphasis on acquiring skill in using these techniques. Arguably they are less important now, but many find that successfully using them to find a difficult antiderivative is a satisfying experience.

8.1 Integration by Parts and by Substitution

Two immensely important rules for finding antiderivatives correspond, respectively, to the rule for differentiation of a product and the chain rule, when these are inverted by means of the fundamental theorem.

Proposition 8.1 (Integration by parts) *Let f and g be functions defined in the open interval A. Assume that f and g are differentiable and that their derivatives, f' and g', are continuous. Then*

$$\int_a^b fg' = f(b)g(b) - f(a)g(a) - \int_a^b f'g$$

for all a and b in A.

Proof We have that $(fg)' = fg' + f'g$, which is continuous; so the fundamental theorem applies and we have

R. Magnus, *Fundamental Mathematical Analysis*, Springer Undergraduate Mathematics Series, https://doi.org/10.1007/978-3-030-46321-2_8

$$\int_a^b (fg' + f'g) = (fg)\Big|_a^b = f(b)g(b) - f(a)g(a).$$

$\square$

The rule is most frequently used to handle antiderivatives in the following way. If u and v are functions of x and we know an antiderivative for $u'v$, then an antiderivative for uv' is given by

$$\int uv' = uv - \int u'v.$$

Given the task of finding an antiderivative $\int f$, skill and experience may suggest suitable functions u and v, such that $f = uv'$, and an antiderivative for $u'v$ is then easy to find.

Proposition 8.2 (Integration by substitution) *Let A and B be open intervals, let $f : A \to \mathbb{R}$ be continuous, and let $\phi : B \to \mathbb{R}$ be differentiable, with ϕ' continuous. Assume that $\phi(B) \subset A$. Then for all a and b in B we have*

$$\int_a^b f \circ \phi \; \phi' = \int_{\phi(a)}^{\phi(b)} f,$$

or, in Leibniz's notation

$$\int_a^b f(\phi(t))\phi'(t)\,dt = \int_{\phi(a)}^{\phi(b)} f(x)\,dx.$$

The use of distinct variables t and x has no logical significance; they are bound variables. However it supports the usual interpretation of the rule, that the integral on the left is obtained from the integral on the right by means of the substitution $x = \phi(t)$. The substitution replaces dx by $\phi'(t)\,dt$, a replacement that is obtained formally by writing

$$x = \phi(t) \;\Rightarrow\; \frac{dx}{dt} = \phi'(t) \;\Rightarrow\; dx = \phi'(t)\,dt.$$

This piece of Leibniz notation, though illegally separating dx and dt, is very useful in practice.

Proof of the Rule Let F be an antiderivative for f (one exists since f is continuous). The composed function $F \circ \phi$ is defined on the domain B and by the chain rule

$$(F \circ \phi)' = (F' \circ \phi)\phi' = (f \circ \phi)\phi'.$$

We see that $(F \circ \phi)'$ is continuous and therefore

$$\int_{\phi(a)}^{\phi(b)} f = (F \circ \phi)(b) - (F \circ \phi)(a) = \int_a^b (F \circ \phi)' = \int_a^b (f \circ \phi)\phi'.$$

$\square$

We note two interesting points:

(a) It is not necessary for ϕ to be monotonic. It does not even need to map the interval $[a, b]$ on to the interval $[\phi(a), \phi(b)]$, but we do not want $\phi(t)$ to go outside the domain of f.

(b) The rule is often used to transform antiderivatives in the following way. If $x = \phi(t)$ then

$$\int f(x)\, dx = \int f(\phi(t))\phi'(t)\, dt + C. \tag{8.1}$$

The equation in point (b) is interpreted to mean that if F is an antiderivative for f then $F \circ \phi$ is an antiderivative for $(f \circ \phi)\phi'$. Conversely, if we can find an antiderivative for $(f \circ \phi)\phi'$, let us call it G, then $G \circ \phi^{-1}$ is an antiderivative for f. This requires us to deploy the inverse ϕ^{-1}; so we would need to work on an interval in which ϕ is monotonic.

Point (b) indicates two somewhat different ways to apply the rule; they are explored in the next section.

8.1.1 Finding Antiderivatives by Substitution

Let us look at two examples of the use of substitution to find an antiderivative. They illustrate two different ways for applying the rule. In each case we present the calculation, as it would normally be presented, and then its explanation.

(A) Find the integral on the right in (8.1) by solving the integral on the left.

$$\int \sqrt{1 - t^2}\, t\, dt = -\frac{1}{2} \int \left(\sqrt{1 - t^2}\right)(-2t)\, dt$$

$$= -\frac{1}{2} \int \sqrt{x}\, dx = -\frac{1}{3}x\sqrt{x} = -\frac{1}{3}(1 - t^2)\sqrt{1 - t^2}.$$

Explanation. We have $\sqrt{1 - t^2}\, t\, dt = f(\phi(t))\phi'(t)\, dt$ where

$$\phi(t) = 1 - t^2, \quad f(x) = -\frac{1}{2}\sqrt{x}.$$

The antiderivative obtained is valid on the interval $-1 < t < 1$.

(B) Find the integral on the left in (8.1) by solving the integral on the right. This is more complicated than case A because in the final step we have to substitute for t as a function of x. This requires the inverse function ϕ^{-1}.

As an example consider the following calculation. For $-1 < x < 1$ we have

$$\int \sqrt{1 - x^2}\, dx = \int \sqrt{1 - \sin^2 t}\, \cos t\, dt = \int \cos^2 t\, dt$$

$$= \int \frac{1}{2}(1 + \cos 2t)\, dt = \frac{1}{2}t + \frac{1}{4}\sin 2t = \frac{1}{2}t + \frac{1}{2}\sin t \cos t$$

$$= \frac{1}{2}\arcsin x + \frac{1}{2}x\sqrt{1 - x^2}.$$

Explanation. We set $x = \sin t$, $dx = \cos t\, dt$. We have to choose an interval for t to fix an inverse for $\sin t$. The simplest is to keep t in the interval $]-\frac{\pi}{2}, \frac{\pi}{2}[$. Then $\sin t$ is increasing and maps this interval on to the interval $]-1, 1[$. Moreover $\sqrt{1 - \sin^2 t} = \cos t$ (and not $-\cos t$, owing to the interval chosen for t). Finally x is reintroduced. Since $-\frac{\pi}{2} < t < \frac{\pi}{2}$ we have $t = \arcsin x$ and $\cos t = \sqrt{1 - x^2}$ (and not $-\sqrt{1 - x^2}$, again, thanks to the choice of interval).

Traditionally, finding an antiderivative of a function f by the techniques of this chapter was called *solving the integral $\int f(x)\, dx$*. Every integral solved can be added to a catalogue and used to solve further integrals.

8.1.2 Exercises

1. Solve the following integrals:

(a) $\int xe^x\, dx$

(b) $\int x^2 \cos x\, dx$

(c) $\int e^x \cos x\, dx$

    Hint. Call the integral I. Integrating twice by parts leads to $I = e^x \sin x + e^x \cos x - I$. Look out for other opportunities to use this trick.

(d) $\int xe^x \cos x\, dx$

(e) $\int \ln x\, dx$, $(x > 0)$

(f) $\int x \ln x\, dx$, $(x > 0)$

(g) $\int \frac{\ln x}{x^2}\, dx$, $(x > 0)$

(h) $\int x(\ln x)^2\, dx$, $(x > 0)$.

2. Solve the following useful integrals, where a is a positive constant:

(a) $\displaystyle\int \frac{1}{\sqrt{a^2 - x^2}}\, dx, \quad (-a < x < a)$

(b) $\displaystyle\int \frac{1}{\sqrt{a^2 + x^2}}\, dx$

(c) $\displaystyle\int \frac{1}{\sqrt{x^2 - a^2}}\, dx, \quad (x > a)$

Note. In this integral, as in some others in the following exercises, a different domain from the one specified is possible, in which case a different formula may be needed for the antiderivative. See the discussion of $\int \frac{1}{x}\, dx$ in Sect. 7.1.1.

(d) $\displaystyle\int \frac{1}{a^2 + x^2}\, dx.$

3. Solve the following integrals:

(a) $\displaystyle\int x\sqrt{1 - x^2}\, dx, \quad (-1 < x < 1)$

(b) $\displaystyle\int \frac{x}{\sqrt{1 + x^2}}\, dx$

(c) $\displaystyle\int \frac{x}{\sqrt{x^4 - 1}}\, dx, \quad (x > 1)$

(d) $\displaystyle\int \frac{x}{1 + x^2}\, dx.$

4. Solve the following integrals:

(a) $\displaystyle\int \frac{\cos x}{1 + \sin^2 x}\, dx$

(b) $\displaystyle\int \frac{\cos x}{\sqrt{1 + \sin^2 x}}\, dx$

(c) $\displaystyle\int x e^{x^2}\, dx$

(d) $\displaystyle\int \tan x\, dx, \quad \left(-\frac{\pi}{2} < x < \frac{\pi}{2}\right)$

(e) $\displaystyle\int \sec^4 x\, dx, \quad \left(-\frac{\pi}{2} < x < \frac{\pi}{2}\right).$

5. Solve the following integrals:

(a) $\displaystyle\int \sin^2 x\, dx$

Hint. The standard trick for this and for $\int \cos^2 x\, dx$ is to use the duplication formulas $\cos 2x = 2\cos^2 x - 1 = 1 - 2\sin^2 x$, as in example B of the preceding section.

(b) $\displaystyle\int \cos^2 x \sin^2 x\, dx$

(c) $\displaystyle\int \cos^3 x \sin^2 x\, dx$

(d) $\displaystyle\int \sin^4 x \, dx$

(e) $\displaystyle\int \cosh^2 x \, dx$

Hint. The same trick as in item (a) but using the duplication formula for hyperbolic cosine (Sect. 7.2 Exercise 19). This also works for $\int \sinh^2 x \, dx$.

(f) $\displaystyle\int \cos ax \cos bx \, dx$ where a and b are real numbers.

Hint. Use the addition formulas for the circular functions. Variants in which one or both factors are replaced by the sine are treated similarly.

6. Solve the following integrals:

(a) $\displaystyle\int \sqrt{x^2 + 1} \, dx$

(b) $\displaystyle\int \sqrt{x^2 - 1} \, dx, \quad (x > 1)$

Hint. The integrals (a) and (b) occur often. A trigonometric or hyperbolic substitution will work, but one can also integrate by parts. The similar integral $\int \sqrt{1 - x^2} \, dx$ was worked out in the text.

(c) $\displaystyle\int \frac{x^2}{\sqrt{x^2 - 1}} \, dx, \quad (x > 1).$

7. The integral $\int \sec x \, dx$ arises frequently in the course of solving other integrals. Solve it by writing

$$\sec x = \frac{\sec^2 x + \sec x \tan x}{\sec x + \tan x}$$

and consulting the derivatives of the trigonometric functions listed in Sect. 7.1. The most convenient domain is $\left]-\frac{\pi}{2}, \frac{\pi}{2}\right[$ since $\sec x + \tan x$ is positive there.

8. Solve the following integrals:

(a) $\displaystyle\int \sec^3 x \, dx, \quad \left(-\frac{\pi}{2} < x < \frac{\pi}{2}\right)$

(b) $\displaystyle\int \sqrt{1 + e^x} \, dx$

(c) $\displaystyle\int e^{\sqrt{x}} \, dx, \quad (x > 0)$

(d) $\displaystyle\int \ln(1 + x^2) \, dx$

(e) $\displaystyle\int \sqrt{\frac{x - 1}{x + 1}} \, dx, \quad (x > 1)$

(f) $\displaystyle\int \sqrt{\frac{x - 1}{x + 1}} \frac{1}{x^2} \, dx, \quad (x > 1)$

(g) $\displaystyle\int \sqrt{1 + \sqrt{x}} \, dx, \quad (x > 0).$

9. Show that the function $F(x) = \int_0^x \cos(1/t)\,dt$ is differentiable at 0 and compute $F'(0)$. This is so, despite the fact that the integrand is discontinuous at 0. *Hint.* Integrate by parts in a cunning way.

10. Let f be continuous in an interval A and let a and b be points in A. Show that

$$\int_a^b f = \int_a^b f(a+b-x)\,dx.$$

Can you prove this without assuming continuity, given that $a < b$ and f is integrable on $[a, b]$?

11. The length of the upper arc of the ellipse $x^2/a^2 + y^2/b^2 = 1$ (with $a > b > 0$), from $x = 0$ to $x = c$ (where $0 < c < a$), is given by the integral (asked for in Sect. 6.7 Exercise 2):

$$\int_0^c \sqrt{\frac{a^2 - e^2x^2}{a^2 - x^2}}\,dx$$

where e, the eccentricity (not necessarily 2.718...), is given by

$$e = \sqrt{1 - \frac{b^2}{a^2}}.$$

Find a substitution that converts this integral to

$$a \int_0^{\arcsin(c/a)} \sqrt{1 - e^2 \sin^2\theta}\,d\theta.$$

Note. The integral obtained is one of the three standard forms that Legendre gave for elliptic integrals.

12. Let f have a continuous derivative on an interval A and let a and b be integers in A with $a < b$. For each real x, let $[x]$ be the highest integer less than or equal to x. Prove that

$$\sum_{n=a}^b f(n) = \int_a^b f + \frac{f(a) + f(b)}{2} + \int_a^b \left(x - [x] - \frac{1}{2}\right) f'(x)\,dx.$$

Hint. Begin by computing the second integral.

Note. This formula is the simplest instance of the Euler–Maclaurin summation formula, and is the first step in proving it. This will be taken up in an exercise in Chap. 12.

13. Let f be continuous in the open interval A and possess an inverse function f^{-1}. Suppose that an antiderivative F is known for f. Show how to express an antiderivative for f^{-1} in terms of the functions x, f^{-1} and F. *Hint.* This is easy if f has a continuous derivative, since one can write

$$\int f^{-1} = \int 1.f^{-1}$$

and integrate by parts. This produces a formula for an antiderivative of f^{-1}. Then one can try to show that the same formula gives an antiderivative of f^{-1} even when f is not assumed to be differentiable. Note that f is strictly monotonic by Sect. 4.4 Exercise 4.

Note. The result is connected with the Legendre transform, Sect. 5.10 Exercise 14.

14. Use the method of the previous exercise to solve the following integrals:

   (a) $\displaystyle\int \ln x \, dx$

   (b) $\displaystyle\int \arcsin x \, dx$

   (c) $\displaystyle\int \arctan x \, dx$

   (d) $\displaystyle\int \cosh^{-1} x \, dx.$

15. Let the functions f and g be continuous in an open interval A. Suppose that f is differentiable, and that f' is continuous and either entirely non-positive or entirely non-negative. Show that for all a and b in A there exists ξ between a and b, such that

$$\int_a^b fg = f(a) \int_a^\xi g + f(b) \int_\xi^b g.$$

Hint. Integrate by parts and use the mean value theorem for integrals, Proposition 6.15.

Note. This result, the second mean value theorem for integrals, was proved in greater generality in Sect. 6.8. See also Exercise 17 below.

The remaining exercises in this section are concerned with extending the rule for integration by parts using the notion of primitive. It is a good idea to take another look at the definition of primitive, as used in this text and introduced in Sect. 6.5, and recall Proposition 6.19 which identified the primitives of piece-wise continuous functions. The first, rather mild, extension allows piece-wise continuous integrands and has some practical uses, as such integrands occur often in technology. A theoretical application will occur in Chap. 12. The second extension goes about as far as is possible for the Riemann–Darboux integral.

16. Let f and g be piece-wise continuous functions in an interval A. Let F be a primitive for f and G a primitive for g.

   (a) Show that FG is a primitive for $Fg + fG$.

   (b) Deduce the rule for integration by parts: for each a and b in A we have

$$\int_a^b Fg = F(b)G(b) - F(a)G(a) - \int_a^b fG.$$

17. ($\Diamond$) There is a very general version of the rule for integration by parts that does not rely on derivatives at all. Its proof is based on the approximation of integrable functions in the mean by step functions (Sect. 6.8).
Let f and g be integrable on an interval $[a, b]$. Let F be a primitive of f and G a primitive of g. Then

$$\int_a^b Fg = F(b)G(b) - F(a)G(a) - \int_a^b fG.$$

(a) Prove the formula in the case that f and g are step functions.
(b) Prove the formula in the case that f and g are integrable.
    Hint. Approximate f and g in the mean by step functions.

Note. Applying the result of this exercise we get another version of the second mean value theorem, where, in the notation of Exercise 15, f is a primitive of a positive integrable function and g is integrable. It is still not as general as the version in Sect. 6.8, in which f is merely monotonic.

8.2 Integrating Rational Functions

Every rational function has an antiderivative that may be expressed using elementary functions. The proof of this remarkable fact will occupy the bulk of this section. Some knowledge of algebra is assumed.

A rational function has the form $P(x)/Q(x)$ where P and Q are polynomials. If the degree of P is greater than or equal to the degree of Q we can divide Q into P and obtain quotient and remainder. The quotient is a polynomial and its antiderivative also a polynomial. The remainder is a polynomial with degree less than that of Q.

We can therefore concentrate on the case of a rational function $P(x)/Q(x)$ where the degree of P is lower than the degree of Q. We also assume that Q is a monic polynomial, meaning that the leading coefficient of Q is 1.

We will need two major inputs from algebra:

(a) *The fundamental theorem of algebra for real polynomials.* The polynomial Q has a factorisation:

$$Q(x) = (x - \lambda_1)^{r_1} \ldots (x - \lambda_m)^{r_m} (x^2 + \alpha_1 x + \beta_1)^{s_1} \ldots (x^2 + \alpha_n x + \beta_n)^{s_n}$$

where the real numbers $\lambda_1, \ldots, \lambda_m$ are distinct, the real number pairs $(\alpha_1, \beta_1), \ldots, (\alpha_n, \beta_n)$ are distinct, the exponents, $r_1, \ldots, r_m, s_1, \ldots, s_n$, are positive integers, and the second-degree factors have no real roots (they are, in other words, irreducible over the reals). The factors $x - \lambda_j$ and $x^2 + \alpha_j x + \beta_j$ are called the prime factors of Q. A prime factor is said to be simple if its exponent in the factorisation is 1.

First-degree polynomials and second-degree polynomials that do not factorise over the reals are the prime elements in the ring of real polynomials in one variable. They play a role in polynomial theory very similar to that of the prime integers in number theory.

(b) The rational function $P(x)/Q(x)$ has a *partial fractions decomposition*; see the next section.

Exercise Show that $\sum_{j=1}^{m} r_j + 2 \sum_{j=1}^{n} s_j = \deg Q$ (the degree of the polynomial Q).

8.2.1 Partial Fractions

The partial fractions decomposition of $P(x)/Q(x)$, given that the degree of Q is higher than the degree of P, and that Q has the factorisation as outlined above, is

$$\frac{P(x)}{Q(x)} = \sum_{j=1}^{r_1} \frac{a_{1j}}{(x-\lambda_1)^j} + \cdots + \sum_{j=1}^{r_m} \frac{a_{mj}}{(x-\lambda_m)^j}$$
$$+ \sum_{j=1}^{s_1} \frac{b_{1j}x + c_{1j}}{(x^2 + \alpha_1 x + \beta_1)^j} + \cdots + \sum_{j=1}^{s_n} \frac{b_{nj}x + c_{nj}}{(x^2 + \alpha_n x + \beta_n)^j}.$$

It is guaranteed that the coefficients a_{ij}, $(1 \le i \le m,\ 1 \le j \le r_i)$, b_{ij} and c_{ij}, $(1 \le i \le n, 1 \le j \le s_i)$, can be found uniquely by solving *simultaneous linear equations*. It is only necessary that the form of the fractions is correctly set down. A foolproof method to find the coefficients is to clear the denominators by multiplying through by $Q(x)$ and then equate coefficients of like powers of x.

8.2.2 Practicalities

We look at some practical hints for finding the partial fractions decomposition of $P(x)/Q(x)$, on the assumption that the real factorisation of Q is already known. Remember that one can always express the problem as a system of linear equations, that is guaranteed a unique set of solutions when the form of the fractions is correct. However, some of the tricks shown below can save labour.

Case A. The prime factors are simple and of first degree. This is the case

$$Q(x) = (x - \lambda_1)(x - \lambda_2)...(x - \lambda_m)$$

where the numbers $\lambda_1, ..., \lambda_m$ are real and distinct.

An example will illustrate a convenient method. Find a, b and c, such that

$$\frac{x}{(x-1)(x-2)(x-3)} = \frac{a}{x-1} + \frac{b}{x-2} + \frac{c}{x-3}.$$

Clearing the denominators we find

$$x = a(x-2)(x-3) + b(x-1)(x-3) + c(x-1)(x-2).$$

Substituting the values $x = 1, 2, 3$ in turn, we get $a = \frac{1}{2}, b = -2, c = \frac{3}{2}$.

Case B. The prime factors are of first degree but are not all simple. This is the case

$$Q(x) = (x - \lambda_1)^{r_1} \dots (x - \lambda_m)^{r_m}$$

where some exponents are greater than 1.

Again we give an illustrative example. Find a, b, c and d, such that

$$\frac{1}{(x-1)^2(x-2)^2} = \frac{a}{x-1} + \frac{b}{(x-1)^2} + \frac{c}{x-2} + \frac{d}{(x-2)^2}.$$

Clearing denominators we find

$$1 = a(x-1)(x-2)^2 + b(x-2)^2 + c(x-1)^2(x-2) + d(x-1)^2.$$

The substitutions $x = 1$ and $x = 2$ give $b = 1$ and $d = 1$. Next we differentiate and obtain

$$0 = a\big((x-2)^2 + 2(x-1)(x-2)\big) + 2(x-2)^2$$
$$+ c\big((2(x-1)(x-2) + (x-1)^2\big) + 2(x-1).$$

Now putting $x = 1$ and $x = 2$ gives $a = -2$ and $c = -2$.

An alternative to differentiating in the second step is to write (having already determined that $b = 1$ and $d = 1$)

$$1 - (x-2)^2 - (x-1)^2 = a(x-1)(x-2)^2 + c(x-1)^2(x-2).$$

Now it will be found that $x - 1$ and $x - 2$ divide the left-hand side and can be cancelled. A further substitution of 1, followed by 2, then reveals a and c.

Case C. Some prime factors are irreducible quadratics but all exponents are 1. This is the case

$$Q(x) = (x - \lambda_1) \dots (x - \lambda_m)(x^2 + \alpha_1 x + \beta_1) \dots (x^2 + \alpha_n x + \beta_n).$$

Consider the example

$$\frac{x^2}{(x-1)(x^2+1)} = \frac{a}{x-1} + \frac{bx+c}{x^2+1}.$$

Clearing denominators we find

$$x^2 = a(x^2+1) + (bx+c)(x-1).$$

Substituting $x = 1$ gives $a = \frac{1}{2}$. Now we write

$$x^2 - \frac{1}{2}(x^2+1) = (bx+c)(x-1).$$

Now $x - 1$ divides the left-hand side and cancelling it we find

$$\frac{1}{2}(x+1) = bx+c$$

giving $b = c = \frac{1}{2}$.

Alternatively, if you know about complex numbers, you can substitute one of the complex roots of $x^2 + 1$ (they are i and $-i$), to find b and c.

8.2.3   Outline of Proof

The existence of the partial fractions decomposition can be proved by linear algebra. We shall sketch a proof, accessible for the reader familiar with vector spaces, in particular the notions of basis and linear independence. The idea is that the partial fractions decomposition is merely a change of basis in a finite-dimensional vector space over the real field. The reader unfamiliar with these ideas can simply skip this section.

For a given positive integer d, the set of all polynomials with degree less than d is vector space over the real field, and it has finite dimension d. In fact, a basis for it is provided by the set

$$\{1, x, x^2, ..., x^{d-1}\}$$

comprising d polynomials.

Our real interest is rational functions. Let Q be a monic polynomial with degree d. We shall denote by R_Q the space of all rational functions, expressible in the form $P(x)/Q(x)$ for some polynomial P with degree less than d. It is a vector space of dimension d. As a basis for R_Q we can indicate the set

$$\left\{ \frac{1}{Q(x)}, \frac{x}{Q(x)}, \frac{x^2}{Q(x)}, ..., \frac{x^{d-1}}{Q(x)} \right\}.$$

The polynomial Q has a factorisation as described in item (a) above (by the fundamental theorem of algebra). Deploying the constants $\lambda_1, \ldots, \lambda_m$, the pairs (α_1, β_1), $\ldots$, (α_n, β_n) and the exponents $r_1, \ldots, r_m, s_1, \ldots, s_n$, we consider the functions in the ensuing list:

$$\frac{1}{(x - \lambda_k)^p}, \quad (1 \le p \le r_k, \quad 1 \le k \le m),$$

$$\frac{x}{(x^2 + \alpha_k x + \beta_k)^q}, \quad (1 \le q \le s_k, \quad 1 \le k \le n),$$

$$\frac{1}{(x^2 + \alpha_k x + \beta_k)^q}, \quad (1 \le q \le s_k, \quad 1 \le k \le n).$$

The functions in the list all belong to the vector space R_Q, as they can obviously be expressed with Q as denominator and with a numerator of degree less than d. The number of functions in the list (the reader is invited to count them) is the same as the degree of Q, that is, it is the same as the dimension of R_Q. We can conclude that they form another basis for R_Q, provided they can be shown to be linearly independent. We omit this step, which is most easily accomplished by exploiting the roots of Q, including the complex roots of the irreducible quadratic factors. Complex numbers will be considered in Chap. 9.

Having shown that the functions in the list constitute a basis for R_Q, it follows that every element of R_Q can be expressed in a unique fashion as a linear combination of them. This is, of course, just the partial fractions decomposition.

Exercise To get an idea of how an independence proof might proceed, the reader can try to prove that the six functions

$$\frac{1}{x - 1}, \quad \frac{1}{(x - 1)^2}, \quad \frac{1}{x^2 + x + 1}, \quad \frac{x}{x^2 + x + 1}, \quad \frac{1}{(x^2 + x + 1)^2}, \quad \frac{x}{(x^2 + x + 1)^2}$$

are linearly independent. This is the case $Q(x) = (x - 1)^2(x^2 + x + 1)^2$.

Hint. One has to show that if a relation

$$\frac{a_1}{x - 1} + \frac{a_2}{(x - 1)^2} + \frac{b_1 + b_2 x}{x^2 + x + 1} + \frac{b_3 + b_4 x}{(x^2 + x + 1)^2} = 0$$

holds (for all x), then the coefficients a_1, a_2, b_1, b_2, b_3 and b_4 are all 0. One way to start is to multiply by $(x - 1)^2$ and set $x = 1$.

8.2.4   How to Integrate the Fractions

The problem of integrating $P(x)/Q(x)$ reduces to integrating each term in the partial
fractions decomposition.

Two of the fractions are easily dealt with:

$$\int \frac{1}{x-\lambda}\, dx = \ln(x-\lambda) \quad (\text{or} \ln(\lambda-x) \text{ if } x < \lambda)$$

and

$$\int \frac{1}{(x-\lambda)^r}\, dx = \frac{1}{(1-r)(x-\lambda)^{r-1}}, \quad (r \neq 1).$$

Next we write

$$\frac{bx+c}{(x^2+\alpha x+\beta)^s} = \frac{\frac{b}{2}(2x+\alpha)}{(x^2+\alpha x+\beta)^s} + \frac{\frac{1}{2}(2c-b\alpha)}{(x^2+\alpha x+\beta)^s}.$$

The first fraction on the right is treated to the substitution $u = x^2 + \alpha x + \beta$, which
leads, in the case $s \neq 1$, to

$$\int \frac{2x+\alpha}{(x^2+\alpha x+\beta)^s}\, dx = \frac{1}{(1-s)u^{s-1}} = \frac{1}{(1-s)(x^2+\alpha x+\beta)^{s-1}},$$

and, in the case $s = 1$, to

$$\int \frac{2x+\alpha}{x^2+\alpha x+\beta}\, dx = \ln u = \ln(x^2+\alpha x+\beta).$$

There remains the more difficult task of solving the integral

$$\int \frac{1}{(x^2+\alpha x+\beta)^s}\, dx.$$

First we write

$$x^2 + \alpha x + \beta = \left(x + \frac{\alpha}{2}\right)^2 + \beta - \frac{\alpha^2}{4}.$$

This is the familiar operation of completing the square. Set $u = x + (\alpha/2)$. The
constant $\beta - (\alpha^2/4)$ is positive (because the polynomial has no real root) and we set
$\gamma^2 = \beta - (\alpha^2/4)$. Now we have the integral

$$I_s := \int \frac{1}{(u^2+\gamma^2)^s}\, du.$$

Using integration by parts we find

$$I_s = \frac{u}{(u^2 + \gamma^2)^s} + 2s \int \frac{u^2}{(u^2 + \gamma^2)^{s+1}}\, du$$

$$= \frac{u}{(u^2 + \gamma^2)^s} + 2s I_s - 2s\gamma^2 I_{s+1}$$

and collecting multiples of I_s we obtain

$$I_{s+1} = \frac{u}{2s\gamma^2(u^2 + \gamma^2)^s} + \frac{2s - 1}{2s\gamma^2} I_s.$$

This is an example of a *reduction formula*. It is convenient for purposes of calculation to knock 1 off s and write it as

$$I_s = \frac{u}{2(s - 1)\gamma^2(u^2 + \gamma^2)^{s-1}} + \frac{2s - 3}{2(s - 1)\gamma^2} I_{s-1}, \quad s = 2, 3, 4, \ldots$$

After a finite number of applications we come down to the integral

$$I_1 = \int \frac{1}{u^2 + \gamma^2}\, du = \frac{1}{\gamma} \arctan \frac{u}{\gamma} = \frac{1}{\gamma} \arctan \frac{x + \alpha}{\gamma}.$$

We summarise the conclusions concerning the form of the antiderivatives just derived.

Proposition 8.3 *The antiderivative of a rational function $P(x)/Q(x)$ can be expressed as the sum of three functions $G_1(x) + G_2(x) + G_3(x)$ (some of which may be 0), where*

(1) *$G_1(x)$ is a rational function.*
(2) *$G_2(x)$ is a linear combination, with real coefficients, of logarithms of first or second-degree polynomials.*
(3) *$G_3(x)$ is a linear combination, with real coefficients, of arctangents of first-degree polynomials.*

We often distinguish the different parts of the integral by calling $G_1(x)$ the rational part and $G_2(x) + G_3(x)$ the transcendental part.

8.2.5 Integrating Rational Functions of $\sin \theta$ and $\cos \theta$

We are going to solve the integral

$$\int R(\cos \theta, \sin \theta)\, d\theta \tag{8.2}$$

where $R(x, y) = f(x, y)/g(x, y)$ is a rational function of x and y, that is, the quotient of two polynomials in the two variables x and y. This can always be accomplished by the *half-angle substitution* $t = \tan(\theta/2)$. Its efficacy can be explained by the fact that the circle $x^2 + y^2 = 1$ has the rational parametrisation:

$$x = \frac{1 - t^2}{1 + t^2}, \quad y = \frac{2t}{1 + t^2}.$$

Proposition 8.4 *The substitution* $t = \tan(\theta/2)$ *transforms the integral* (8.2) *into the integral of a rational function of* t.

Proof The perfectly explicit proof is the series of elementary calculations:

$$\sin \theta = 2 \sin \frac{\theta}{2} \cos \frac{\theta}{2} = \frac{2 \sin \frac{\theta}{2} \cos \frac{\theta}{2}}{\cos^2 \frac{\theta}{2} + \sin^2 \frac{\theta}{2}} = \frac{2t}{1 + t^2}$$

$$\cos \theta = \cos^2 \frac{\theta}{2} - \sin^2 \frac{\theta}{2} = \frac{\cos^2 \frac{\theta}{2} - \sin^2 \frac{\theta}{2}}{\cos^2 \frac{\theta}{2} + \sin^2 \frac{\theta}{2}} = \frac{1 - t^2}{1 + t^2}$$

$$dt = \frac{1}{2 \cos^2 \frac{\theta}{2}} d\theta = \frac{\cos^2 \frac{\theta}{2} + \sin^2 \frac{\theta}{2}}{2 \cos^2 \frac{\theta}{2}} d\theta = \frac{1 + t^2}{2} d\theta$$

all of which leads to

$$\int R(\cos \theta, \sin \theta) \, d\theta = \int R\left(\frac{1 - t^2}{1 + t^2}, \frac{2t}{1 + t^2}\right) \frac{2}{1 + t^2} \, dt.$$

$\square$

8.2.6   Further Useful Reduction Formulas

We list some more reduction formulas that the reader is likely to encounter. They all come in useful for finding antiderivatives. Yet others are explored in the exercises.

(i) $\displaystyle \int (\ln x)^n \, dx = x (\ln x)^n - n \int (\ln x)^{n-1} \, dx$

(ii) $\displaystyle \int x^n e^x \, dx = x^n e^x - n \int x^{n-1} e^x \, dx$

(iii) $\displaystyle \int \sin^n x \, dx = -\frac{1}{n} \sin^{n-1} x \cos x + \frac{n-1}{n} \int \sin^{n-2} x \, dx$

(iv) $\displaystyle \int \cos^n x \, dx = \frac{1}{n} \cos^{n-1} x \sin x + \frac{n-1}{n} \int \cos^{n-2} x \, dx.$

The last two rules lead to *Wallis' integrals*:

$$\int_0^{\frac{\pi}{2}} \sin^n x \, dx = \int_0^{\frac{\pi}{2}} \cos^n x \, dx = \begin{cases} \dfrac{n-1}{n}\dfrac{n-3}{n-2}\cdots\dfrac{1}{2}\dfrac{\pi}{2} & \text{if } n \text{ is even} \\[2mm] \dfrac{n-1}{n}\dfrac{n-3}{n-2}\cdots\dfrac{2}{3} & \text{if } n \text{ is odd.} \end{cases}$$

8.2.7 Exercises

In the case of a rational integrand the precise form of the antiderivative may depend on which interval, whose endpoints are successive zeros of the denominator, is being considered. If this is the case it is simplest to assume that x is higher than all roots of the denominator. It is then easy to adjust the antiderivative thus obtained for other intervals separated by roots of the denominator.

1. Solve the following integrals:

(a) $\displaystyle\int \frac{x^2}{x^2+1}\,dx$

(b) $\displaystyle\int \frac{x}{(x^2-1)(x^2-2)}\,dx$

(c) $\displaystyle\int \frac{1}{(x^2+1)(x^2+2)}\,dx$

(d) $\displaystyle\int \frac{x}{(x^2-1)(x^2+2)}\,dx$

(e) $\displaystyle\int \frac{x^2}{(x-1)^2(x^2+1)}\,dx$

(f) $\displaystyle\int \frac{1}{(x-1)^2(x^2+1)^2}\,dx.$

2. (a) Evaluate the integral

$$\int_0^1 \frac{x^4(1-x)^4}{x^2+1}\,dx.$$

(b) Using the estimate $\frac{1}{2} < (x^2+1)^{-1} < 1$ obtain the inequalities

$$\frac{22}{7}-\frac{1}{630} < \pi < \frac{22}{7}-\frac{1}{1260}.$$

Note. The approximation $22/7$ to π was known to Archimedes. The result shows that $1979/630$, or $3.14126..$, is an approximation to π with error less than $1/1000$. The actual error is around $3/10000$.

3. Solve the integral

$$\int \frac{1}{a+\sin\theta}\,d\theta$$

where a is a non-zero constant. Distinguish carefully between the cases $a^2 < 1$, $a^2 = 1$ and $a^2 > 1$.

4. Prove the reduction formulas listed in this section.
5. Verify the formulas known as Wallis' integrals.
6. Obtain a reduction formula for $\int \sec^n x \, dx$. This is useful because integrals involving powers of secant often arise from using the substitution $x = \tan t$ in integrals involving the factor $\sqrt{1 + x^2}$.
7. To exploit the result of the previous exercise one needs the useful integral $\int \sec x \, dx$. This was considered in Sect. 8.1 Exercise 7. Solve it again by using the half-angle substitution $t = \tan(x/2)$. Reconcile the result with that of Sect. 8.1 Exercise 7.
8. Using the previous two exercises and the substitution $x = \tan t$ solve the integral $\int (1 + x^2)^{3/2} \, dx$.
9. Obtain a reduction formula for $\int x^n e^{ax^2} \, dx$. The results should show that this integral can be expressed by elementary functions when n is odd, including the case of negative n.
10. Show that the integral $\int \sin^m x \cos^n x \, dx$, where m and n are non-negative integers, reduces to one of the following easily solvable integrals:

$$\int f(\sin x) \cos x \, dx, \quad \int f(\cos x) \sin x \, dx, \quad \int f(\sin x) \, dx$$

where f is a polynomial.

11. Obtain reduction formulas for the integrals

(a) $\displaystyle\int \tan^n x \, dx$

(b) $\displaystyle\int \tan^n x \sec x \, dx$

(c) $\displaystyle\int \frac{x^n}{\sqrt{x^2 + 1}} \, dx$.

12. Some integrals involving fractional powers can be reduced to the integral of a rational function by a suitable substitution, and the resulting integral solved by the methods of this chapter. Try to do this (or at least the reduction part) for the following integrals:

(a) $\displaystyle\int \frac{1}{x^{1/3} + x^{1/5}} \, dx$

(b) $\displaystyle\int \sqrt{\tan x} \, dx$

(c) $\displaystyle\int \sqrt[3]{\frac{x}{x+1}} \, dx$

(d) $\displaystyle\int \frac{1}{(x-a)^{3/2} + (x+a)^{3/2}} \, dx$.

13. Solve the following integrals:

(a) $\displaystyle\int \frac{1}{x^3 - 1}\, dx$

(b) $\displaystyle\int \frac{1}{x^4 + 1}\, dx$

(c) $\displaystyle\int \frac{1}{x^6 + 1}\, dx.$

Hint. Try to find the irreducible quadratic factors using inspired guesswork. Complex numbers can help in factorising the denominators; see the next chapter.

8.3 ($\Diamond$) Ostrogradski's Method

In order to find the partial fractions decomposition of a rational function P/Q, prior to integrating it, one must know the roots of the denominator. This may be a hard task. In this nugget we shall explore what can be discovered about the integral $\int P/Q$ without finding the roots of the denominator. In particular it turns out that the rational part of the integral, see Proposition 8.3, may be found by linear algebra alone. This is variously known as Ostrogradski's method or Hermite's method (the former seems to have priority but the latter is better known). Some knowledge of algebra will be assumed.

We recall the fundamental theorem of algebra, according to which a real non-constant polynomial Q has a factorisation into real prime factors

$$Q = g_1^{k_1} g_2^{k_2} \dots g_\ell^{k_\ell}.$$

The prime factors g_j, all distinct, are first-degree polynomials, or second-degree polynomials without real roots. The exponents are positive integers. Recall that a prime factor g_j of Q is called simple if $k_j = 1$.

The polynomial Q is called square-free if all the exponents k_j equal 1. The usage is the same as for the factorisation of integers into primes. Square-free indicates that it is not divisible by any perfect square (in which context constant polynomials do not count).

Prime polynomials (for our purposes this means first-degree polynomials or second-degree polynomials without real roots) have the following important property: if P and Q are polynomials and a prime polynomial g divides PQ, then g divides either P or Q.

The role that the following, rather mysterious, lemma plays will be apparent shortly. The reader may prefer to look ahead and see how the lemma is used before reading its proof.

Lemma 8.1 *Suppose that P_1, P_2, Q_1 and Q_2 are polynomials that satisfy the conditions:*

(i) $\displaystyle\left(\frac{P_1}{Q_1}\right)' + \frac{P_2}{Q_2} = 0$

(ii)   All prime factors common to both Q_1 and Q_2 are simple factors of Q_2.

Then Q_1 divides P_1 and Q_2 divides P_2.

Proof Our strategy for proving this is to show that all prime factors of the denominators Q_1 and Q_2 can be cancelled against the numerators P_1 and P_2, using a process of descent (which is really the same as induction). There are two possible cases in each step of the descent.

(A) The polynomial g is a prime factor of Q_2 but not of Q_1.

In this case we show that g divides P_2. By condition (i) we have

$$Q_1 Q_2 P_1' - Q_1' Q_2 P_1 + P_2 Q_1^2 = 0. \tag{8.3}$$

Since g divides Q_2 it must divide $P_2 Q_1^2$ also. But g does not divide Q_1, and therefore g divides P_2.

(B) The polynomial g is a prime factor of Q_1.

In this case we show that g divides P_1. To see this we first write $Q_1 = g^r \psi_1$ and $Q_2 = g^s \psi_2$, where ψ_1 and ψ_2 are polynomials not divisible by g, the exponent r is a positive integer, and, according to assumption (ii), the exponent s is 0 or 1. Substituting into (8.3) and cancelling powers of g we find

$$g \psi_1 \psi_2 P_1' - r g' \psi_1 \psi_2 P_1 - g \psi_1' \psi_2 P_1 + g^{r-s+1} \psi_1^2 = 0.$$

We see that $r \geq 1$ and $r - s + 1 \geq 1$, so that g divides $g' \psi_1 \psi_2 P_1$. But g does not divide g', nor by assumption does it divide ψ_1 or ψ_2. It follows that g divides P_1.

In case A we replace P_2 by P_2/g and Q_2 by Q_2/g. In case B we replace P_1 by P_1/g and Q_1 by Q_1/g. We proceed to eliminate all prime factors of Q_1 and Q_2, at each step applying case A or B as appropriate. A simple way to implement this is to apply case B until all prime factors of Q_1 have been stripped, then to apply case A to do the same to Q_2. It is clear that this process eventually clears both denominators of all their prime factors, and leads to the conclusion that Q_1 divides P_1 and Q_2 divides P_2. $\qquad\square$

Suppose as before that Q has the prime factorisation

$$Q = g_1^{k_1} g_2^{k_2} \cdots g_\ell^{k_\ell}. \tag{8.4}$$

We define the polynomials

$$Q_1 = g_1^{k_1-1} g_2^{k_2-1} \cdots g_\ell^{k_\ell-1}, \quad Q_2 = g_1 g_2 \cdots g_\ell. \tag{8.5}$$

Exercise Show that the polynomial Q_1 is the highest common factor of Q and Q', and that $Q_2 = Q/Q_1$. Show also that Q_1 and Q_2 satisfy condition (ii) of lemma 8.1.

We will also need the formula

$$\frac{Q_1'}{Q_1} = \sum_{j=1}^{\ell} \frac{(k_j - 1)g_j'}{g_j}. \tag{8.6}$$

Exercise Prove this formula.

We will need the following input from algebra. The highest common factor of two polynomials can be computed without factorising them, using the Euclidean algorithm (this was also referred to in the nugget "Multiplicity"). The reader unfamiliar with this should consult a book of algebra. The significance here is that the polynomials Q_1 and Q_2 can be found without factorising Q, since Q_1 is the highest common factor of Q and Q'.

The following decomposition of P/Q is the basis of Ostrogradski's method.

Lemma 8.2 *Suppose that* $\deg P < \deg Q$ *and let* Q_1 *and* Q_2 *be defined by* (8.4) *and* (8.5). *Then there exist unique polynomials* P_1 *and* P_2, *such that* $\deg P_1 < \deg Q_1$ *and* $\deg P_2 < \deg Q_2$, *satisfying*

$$\frac{P}{Q} = \left(\frac{P_1}{Q_1}\right)' + \frac{P_2}{Q_2}. \tag{8.7}$$

Moreover P_1 *and* P_2 *can be found by linear algebra, more precisely by the method of undetermined coefficients.*

Outline of Proof We first show that if $\deg P_1 < \deg Q_1$ and $\deg P_2 < \deg Q_2$, and if

$$\left(\frac{P_1}{Q_1}\right)' + \frac{P_2}{Q_2} = 0,$$

then $P_1 = P_2 = 0$. This follows from lemma 8.1, according to which we can conclude that Q_1 divides P_1 and Q_2 divides P_2. Since $\deg P_1 < \deg Q_1$ and $\deg P_2 < \deg Q_2$ it follows that P_1 and P_2 are both 0.

The rest of the argument is linear algebra. We define two vector spaces. Firstly we need the vector space R_Q of all rational functions expressible in the form $P(x)/Q(x)$ with $\deg P < \deg Q$ (also used in the discussion of partial fractions in Sect. 8.2). Secondly we need the vector space W comprising all pairs of polynomials (P_1, P_2), such that $\deg P_1 < \deg Q_1$ and $\deg P_2 < \deg Q_2$. The first space has dimension $\deg Q$ and the second $\deg Q_1 + \deg Q_2$, but these are equal since $Q = Q_1 Q_2$.

We introduce a linear mapping $T : W \to R_Q$ defined by

$$T(P_1, P_2) = \left(\frac{P_1}{Q_1}\right)' + \frac{P_2}{Q_2}.$$

That the right-hand side really belongs to the vector space R_Q is easily seen by expanding the derivative and using (8.6). The reader should check this point.

The argument at the beginning of the proof and based on the lemma tells us that the kernel of T contains only the zero vector in W. Hence T is injective. Since the domain and codomain of T are vector spaces of the same dimension, it follows by linear algebra that it is also surjective, and hence bijective. This establishes the required decomposition.   □

As a corollary of the decomposition (8.7) we obtain

$$\int \frac{P}{Q} = \frac{P_1}{Q_1} + \int \frac{P_2}{Q_2}.$$

The integral on the right-hand side is a transcendental function since the denominator Q_2 is square-free and the numerator has lower degree than the denominator. The first term on the right-hand side P_1/Q_1 is the rational part of the integral.

8.3.1   Exercises

1. Let ψ be a square-free polynomial of degree d and let P have degree less than $2d$ and no prime factor in common with ψ. Show that the integral $\int P/\psi^2$ is rational if and only if ψ divides $P'\psi' - P\psi''$.

2. Show that the integral $\displaystyle\int \frac{4x^3 - 3x^2 - 2}{(x^3 + 1)^2}\, dx$ is rational and evaluate it.

3. Solve the integral $\displaystyle\int \frac{1}{(x^4 + 1)^2}\, dx$.

8.3.2   Pointers to Further Study

→ Symbolic integration

8.4   (◇) Numerical Integration

If an antiderivative for f is not forthcoming, it is often the case that to calculate the integral $\int_a^b f$ one has to approximate it.

We have seen how Riemann sums furnish an approximation, but one that yields few extra decimal digits with each improvement step. More exact methods exist that build on the same idea. Points t_k are chosen in the interval $[a, b]$ and the sum $\sum \alpha_k f(t_k)$ is formed with coefficients α_k that satisfy $\sum \alpha_k = b - a$. There are a number of different prescriptions that have been developed. They are called numerical integration rules. By choosing the points and the coefficients appropriately it is possible to reach some remarkably accurate approximations.

8.4.1 Trapezium Rule[1]

The interval $[a, b]$ is divided into n equal parts. Let

$$h = \frac{b - a}{n}, \quad t_0 = a, \quad t_k = a + kh, \ (k = 1, 2, ..., n - 1), \quad t_n = b.$$

The trapezium approximation is

$$T_n = \frac{h}{2} \left(f(a) + 2 \sum_{k=1}^{n-1} f(t_k) + f(b) \right). \tag{8.8}$$

It is possible to bound the error. Assume that f is twice differentiable and that $|f''(x)| < M$ in $[a, b]$. Then

$$\left| \int_a^b f - T_n \right| < \frac{M(b - a)^3}{12n^2}.$$

A convenient approach is to double n in each improvement step, as then previous points can be used again. We can roughly say that each improvement step yields on average at least 0.6 (being near to $\log_{10} 4$) extra correct digits after the decimal point, as opposed to 0.3 digits for Riemann sums.

8.4.2 Midpoint Rule

The interval $[a, b]$ is divided into n equal parts. Let

$$h = \frac{b - a}{n}, \quad t_0 = a, \quad t_k = a + kh, \ (k = 1, 2, ..., n - 1), \quad t_n = b.$$

The midpoint rule is the approximation

$$M_n = h \sum_{k=0}^{n-1} f\left(\frac{t_k + t_{k+1}}{2} \right). \tag{8.9}$$

This is just a Riemann sum where in each subinterval we choose to evaluate f at the midpoint.

---

[1] Also known as the trapezoidal rule.

8.4.3 Simpson's Rule

The interval is partitioned into $2n$ equal subintervals. Set

$$h = \frac{b-a}{2n}, \quad t_0 = a, \quad t_k = a + kh, \; (k = 1, 2, ..., 2n-1), \quad t_{2n} = b.$$

Simpson's rule is the approximation

$$S_n = \frac{h}{3}\left(f(a) + 4\sum_{k=1}^{n} f(t_{2k-1}) + 2\sum_{k=1}^{n-1} f(t_{2k}) + f(b) \right). \tag{8.10}$$

The error is much improved over that for the trapezium rule. Assume that f is four times differentiable and that $|f^{(4)}(x)| < M$ in the interval $[a, b]$. Then

$$\left| \int_a^b f - S_n \right| < \frac{M(b-a)^5}{2880n^4}.$$

Each doubling of n contributes, at a rough estimate, a further 1.2 (near to $\log_{10} 16$) correct decimal digits. Note also that the rule is exact for third-degree polynomials, as is clear from the error estimate.

The rule is easy to remember in the form:

$$\int_a^b f \approx \frac{h}{3}\big[\text{initial} + \text{twice even} + \text{four times odd} + \text{final}\,\big].$$

8.4.4 Proof of the Error Estimate

We will prove the error estimate for Simpson's rule in some detail. Error estimates for the trapezium rule and the midpoint rule are simpler and are left to the reader as exercises.

The interval is partitioned into an even number of intervals so we first estimate the error for an interval partitioned into two subintervals. It is convenient to take the interval as $[-c, c]$, with partition points

$$t_0 = -c, \quad t_1 = 0, \quad t_2 = c,$$

so that $h = c$. Simpson's approximation is

$$\frac{c}{3}\Big(f(-c) + 4f(0) + f(c) \Big).$$

We set

$$E(t) = \int_{-t}^{t} f - \frac{t}{3}\Big(f(-t) + 4f(0) + f(t)\Big), \quad (0 \le t \le c).$$

Now we differentiate repeatedly with respect to t. The reader should check the algebra in the following steps:

$$E'(t) = \frac{2}{3}f(t) + \frac{2}{3}f(-t) - \frac{4}{3}f(0) + \frac{t}{3}f'(-t) - \frac{t}{3}f'(t)$$

$$E''(t) = \frac{1}{3}f'(t) - \frac{1}{3}f'(-t) - \frac{t}{3}f''(-t) - \frac{t}{3}f''(t)$$

$$E'''(t) = \frac{t}{3}f'''(-t) - \frac{t}{3}f'''(t).$$

Recall now that we are assuming that $|f^{(4)}(x)| < M$ for all x. By the mean value theorem we therefore have

$$|f'''(s) - f'''(-s)| < 2Ms$$

for $s > 0$, so that

$$-\frac{2M}{3}s^2 < E'''(s) < \frac{2M}{3}s^2, \quad (s > 0).$$

Integrate these inequalities repeatedly from 0 to t; the correct values for the middle term at $t = 0$ can be read from the derivatives calculated above. After the third integration we find

$$-\frac{M}{90}t^5 < E(t) < \frac{M}{90}t^5,$$

which gives

$$|E(c)| < \frac{M}{90}c^5.$$

We apply this to the partition of $[a, b]$ into $2n$ intervals of length h. We have $h = (b - a)/2n$ and the error for each pair of consecutive subintervals is bounded by

$$\frac{M}{90}h^5 = \frac{M(b-a)^5}{(90 \times 2^5)n^5} = \frac{M(b-a)^5}{2880n^5}.$$

Adding together over the n pairs of intervals we get the error bound

$$\frac{M(b-a)^5}{2880n^4}$$

for Simpson's rule.

8.4.5 Exercises

1. Show that Simpson's approximation for the two interval partition is the integral from a to b of the second-degree polynomial whose graph (a parabola) passes through the three points

$$\left(a, f(a)\right), \quad \left(\frac{a+b}{2}, f\left(\frac{a+b}{2}\right)\right), \quad \left(b, f(b)\right).$$

   This makes it obvious that Simpson's approximation is exact for a parabola, but does not reveal why it should be exact for a cubic.
   Hint. It's easy to do it for the case $a = -b$.

2. Prove the error estimate for the trapezium rule.
   Hint. Do it first for a partition with only one subinterval. You can take the interval as $[-c, c]$ so the trapezium approximation is $c(f(-c) + f(c))$ and $h = 2c$. Estimate the error by copying the method used above for Simpson's rule; only it's much easier.

3. Develop an error estimate for the midpoint rule. Compare it with the error in the trapezium rule. Which is more accurate?

4. There is another way to write the error that sometimes gives more information, for example it can tell you whether the approximation lies above or below the true value. The treatment suggested here, and in the next two exercises, is closely related to Taylor's theorem (to be studied in Chap. 11).
   For the trapezium approximation show that if f is twice differentiable, the error (the integral minus the approximation) is

$$-\frac{(b-a)^3}{12n^2} f''(\xi),$$

   for some ξ in $]a, b[$. This gives the intuitive result that for a convex function the integral is below the approximation.
   Hint. Do it first for one interval, letting the interval be $[-c, c]$ and $h = 2c$. Let

$$E(x) = \int_{-x}^{x} f - x\left(f(-x) + f(x)\right),$$

   so that the error is $E(c)$. Apply Cauchy's form of the mean value theorem repeatedly to the quotient $E(x)/x^3$. You will need the fact that derivatives have the intermediate value property (see Sect. 5.6 Exercise 8).

5. Develop a similar result for the midpoint rule and compare it with the trapezium rule as regards magnitude and sign.

6. A similar result is also available for Simpson's rule, but it is more complicated. If f is four times differentiable then

$$\int_a^b f - \text{Simpson's approx.} = -\frac{(b-a)^5}{2880n^4} f^{(4)}(\xi)$$

for some ξ in $]a, b[$. A proof is outlined in the following steps:

(a) Obtain the result in the case of a Simpson approximation with two subintervals.

Hint. You can take the interval $[-c, c]$ subdivided at 0, so that $h = c$. Let

$$E(x) = \int_{-x}^x f - \frac{x}{3}\left(f(-x) + 4f(0) + f(x)\right).$$

Apply Cauchy's form of the mean value theorem repeatedly to the quotient $E(x)/x^5$. You will again need the fact that derivatives satisfy the intermediate value property.

(b) Obtain the result for $2n$ subintervals. Again you will need the intermediate value property.

7. Let I_n denote the approximation to $\int_a^b f$ obtained from applying Simpson's rule with $2n$ subintervals. Define a new approximation

$$J_n = \frac{16I_{2n} - I_n}{15}.$$

Find a formula for the error $\int_a^b f - J_n$. The result should indicate that J_n is a slight improvement over Simpson's rule.

8.4.6 Pointers to Further Study

→ Numerical integration
→ Gauss quadrature
→ Numerical analysis

Chapter 9
Complex Numbers

> *The shortest path between two truths in the real domain passes through the complex domain*
>
> J. Hadamard

Negative real numbers have no square root. This apparently regrettable fact is alleviated by extending the real number field $\mathbb{R}$ to the complex number field $\mathbb{C}$. In it all numbers have a square root.

9.1 The Complex Number Field

Formally the complex number field is a field $\mathbb{C}$, that is, a set of elements satisfying axioms A1–A6 of Chap. 1, that includes the real numbers, contains an element i, necessarily non-real, that satisfies $i^2 = -1$, and contains no other non-real elements except those that it has to in virtue of being a field. Its non-real elements must include all the elements of the form $a + bi$, where a and b are real numbers, but plainly we do not have to include especially elements such as $1 + 2i + 3i^2 + 4i^3 + 5i^5$, involving second and higher powers of i, since they can be expressed in the form $a + bi$ by using the property $i^2 = -1$, and therefore are already there.

Interestingly we do not have to include especially the reciprocals of elements (although fields must contain reciprocals of all non-zero elements). They are expressible in the form $a + bi$ and are already included. The reciprocal of the complex number $z = a + bi$, assuming that $z \neq 0$, is given by

$$z^{-1} = \frac{a}{a^2 + b^2} - \frac{b}{a^2 + b^2}i,$$

R. Magnus, *Fundamental Mathematical Analysis*, Springer Undergraduate Mathematics Series, https://doi.org/10.1007/978-3-030-46321-2_9

as the reader should check.

The important identity

$$(a + bi)(a - bi) = a^2 + b^2$$

is used very frequently. As a simple instance suppose that $a + bi = 0$ (where a and b are real numbers). We deduce

$$0 = (a + bi)(a - bi) = a^2 + b^2,$$

so that $a = b = 0$. This implies that the expression of a given complex number in the form $a + bi$, with real a and b, is unique. This allows us to define the functions

$$\mathrm{Re}\, z = x, \quad \mathrm{Im}\, z = y, \quad (z = x + yi,\ x,\ y \text{ real})$$

called the real part of z and the imaginary part of z.

We often give ourselves a simple geometric picture of the complex numbers. Just as we think of the real numbers as a line, we think of the complex numbers as forming a plane, by identifying the number $z = x + yi$ with the point (x, y) of the coordinate plane. In this way the real numbers are identified with the x-axis, called the real axis, and the complex numbers of the form yi with the y-axis, called the imaginary axis. We sometimes identify z with the vector joining $(0, 0)$ to (x, y), instead of simply with the point (x, y). These identifications sometimes give a way of proving things in Euclidean geometry using algebraic operations on complex numbers.

Extending a field by joining a new element to serve as a root of some equation is one of the most elementary operations of field theory. For example we can start with the field $\mathbb{Q}$ of rational numbers. We saw that the equation $x^2 - 2 = 0$ has no root in $\mathbb{Q}$ since there can be no rational square root of 2. Let us join a root, denoted by $\sqrt{2}$, to $\mathbb{Q}$. We obtain a new field, denoted by $\mathbb{Q}(\sqrt{2})$, consisting of all elements of the form $a + b\sqrt{2}$ with rational a and b. Again we can simplify an expression such as $1 + 2\sqrt{2} + 3(\sqrt{2})^5$ to the form $a + b\sqrt{2}$. And the reciprocal of $a + b\sqrt{2}$ is given by

$$(a + b\sqrt{2})^{-1} = \frac{a}{a^2 - 2b^2} - \frac{b\sqrt{2}}{a^2 - 2b^2}.$$

Note how the denominators fail to be 0; because the square root of 2 is not rational.

This procedure gives us another way to enlarge the field of rational numbers to include the square root of two, quite different from that of Chap. 1. And it seems neither more nor less acceptable than conjuring up an element i to serve as a square root of -1. But there is a big problem. The field $\mathbb{Q}(\sqrt{2})$ takes us only part of the way from $\mathbb{Q}$ to $\mathbb{R}$. There is much still missing; for example there is no square root of 3. We can join it, and then one by one, all square roots that are not so far expressible, then all cube roots and so on. After that we can join roots of polynomials that are still stubbornly unfactorisable. We obtain a whole sequence of fields intermediate between $\mathbb{Q}$ and $\mathbb{R}$. However we can never reach $\mathbb{R}$ by this piecemeal means; there are some irrationals that are not even expressible as a root of a polynomial equation with rational coefficients. These are the transcendental numbers. They include e and π for example.

So we have to arrive at $\mathbb{R}$ by a big leap, accomplished by means of the axiom of completeness. From $\mathbb{R}$ to $\mathbb{C}$ is but a small further step, joining a non-real root of $x^2 + 1 = 0$. But the remarkable thing is that the process ends here. Every polynomial equation with coefficients in $\mathbb{C}$ has a root in $\mathbb{C}$; no further extensions are needed.

The fact just mentioned is called the fundamental theorem of algebra and it was mentioned in the previous chapter in connection with the partial fractions decomposition of a rational function, though without mentioning complex numbers. The fact stated there is a simple corollary of the fundamental theorem of algebra.

9.1.1 Square Roots

We will take a little detour to examine the claim made before that every complex number has a square root. As a corollary we solve any quadratic equation. We will use purely algebraic arguments. Later we will examine radicals (cube roots, fourth roots, etc.) by different methods.

Suppose we are given the complex number $a + bi$ and we wish to find its square root $x + yi$. We assume that x, y, a and b are real and in future will often not make such assumptions explicit. If $x + yi$ is a square root of $a + bi$ then

$$(x + yi)^2 = x^2 - y^2 + 2xyi = a + bi,$$

and so

$$x^2 - y^2 = a, \quad 2xy = b.$$

Using the identity $(x^2 + y^2)^2 = (x^2 - y^2)^2 + 4x^2y^2$ we find that

$$x^2 + y^2 = \sqrt{a^2 + b^2}.$$

Solving for x^2 we find

$$x^2 = \frac{a + \sqrt{a^2 + b^2}}{2}$$

and since $a + \sqrt{a^2 + b^2} \geq 0$ we obtain

$$x = \pm\sqrt{\frac{a + \sqrt{a^2 + b^2}}{2}}.$$

For each of the two values of x we find y by the relation $y = b/2x$, except that when $a = b = 0$ we take $x = y = 0$.

We thus get two square roots (except for when $a = b = 0$) given by

$$x = \sqrt{\frac{a + \sqrt{a^2 + b^2}}{2}} + \frac{bi}{2} \bigg/ \sqrt{\frac{a + \sqrt{a^2 + b^2}}{2}}$$

and its negative.

It will be rightly objected that we have not proved the existence of any square root, merely shown that if a square root exists it must be given by this formula. The reader is therefore invited to check that this is indeed the square root of $a + bi$ by squaring it.

Now we have the square root we can solve any quadratic equation, using the usual formula.

Proposition 9.1 *The quadratic equation $ax^2 + bx + c = 0$, with coefficients a, b, c in $\mathbb{C}$ and with $a \neq 0$, has a root in $\mathbb{C}$. Its roots are found by the usual formula. More precisely, let $d = b^2 - 4ac$ and let ω be a square root of d. Then the roots are $\frac{1}{2a}(-b \pm \omega)$.*

The reader is invited to check that these are the roots. They are the only ones; this can be seen by the usual method of completing the square, based on observing that the equation $ax^2 + bx + c = 0$ is equivalent to

$$\left(x + \frac{b}{2a}\right)^2 = \frac{1}{4a^2}(b^2 - 4ac).$$

We can also appeal to a simple fact of field theory, that a polynomial with degree n cannot have more than n roots.

9.1.2   Modulus and Conjugate

Let z be a complex number and let $z = x + yi$, where x and y are real numbers. The number

$$\bar{z} := x - yi$$

is called the complex conjugate of z, or just conjugate for short. The real number

$$|z| := \sqrt{x^2 + y^2}$$

is called the modulus of z, but is often called its absolute value. It has a geometric interpretation. If you think of z as the coordinate vector (x, y) (if you like, the vector joining $(0, 0)$ to the point (x, y)), then $|z|$ is its Euclidean length.

The real part $\operatorname{Re} z = x$ and the imaginary part $\operatorname{Im} z = y$ have already been mentioned. Here is a list of important properties of these operations.

$$
\begin{array}{llll}
(1) & \overline{\overline{z}} = z & (2) & \overline{z + w} = \overline{z} + \overline{w} \\
(3) & \operatorname{Re} z = \frac{1}{2}(z + \overline{z}) & (4) & \operatorname{Im} z = \frac{1}{2i}(z - \overline{z}) \\
(5) & \overline{zw} = \overline{z}\,\overline{w} & (6) & \overline{z^{-1}} = \overline{z}^{-1} \\
(7) & |z|^2 = z\overline{z} & (8) & |zw| = |z|\,|w| \\
(9) & \operatorname{Re} z \le |z| & (10) & \operatorname{Im} z \le |z|.
\end{array}
$$

Some indications of the proofs. Rules 1–6 are obvious consequences of the definitions. Rule 7 is the identity $x^2 + y^2 = (x + yi)(x - iy)$. Rule 8 follows from rule 7 using the calculation

$$
|zw|^2 = zw\overline{zw} = z\overline{z}w\overline{w} = |z|^2|w|^2.
$$

Rule 9 (and similarly rule 10) is the inequality $x \le \sqrt{x^2 + y^2}$.

The following result (more precisely the first inequality) is called the triangle inequality and is immensely important.

Proposition 9.2 *For all complex z and w we have*

(1) $|z + w| \le |z| + |w|$ *(triangle inequality)*
(2) $\big||z| - |w|\big| \le |z - w|$.

Proof We calculate, applying at least seven of the rules listed above:

$$
\begin{aligned}
|z + w|^2 &= (z + w)(\overline{z} + \overline{w}) \\
&= z\overline{z} + z\overline{w} + w\overline{z} + w\overline{w} = |z|^2 + 2\operatorname{Re}(z\overline{w}) + |w|^2 \\
&\le |z|^2 + 2|z\overline{w}| + |w|^2 = |z|^2 + 2|z||w| + |w|^2 \\
&= (|z| + |w|)^2.
\end{aligned}
$$

This proves the inequality in item 1.

Now $z = z - w + w$, so by item 1 we find $|z| \le |z - w| + |w|$, or

$$
|z| - |w| \le |z - w|.
$$

Interchange z and w. This makes no difference to the right-hand side. We find

$$
|w| - |z| \le |z - w|.
$$

In one of the displayed inequalities the left-hand side is $\big||z| - |w|\big|$. $\square$

We see from the calculation proving the triangle inequality that equality holds in it if and only if $\operatorname{Re}(z\overline{w}) = |z\overline{w}|$, which is equivalent to saying $\operatorname{Im}(z\overline{w}) = 0$. If $z = a + bi$ and $w = c + di$ then

$$\text{Im}\,(z\overline{w}) = bc - ad = \begin{vmatrix} b & a \\ d & c \end{vmatrix}$$

so the equality $\text{Re}\,(z\overline{w}) = |z\overline{w}|$ says that the vectors (a, b) and (c, d) are linearly dependent. Putting it differently it says that equality holds in the triangle inequality if and only if either one or both of z and w are 0, or there is a non-zero real number λ, such that $z = \lambda w$.

There is a more geometric interpretation of $\text{Im}\,(z\overline{w})$ that derives from its determinantal form. Its absolute value is the area of the parallelogram with vertices 0, z, w and $z + w$, or twice the area of the triangle with vertices 0, z and w. Even $\text{Re}\,(z\overline{w})$ has a geometric interpretation. It is the scalar product of the vectors z and w.

9.1.3 Exercises

1. Check the claim that the reciprocal of $a + bi$ is $(a - bi)/(a^2 + b^2)$.
2. Check that $\text{Re}\,(z\overline{w})$ is the scalar product of the vectors z and w, where the complex numbers are identified with plane vectors.
3. Check that the formula

$$x = \sqrt{\frac{a + \sqrt{a^2 + b^2}}{2}} + \frac{bi}{2} \Big/ \sqrt{\frac{a + \sqrt{a^2 + b^2}}{2}}$$

   really does provide a square root of $a + bi$.
4. Show that the area of the triangle with vertices z_1, z_2 and z_3 is the absolute value of

$$\frac{1}{2}\text{Im}\,(\overline{z_1}z_2 + \overline{z_2}z_3 + \overline{z_3}z_1).$$

5. Prove the inequality

$$|z| \leq \sqrt{2}\,\max(|\text{Re}\,z|, |\text{Im}\,z|).$$

   Show that $\sqrt{2}$ cannot be replaced by a smaller number.
6. Let

$$f(x) = a_n x^n + a_{n-1} x^{n-1} + \cdots + a_1 x + a_0$$

   be a polynomial with real coefficients. Show that for all complex z we have

$$\overline{f(z)} = f(\overline{z}).$$

   Deduce that if w is a complex root of $f(z) = 0$ so is $\overline{w}$.

9.2 Algebra in the Complex Plane

The identification of $\mathbb{C}$ with the plane produces a valuable geometric picture of the algebra of the complex numbers. We can, for example, introduce polar coordinates in the plane,

$$x = r \cos \theta, \quad y = r \sin \theta, \quad (r \geq 0, \; -\infty < \theta < \infty)$$

(we allow here the cases $r = 0$ and do not restrict θ to any particular interval) and then write

$$x + yi = r(\cos \theta + i \sin \theta).$$

Here we have $r = |x + yi|$, but θ (called the argument of z) is not uniquely determined by z, as we may always add to it an integer multiple of 2π. The point 0 is an exception, as it does not have an argument.

If we agree to choose θ in the interval $]-\pi, \pi]$ then it is uniquely determined by z (if $z \neq 0$). It is then called the principal argument and is denoted by $\operatorname{Arg} z$ (always with upper case "A" as befits its privileged status). The principal argument of z is the polar angular coordinate according to the most common convention.

The following proposition describes complex multiplication in terms of polar coordinates.

Proposition 9.3 *Let $z_1 = r_1(\cos \theta_1 + i \sin \theta_1)$ and $z_2 = r_2(\cos \theta_2 + i \sin \theta_2)$. Then*

$$z_1 z_2 = r_1 r_2 \big(\cos(\theta_1 + \theta_2) + i \sin(\theta_1 + \theta_2) \big).$$

Proof We multiply and use the addition formulas for sine and cosine:

$$r_1(\cos \theta_1 + i \sin \theta_1) \, r_2(\cos \theta_2 + i \sin \theta_2)$$
$$= r_1 r_2 \Big((\cos \theta_1 \cos \theta_2 - \sin \theta_1 \sin \theta_2) + i (\sin \theta_1 \cos \theta_2 + \cos \theta_1 \sin \theta_2) \Big)$$
$$= r_1 r_2 \big(\cos(\theta_1 + \theta_2) + i \sin(\theta_1 + \theta_2) \big).$$

$\square$

A special case is the rule known as de Moivre's theorem,

$$(\cos \theta + i \sin \theta)^n = \cos n\theta + i \sin n\theta, \quad (n \in \mathbb{N}).$$

9.2.1 n^{th} Root of a Complex Number

Using de Moivre's theorem we can show that every complex number has an nth root. It is noteworthy that to express it we use transcendental functions, albeit elementary ones, seemingly moving outside the realm of algebra.

Fig. 9.1 The 5th roots of 1

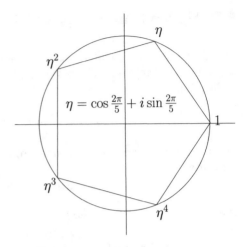

Let $w = r(\cos\theta + i\sin\theta)$ where $r = |w|$. Then one n^{th} root of w is

$$\alpha = r^{\frac{1}{n}}\left(\cos\left(\frac{\theta}{n}\right) + i\sin\left(\frac{\theta}{n}\right)\right).$$

Others are

$$r^{\frac{1}{n}}\left(\cos\left(\frac{\theta + 2\pi k}{n}\right) + i\sin\left(\frac{\theta + 2\pi k}{n}\right)\right) = \alpha\,\eta^k, \quad (k = 1, 2, ..., n-1),$$

where α was the root given first and

$$\eta = \cos\left(\frac{2\pi}{n}\right) + i\sin\left(\frac{2\pi}{n}\right).$$

These n^{th} roots, a total of n of them, are all distinct if $w \neq 0$.

Actually these are all the nth roots of w, as they are roots of the polynomial equation $x^n - w = 0$, and it is known from algebra that a polynomial equation of degree n cannot have more than n roots.

The numbers η^k, $k = 0, 1, ..., n-1$ are the nth roots of 1, and may be visualised geometrically as the vertices of a regular polygon with n sides inscribed in the unit circle as illustrated in Fig. 9.1.

9.2.2 Logarithm of a Complex Number

Properly, functions of a complex variable need a text of their own. The reason for introducing logarithm here, perhaps prematurely, is the light that it throws on the integration of rational functions.

Let z be a complex number, not of the form $x + 0i$ with $x \le 0$. We define

$$\operatorname{Log} z = \ln |z| + i \operatorname{Arg} z.$$

The use of upper case "L" is conventional here, the function being sometimes called the principal logarithm.

Now we allow the differentiation of functions of a real variable with values in the complex plane. This is accomplished in the most obvious way. A function $f : A \to \mathbb{C}$ (where A is an interval of real numbers) has the form

$$f(t) = u(t) + iv(t)$$

where u and v are real-valued functions with domain A. We define

$$f'(t) = u'(t) + iv'(t)$$

for all t at which u and v are differentiable.

As an example we can consider the parametrisation of the unit circle

$$f(t) = \cos t + i \sin t.$$

Differentiation gives

$$f'(t) = -\sin t + i \cos t,$$

and, identifying a complex number with a plane vector, we can interpret this as saying that the velocity vector is normal to the radius and has length (speed) 1.

Now let $a \ne 0$ and consider the function $\operatorname{Log}(x + ia)$ of the real variable x. We have

$$\frac{d}{dx} \operatorname{Log}(x + ia) = \frac{1}{x + ia}, \quad (-\infty < x < \infty). \tag{9.1}$$

Note that there is no need to restrict x to positive values only.

We also have, for a positive integer exponent n,

$$\frac{d}{dx} \frac{1}{(x + ia)^n} = -\frac{n}{(x + ia)^{n+1}}, \quad (-\infty < x < \infty). \tag{9.2}$$

The proofs of these formulas are left to the exercises. They are important because they provide antiderivatives for functions that arise naturally when the fundamental theorem of algebra is applied, in its more usual complex form, to the decomposition of the rational function P/Q into partial fractions. According to the fundamental theorem the polynomial Q (with leading coefficient 1) has a factorisation

$$Q(x) = (x - \alpha_1)^{r_1} ... (x - \alpha_\ell)^{r_\ell}$$

with complex numbers α_j. The fraction P/Q decomposes into

$$\text{polynomial} + \sum_{j=1}^{\ell} \sum_{k=1}^{r_j} \frac{b_{jk}}{(x - \alpha_j)^k}.$$

The transcendental part of the integral can then be expressed as a sum of logarithms of the complex first-degree polynomials $x - \alpha_j$. Of course this implies a connection between arctangent and the principal logarithm. It is left to the reader to show that

$$\arctan x = \frac{1}{2i} \left(\text{Log} (x - i) - \text{Log} (x + i) \right) + \frac{\pi}{2}. \tag{9.3}$$

Complex numbers can also simplify the determination of the coefficients in the partial fractions decomposition. Consider the case of a polynomial $Q(x)$ that has only simple roots, including all its complex roots. Then

$$Q(x) = (x - \alpha_1)(x - \alpha_2)...(x - \alpha_n)$$

with n distinct first-degree prime factors, where the numbers α_j may be complex. Still thinking of x as a real variable, we can differentiate. Leibniz's rule applies (see the exercises) so we find

$$Q'(x) = F_1(x) + F_2(x) + \cdots + F_n(x)$$

where $F_j(x)$ is the polynomial obtained by omitting the factor $x - \alpha_j$ from $Q(x)$. Since $F_k(\alpha_j) = 0$ if $k \neq j$ we have

$$F_j(\alpha_j) = Q'(\alpha_j).$$

Now we have a simple formula for the partial fractions decomposition:

$$\frac{1}{Q(x)} = \sum_{j=1}^{n} \frac{1}{Q'(\alpha_j)} \frac{1}{x - \alpha_j}. \tag{9.4}$$

More generally, if $P(x)$ is another polynomial with degree lower than that of $Q(x)$, we have

$$\frac{P(x)}{Q(x)} = \sum_{j=1}^{n} \frac{P(\alpha_j)}{Q'(\alpha_j)} \frac{1}{x - \alpha_j}. \tag{9.5}$$

A theory of differentiation with respect to a complex variable is not needed for this simple case. We create the polynomial $Q'(x)$ by differentiating $Q(x)$ in the normal way and then substituting α_j for x. If Q has multiple complex roots a similar formulation is possible using higher derivatives of Q. However, to treat this more general case properly it is best to use the theory of complex analytic functions.

As we have seen, the antiderivative of a rational function can be expressed without using arctangents, if we use instead the principal logarithm of complex first-degree polynomials. Such formulations are often thrown up by computer algebra programs that know how to find antiderivatives of elementary functions and understand complex numbers. Thus, given the roots of the polynomial Q in the complex plane, and assuming that they are simple, we have

$$\int \frac{P(x)}{Q(x)} \, dx = \sum_{j=1}^{n} \frac{P(\alpha_j)}{Q'(\alpha_j)} \, \mathrm{Log}\,(x - \alpha_j). \tag{9.6}$$

To ensure the validity of this formula we do not want $x - \alpha_j$ to be a non-positive, real number. This is ensured if $x > \alpha_j$, for all j such that α_j is real. If α_j is real and $x < \alpha_j$ we simply replace $x - \alpha_j$ by $\alpha_j - x$ in the corresponding term.

9.2.3 Exercises

1. Prove the formula

$$\mathrm{Arg}\, z = 2 \arctan\left(\frac{\mathrm{Im}\, z}{|z| + \mathrm{Re}\, z}\right),$$

   given that z is not of the form $x + 0i$ with $x \leq 0$.
   Hint. A proof by geometry is easiest.
2. The real factorisation of $x^4 + 1$ into irreducible quadratic factors is needed to calculate the integral $\int 1/(x^4 + 1)\, dx$. Obtain the factorisation using complex numbers by noting that the roots of $x^4 + 1 = 0$, the four complex numbers

$$w_1 = \frac{1+i}{\sqrt{2}}, \quad w_2 = \frac{-1+i}{\sqrt{2}}, \quad w_3 = \frac{-1-i}{\sqrt{2}}, \quad w_4 = \frac{1-i}{\sqrt{2}},$$

   form the corners of a square, and that the quadratics, $(x - w_1)(x - w_4)$ and $(x - w_2)(x - w_3)$, have real coefficients.
3. Let $\eta = \cos(2\pi/5) + i \sin(2\pi/5)$. The four non-real fifth roots of 1 are η, η^2, η^3 and η^4.

   (a) Show that they are the roots of the polynomial

$$x^4 + x^3 + x^2 + x + 1.$$

   (b) Express the sum $\eta + \eta^2 + \eta^3 + \eta^4$ in terms of $\lambda := \cos(2\pi/5)$.
   (c) Deduce that $\cos(2\pi/5) = \dfrac{\sqrt{5} - 1}{4}$.

(d) Show that

$$x^4 + x^3 + x^2 + x + 1 = (x^2 - 2\lambda x + 1)(x^2 - 2(2\lambda^2 - 1)x + 1).$$

Note. Item (c) shows that the number λ is constructible with straight edge and compass, and therefore the regular pentagon is also constructible. One such construction is described in Euclid (book 4, prop. 11).

4. In the previous exercise an expression was obtained for $\cos(2\pi/5)$ as an algebraic number involving a square root. We now express angles in degrees, defining $x° = x\pi/180$. The formulas $\sin 30° = \frac{1}{2}$, $\sin 45° = 1/\sqrt{2}$ are doubtlessly familiar. We also have $\cos 72° = (\sqrt{5} - 1)/4$ by the previous exercise. Formulas such as these are sometimes said to give "exact" values of the circular functions. They are characterised by including only arithmetic operations on rationals, and radicals, possibly nested.

Find exact values of the following circular functions of the given angles:

   (a) $\sin 36°$
   (b) $\cos 36°$
   (c) $\sin 6°$
   (d) $\cos 6°$
   (e) $\sin 3°$
   (f) $\cos 3°$

Conclude that the sine and cosine of any multiple of $3°$ are expressible exactly with square roots.

5. Derive the rules for differentiation of product and reciprocal

$$\frac{d}{dx}(fg) = f'g + fg', \quad \frac{d}{dx}\left(\frac{1}{f}\right) = -\frac{f'}{f^2}$$

for complex-valued functions f and g of a real variable x.

Hint. It should be obvious just by looking at the real and imaginary parts that $1/f$ is differentiable if f is. Knowing this means that the rule for reciprocal can be obtained with little effort from the rule for product.

6. Derive formulas (9.1) and (9.2).
7. Derive formula (9.3).
   Hint. Differentiate the formula.
8. Express the antiderivative

$$\int \frac{1}{x^5 - 1} \, dx$$

using the principal logarithm of first-degree complex polynomials.

9. Let $w = a + ib$ and let $f(x) = e^{ax}(\cos bx + i \sin bx)$ for real x. Show that $f'(x) = wf(x)$.

Note. We have not yet defined exponentials of complex numbers. When we study the unification of exponential and circular functions in Chap. 11, we will define $e^{(a+ib)x}$ and show that it equals $e^{ax}(\cos bx + i \sin bx)$.

10. We study Cardano's solution of the cubic equation.

(a) Show that $-b - c$ is a root of $x^3 + px + q = 0$ if b and c satisfy

$$bc = -\frac{p}{3}, \quad b^3 + c^3 = q. \tag{9.7}$$

Hint. One way is to use the identity

$$a^3 + b^3 + c^3 - 3abc = (a+b+c)(a^2 + b^2 + c^2 - ab - bc - ac).$$

(b) Show that the solutions to (9.7) are given by

$$b = \sqrt[3]{\frac{q + \sqrt{q^2 + 4p^3/27}}{2}}, \quad c = -\frac{p}{3b} \tag{9.8}$$

for any choice of the two possible square roots and the three possible cube roots.

(c) Show that all solutions to $x^3 + px + q = 0$ are obtained by fixing the square root in (9.8) and using all three cube roots.

(d) Show that the substitution $y = x - a/3$ reduces the general cubic equation $x^3 + ax^2 + bx + c = 0$ to an equation (for y) of the form $y^3 + py + q = 0$, that can then be solved by the preceding method.

11. In this problem we assume that the coefficients p and q in the cubic equation are real. In the formula for the solution in the previous exercise the quantity $D := q^2 + 4p^3/27$ plays a crucial role. Note that $-27D$ is what in algebra is called the discriminant. If D is positive then all three roots can be found by taking the square root of a positive number followed by the cube root of a real number. If D is negative it looks as if we are forced to find the cube root of a complex, non-real number.

(a) Show that $D = 0$ if and only if the graph $y = x^3 + px + q$ is tangent to the x-axis.

(b) Show, by examining the graph for example, that $D > 0$ if and only if there is one real root and two complex, non-real roots.

The third case, $D < 0$, is the case when there are three distinct real roots. This was known as the *casus irreducibilis*. It can be shown that in this case there is in general no way to express the roots using only real radicals (that is, square roots, cube roots etc. of real numbers).

12. We study the trigonometric solution of the cubic equation $x^3 + px + q = 0$ with real coefficients p and q. In the case $D < 0$, (see the previous exercise), there are three real roots and they cannot be found using Cardano's method without taking the cube root of a non-real number. The roots can be found

more easily by exploiting the triplication rule for the cosine function, $\cos 3\theta = 4\cos^3\theta - 3\cos\theta$. In this exercise we assume that $D < 0$. Note that this implies that $p < 0$.

(a) Let $x = \lambda\cos\theta$. Show that x satisfies the cubic equation if

$$\lambda = \sqrt{-4p/3} \quad \text{and} \quad \cos 3\theta = -4q/\lambda^3.$$

Check that $|4q/\lambda^3| < 1$.

(b) Taking any θ that satisfies the second equation, show that the three real solutions of the cubic equation are

$$\lambda\cos\theta, \quad \lambda\cos\left(\theta + \frac{2\pi}{3}\right), \quad \lambda\cos\left(\theta + \frac{4\pi}{3}\right).$$

13. In this series of exercises we look at the Chebyshev polynomials.

(a) Show that for all natural numbers n there exist polynomials $T_n(x)$ and $U_n(x)$, each of degree n, that satisfy

$$T_n(\cos\theta) = \cos n\theta, \qquad U_n(\cos\theta)\sin\theta = \sin(n+1)\theta$$

for all θ, and find explicit expressions for them.

Hint. Use de Moivre's theorem.

(b) Show that in the interval $[-1, 1]$ both T_n and U_n have n distinct roots.

(c) Show that the polynomial T_n satisfies $|T_n(x)| \leq 1$ in the interval $[-1, 1]$, and that it attains the value 1 at n points in $[-1, 1]$.

(d) You may not think the explicit formulas (see item (a)) very appealing as ways to calculate T_n and U_n. However, it is easy to compute them recursively. Show that both sequences of polynomials satisfy the same recurrence relation

$$T_{n+2}(x) = 2x\,T_{n+1}(x) - T_n(x), \quad U_{n+2}(x) = 2x\,U_{n+1}(x) - U_n(x)$$

though with different initial conditions.

Chapter 10
Complex Sequences and Series

I mean the word proof not in the sense of the lawyers, who set two half proofs equal to a whole one, but in the sense of a mathematician, where half proof = 0, and it is demanded for proof that every doubt becomes impossible.

C. F. Gauss

The main task of this chapter is to extend the theory of real sequences and real series to complex sequences and complex series. In Chap. 3 we dealt predominantly with positive series and stopped short of saying anything useful about real series that were not positive. This shortcoming will be amended here.

10.1 The Limit of a Complex Sequence

Let $(z_n)_{n=1}^{\infty}$ be a sequence of complex numbers and let w be a complex number.

Definition The number w is said to be the limit of the sequence $(z_n)_{n=1}^{\infty}$ if the following condition is satisfied:

For every $\varepsilon > 0$, there exists a natural number N, such that $|z_n - w| < \varepsilon$ for all $n \geq N$.

Formally this definition is the same as that for convergence of real sequences; only that the absolute value is reinterpreted as the modulus of a complex number.

A complex sequence $(z_n)_{n=1}^{\infty}$ is said to be convergent if there exists some w, such that $\lim_{n\to\infty} z_n = w$. If there is no such w then the sequence is said to be divergent. We will not make any use of infinite limits in the complex realm.

R. Magnus, *Fundamental Mathematical Analysis*, Springer Undergraduate Mathematics Series, https://doi.org/10.1007/978-3-030-46321-2_10

Let $z_n = a_n + b_n i$ and $w = s + ti$ where a_n, b_n, s and t are real numbers. That is, $a_n = \operatorname{Re} z_n, b_n = \operatorname{Im} z_n, s = \operatorname{Re} w$ and $t = \operatorname{Im} w$. By the rules for complex numbers listed in 9.1 we have

$$|a_n - s| \leq |z_n - w| \quad \text{and} \quad |b_n - t| \leq |z_n - w|,$$

so that if $\lim_{n \to \infty} z_n = w$ then

$$\lim_{n \to \infty} a_n = s \quad \text{and} \quad \lim_{n \to \infty} b_n = t.$$

But the converse of this last statement is also true because

$$|z_n - w|^2 = |a_n - s|^2 + |b_n - t|^2.$$

We have proved the following.

Proposition 10.1 *A complex sequence* $(z_n)_{n=1}^{\infty}$ *satisfies* $\lim_{n \to \infty} z_n = w$ *if and only if* $\lim_{n \to \infty} \operatorname{Re} z_n = \operatorname{Re} w$ *and* $\lim_{n \to \infty} \operatorname{Im} z_n = \operatorname{Im} w$.

This also shows that if a complex sequence has a limit, then the limit is unique (because that is known for real sequences). Another important conclusion is as follows.

Proposition 10.2 *Let* $\lim_{n \to \infty} z_n = w$. *Then* $\lim_{n \to \infty} |z_n| = |w|$.

Proof It follows from the inequality $\big| |z| - |w| \big| \leq |z - w|$ stated in Proposition 9.2.
□

Cauchy's principle of convergence for real sequences (Proposition 3.12) extends almost without change to complex sequences.

Proposition 10.3 (Cauchy's convergence principle) *A complex sequence* $(z_n)_{n=1}^{\infty}$ *is convergent if and only if it satisfies Cauchy's condition: for each* $\varepsilon > 0$ *there exists* N, *such that* $|z_n - z_m| < \varepsilon$ *for all* n *and* m *that satisfy* $n \geq N$ *and* $m \geq N$.

Proof Let $a_n = \operatorname{Re} z_n$ and $b_n = \operatorname{Im} z_n$. Since $(z_n)_{n=1}^{\infty}$ is convergent if and only if $(a_n)_{n=1}^{\infty}$ and $(b_n)_{n=1}^{\infty}$ are both convergent real sequences, it suffices to show that $(z_n)_{n=1}^{\infty}$ satisfies Cauchy's condition for complex sequences if and only if $(a_n)_{n=1}^{\infty}$ and $(b_n)_{n=1}^{\infty}$ satisfy Cauchy's condition for real sequences. But that is obvious in virtue of the inequalities (all of which appeared in Sect. 9.1):

$$|a_n - a_m| \leq |z_n - z_m|, \qquad |b_n - b_m| \leq |z_n - z_m|$$

$$|z_n - z_m| \leq \sqrt{2} \max(|a_n - a_m|, |b_n - b_m|).$$

□

10.2 Complex Series

A series of complex numbers $\sum_{k=1}^{\infty} z_k$ is said to be convergent, and its sum is the complex number w, when the sequence $(s_n)_{n=1}^{\infty}$, given by $s_n = \sum_{k=1}^{n} z_k$, is convergent and $\lim_{k \to \infty} s_n = w$.

Proposition 10.4 *A complex series $\sum_{k=1}^{\infty} z_k$ is convergent if and only if it satisfies the following condition: for each $\varepsilon > 0$ there exists N, such that for all m and n that satisfy $N \leq m \leq n$ we have*

$$\left| \sum_{k=m}^{n} z_k \right| < \varepsilon.$$

Proof Since $s_n - s_{m-1} = \sum_{k=m}^{n} z_k$, the condition of the proposition is equivalent to Cauchy's condition (Proposition 10.3) applied to the sequence $(s_n)_{n=1}^{\infty}$. $\qquad\square$

The condition of the proposition may also be written as follows: for each $\varepsilon > 0$ there exists N, such that for all m that satisfy $m \geq N$, and for every natural number p, we have $\left| \sum_{k=m}^{m+p} a_k \right| < \varepsilon$. This throws into relief the point that there is no upper limit placed on the separation of m and n.

10.2.1 Absolutely Convergent Series

Proposition 10.5 *Let $\sum_{k=1}^{\infty} z_k$ be a complex series and assume that the positive series $\sum_{k=1}^{\infty} |z_k|$ is convergent. Then the series $\sum_{k=1}^{\infty} z_k$ is convergent and we have*

$$\left| \sum_{k=1}^{\infty} z_k \right| \leq \sum_{k=1}^{\infty} |z_k|.$$

Proof Let $\varepsilon > 0$. Choose N, such that for all m and n that satisfy $N \leq m \leq n$ we have $\sum_{k=m}^{n} |z_k| < \varepsilon$. Now if $N \leq m \leq n$ we have

$$\left| \sum_{k=m}^{n} z_k \right| < \sum_{k=m}^{n} |z_k| < \varepsilon$$

and so, by Proposition 10.4, the series $\sum_{k=m}^{n} z_k$ is convergent.
Moreover, we have

$$\left| \sum_{k=1}^{n} z_k \right| \leq \sum_{k=1}^{n} |z_k| \leq \sum_{k=1}^{\infty} |z_k|,$$

and therefore going to the limit and applying Proposition 10.2, we obtain

$$\left| \sum_{k=1}^{\infty} z_k \right| = \left| \lim_{n \to \infty} \sum_{k=1}^{n} z_k \right| = \lim_{n \to \infty} \left| \sum_{k=1}^{n} z_k \right| \leq \sum_{k=1}^{\infty} |z_k|.$$

$\square$

A series $\sum_{k=1}^{\infty} z_k$, which is such that the positive series $\sum_{k=1}^{\infty} |z_k|$ is convergent, is said to be *absolutely convergent*. As we have just seen, it is then convergent. The inequality in Proposition 10.5 may be viewed as an infinite version of the triangle inequality.

10.2.2 Cauchy's Root Test

Convergence tests for positive series can be applied to the series of moduli of a complex series; they are therefore also tests for absolute convergence of complex series. The most important of these tests were covered in Chap. 3. We include another here, Cauchy's test, also known as the root test.

Proposition 10.6 *Let* $\sum_{k=1}^{\infty} a_k$ *be a positive series and suppose that the limit* $\lim_{n \to \infty} a_n^{1/n}$ *exists and equals* t. *The following then hold:*

(1) If $t < 1$ *the series is convergent.*
(2) If $t > 1$ *the series is divergent.*

There is no conclusion if $t = 1$.

Proof If $t < 1$ we choose s, such that $t < s < 1$, and choose N, such that $a_n^{1/n} < s$ for all $n \geq N$. Then we have $a_n < s^n$ for all $n \geq N$, and the series is convergent by comparison with the geometric series $\sum s^n$. If $t > 1$ then $a_n^{1/n} > 1$ when n is sufficiently high, and then $a_n > 1$ and cannot tend to 0. The series $\sum_{k=1}^{\infty} a_k$ then diverges.

The lack of a conclusion if $t = 1$ is illustrated by the series $\sum_{n=1}^{\infty} n^{-1}$ and $\sum_{n=1}^{\infty} n^{-2}$. $\square$

We showed in Proposition 3.25 (in the nugget on limits inferior and superior), that if $\lim_{n \to \infty} a_{n+1}/a_n = t$ (with all denominators strictly positive), then we also have $\lim_{n \to \infty} a_n^{1/n} = t$. Hence, if a conclusion can be obtained from the ratio test, it can also be obtained from Cauchy's test. However, there are cases when Cauchy's test works, but the ratio test gives no conclusion.

10.2.3 Extended Forms of the Ratio and Cauchy's Tests

In this section the series $\sum_{n=1}^{\infty} a_n$ is a positive series. The results are therefore applicable to proving that a complex series is absolutely convergent. We list some versions, occasionally useful, of the ratio test and Cauchy's root test with weaker conditions than the usual ones. They are weaker in the sense that they do not require a limit, only an inequality. The proofs, all easy, are left to the reader. Some of them can be expressed using limit superior; it is left to the reader to see how.

Ratio Test

(a) Assume that $a_n \neq 0$ for all n. Instead of requiring that $\lim_{n\to\infty} a_{n+1}/a_n = t$ and $t < 1$, it suffices for convergence to assume there exist N and $s < 1$, such that $a_{n+1}/a_n < s$ for all $n \geq N$.
(b) Assume that $a_n \neq 0$ for all n. Instead of requiring that $\lim_{n\to\infty} a_{n+1}/a_n = t$ and $t > 1$, it suffices for divergence to assume that there exists N, such that $a_{n+1}/a_n \geq 1$ for all $n \geq N$.

Root Test

(c) Instead of requiring $\lim_{n\to\infty} a_n^{1/n} = t$ and $t < 1$, it suffices for convergence to assume that there exist N and $s < 1$, such that $a_n^{1/n} < s$ for all $n \geq N$.
(d) Instead of requiring $\lim_{n\to\infty} a_n^{1/n} = t > 1$, it suffices for divergence to assume that $a_n^{1/n} \geq 1$ for infinitely many n.

10.2.4 Conditional Convergence: Leibniz's Test

A series that is convergent but not absolutely convergent is said to be *conditionally convergent* . An example of such a series is

$$1 - \frac{1}{2} + \frac{1}{3} - \frac{1}{4} + \frac{1}{5} - \cdots = \sum_{n=1}^{\infty} \frac{(-1)^{n-1}}{n}$$

the sum of which is ln 2, as we shall see. This is an example of an *alternating series*, meaning that the terms are alternately positive and negative.

The following test is called Leibniz's test, or, the alternating series test.

Proposition 10.7 *Let $(a_n)_{n=1}^{\infty}$ be a sequence of positive numbers, that are decreasing and tend to 0. The following conclusions hold:*

(1) The series $\sum_{n=1}^{\infty}(-1)^{n-1}a_n$ is convergent.
(2) Let $s_n = \sum_{k=1}^{n}(-1)^{k-1}a_k$ and $s = \sum_{k=1}^{\infty}(-1)^{k-1}a_k$. Then s_{2n-1} tends to s from above and s_{2n} tends to s from below.
(3) We have the error estimate $|s - s_n| < a_{n+1}$.

Proof The reader is invited to supply the proof, by induction, that

$$s_2 < s_4 < \cdots < s_{2n} < s_{2n-1} < \cdots < s_3 < s_1 \qquad (n = 1, 2, 3, \ldots).$$

It follows that the sequence s_{2n} is increasing and bounded above, whilst the sequence s_{2n-1} is decreasing and bounded below. We conclude that both sequences are convergent. But $s_{2n-1} - s_{2n} = a_{2n} \to 0$, so that both have the same limit s, which is then the limit $\lim_{n\to\infty} s_n$. This proves conclusion 1.

We next note that

$$s_{2n} < s < s_{2n+1} < s_{2n-1},$$

which gives

$$s - s_{2n} < s_{2n+1} - s_{2n} = a_{2n+1}$$

and

$$s_{2n-1} - s < s_{2n-1} - s_{2n} = a_{2n}.$$

This proves conclusions 2 and 3.   □

Now we are going to study the series $\sum_{n=1}^{\infty} (-1)^{n-1}/n$ which, as we said before, has the sum $\ln 2$. We are going to change the order of the terms.

We separate the terms of the series into two sequences: the positive terms form the sequence $(1/(2n-1))_{n=1}^{\infty}$, and the negative terms the sequence $(-1/2n)_{n=1}^{\infty}$. From these two sequences we shall build a new series that has exactly the same terms as the original series, but presented in a different order.

We take the first positive term, then the first and second negative terms, then the next positive term, then the next two negative terms and so on, always taking one positive term followed by two negative ones. Proceeding rather recklessly in the spirit of the mathematicians of the eighteenth century, we write this down and calculate:

$$1 - \frac{1}{2} - \frac{1}{4} + \frac{1}{3} - \frac{1}{6} - \frac{1}{8} + \frac{1}{5} - \frac{1}{10} - \frac{1}{12} + \frac{1}{7} - \frac{1}{14} + \cdots$$

$$= \left(1 - \frac{1}{2}\right) - \frac{1}{4} + \left(\frac{1}{3} - \frac{1}{6}\right) - \frac{1}{8} + \left(\frac{1}{5} - \frac{1}{10}\right) - \frac{1}{12} + \left(\frac{1}{7} - \frac{1}{14}\right) + \cdots$$

$$= \frac{1}{2} - \frac{1}{4} + \frac{1}{6} - \frac{1}{8} + \frac{1}{10} - \frac{1}{12} + \frac{1}{14} - \cdots$$

$$= \frac{\ln 2}{2}.$$

$$(10.1)$$

The striking conclusion that we wish to draw is that the sum of the series in line 1 of (10.1) is different from that of the series $\sum_{n=1}^{\infty} (-1)^{n-1}/n$, although both series have the same terms, but presented in a different order. However, some doubt may linger over the validity of the first equals sign. We have not shown that the series in line 1 is convergent. Although the series in line 2 is convergent (being $\sum_{n=1}^{\infty} (-1)^{n-1}/2n$), it is not the same series as in line 1.

A more rigorous argument might proceed as follows. Let s_n be the sum of the first n terms of the series in line 1 of (10.1). Taking three terms at a time we have

$$s_{3n} = \frac{1}{2} - \frac{1}{4} + \frac{1}{6} - \frac{1}{8} + \cdots + \frac{1}{4n-2} - \frac{1}{4n} = \frac{1}{2} \sum_{k=1}^{2n} \frac{(-1)^{k-1}}{k}.$$

Hence $s_{3n} \to \ln 2/2$. Now we note that the nth term of the series in line 1 tends to zero as n tends to infinity. Hence we conclude, by an easy argument left to the reader, that $s_n \to \ln 2/2$.

It may seem a contradiction; by changing the order of the terms we obtain half the original sum. But it simply means that the commutative rule that holds for finite sums does not hold for infinite ones; and, after all, an infinite sum is really a limit. For absolutely convergent series things are much tamer.

10.2.5 Rearrangements of Absolutely Convergent Series

A permutation of a set A is a bijective mapping $\phi : A \to A$. It is often convenient to picture a permutation as a table. For example

$$\begin{array}{c|ccccccc} n & 1, & 2 & 3 & 4 & 5 & 6 & 7 \\ \hline \phi(n) & 2 & 7 & 1 & 3 & 5 & 4 & 6 \end{array}$$

describes a permutation of the set $\{1, 2, 3, 4, 5, 6, 7\}$. An infinite example might be

$$\begin{array}{c|ccccccccccc} n & 1 & 2 & 3 & 4 & 5 & 6 & 7 & 8 & 9 & 10 & 11 & \dots \\ \hline \phi(n) & 1 & 2 & 4 & 3 & 6 & 8 & 5 & 10 & 12 & 7 & 14 & \dots \end{array}$$

This is a permutation of $\mathbb{N}_+$. Is it obvious how to go on?

In fact a permutation of $\mathbb{N}_+$ may be viewed as a sequence of positive integers in which every positive integer appears exactly once. To describe one a table may be impracticable. In some cases a formula for $\phi(n)$ may be available. However, it may not always be convenient to specify $\phi(n)$ by a formula; a verbal description may be deemed enough to define it. The main thing is to ensure that every number in the upper row of the table occurs exactly once in the lower row.

A permutation ϕ of the infinite set $\mathbb{N}_+$ produces a so-called *rearrangement* of the series $\sum_{n=1}^{\infty} a_n$. This is the new series $\sum_{n=1}^{\infty} a_{\phi(n)}$.

Proposition 10.8 *Let $\sum_{n=1}^{\infty} a_n$ be an absolutely convergent series of real or complex terms. Let ϕ be a permutation of $\mathbb{N}_+$. Then the rearranged series $\sum_{n=1}^{\infty} a_{\phi(n)}$ is absolutely convergent and $\sum_{n=1}^{\infty} a_{\phi(n)} = \sum_{n=1}^{\infty} a_n$.*

Proof Let $\varepsilon > 0$. Write $b_n = a_{\phi(n)}$. By Cauchy's principle, there exists N, such that $\sum_{k=n}^{m} |a_k| < \varepsilon$ for all m and n that satisfy $N \le m \le n$. More than this, because the terms are positive, the sum of a finite number of terms $|a_k|$, all of which have place numbers $k \ge N$, is less than ε. The place numbers here do not have to be consecutive.

There exists N_1, such that the numbers $\phi(1), ..., \phi(N_1)$ include all the numbers $1, ..., N$. This is so, because, as we recall, every number in the first row of the table representing the permutation appears exactly once in the second. If $N_1 \leq m \leq n$ then all the numbers $\phi(m), \phi(m + 1), ..., \phi(n)$ are higher than or equal to N. But then we have

$$\sum_{k=m}^{n} |b_k| = \sum_{k=m}^{n} |a_{\phi(k)}| < \varepsilon.$$

We conclude, by Cauchy's principle, that the series $\sum_{k=1}^{\infty} |b_k|$ is convergent.

Let $t = \sum_{n=1}^{\infty} a_n$ and $s = \sum_{n=1}^{\infty} b_n$. We wish to show that $s = t$. Let $\varepsilon > 0$. Choose N and N_1 as we did above. If $n \geq N_1$ then the numbers $\phi(1), ..., \phi(n)$ include all the numbers $1, ..., N$, and so

$$\left| \sum_{k=1}^{n} b_k - \sum_{k=1}^{N} a_k \right| = \left| \sum_{k=1}^{n} a_{\phi(k)} - \sum_{k=1}^{N} a_k \right| < \varepsilon,$$

as all the terms in the sum $\sum_{k=1}^{N} a_k$ get cancelled. Letting $n \to \infty$ we conclude that $|s - \sum_{k=1}^{N} a_k| \leq \varepsilon$. But now we find, applying the triangle inequality:

$$|s - t| = \left| s - \sum_{k=1}^{\infty} a_k \right| \leq \left| s - \sum_{k=1}^{N} a_k \right| + \left| \sum_{k=N+1}^{\infty} a_k \right| \leq 2\varepsilon.$$

Since this holds for all $\varepsilon > 0$ we must have $s = t$. $\square$

10.2.6 Exercises

1. Test the following series for convergence. In each case determine whether the series is absolutely convergent, conditionally convergent or divergent.

   (a) $\displaystyle\sum_{n=1}^{\infty} \frac{(-1)^n}{\sqrt{n}}$

   (b) $\displaystyle\sum_{n=1}^{\infty} \frac{(-1)^n n(n + 1)}{(n + 2)(n + 3)}$

   (c) $\displaystyle\sum_{n=1}^{\infty} \frac{(-1)^n n(n + 1)}{(n + 2)(n + 3)(n + 4)}$

   (d) $\displaystyle\sum_{n=1}^{\infty} \frac{(-1)^n n(n + 1)}{(n + 2)(n + 3)(n + 4)(n + 5)}$

   (e) $\displaystyle\sum_{n=1}^{\infty} \frac{(-1)^n (n + a)(n + b)}{(n + c)(n + d)(n + e)},$

where a, b, c, d and e are real numbers and none of c, d or e is a negative integer (but e is not necessarily the base of the natural logarithm).

(f) $\displaystyle\sum_{n=1}^{\infty} \frac{n+a}{(n+b)(n+c)(n+d)}$,

where a, b, c and d are non-real *complex numbers*.

(g) $\displaystyle\sum_{n=1}^{\infty} \frac{1}{n+i}$

(h) $\displaystyle\sum_{n=1}^{\infty} \frac{n+a}{(n+b)(n+c)}$,

where a, b and c are non-real *complex numbers*.

2. Test the following series for convergence. In each case determine whether the series is absolutely convergent, conditionally convergent or divergent.

(a) $\displaystyle\sum_{n=1}^{\infty} (-a_n)^n$,

where $a_n = \frac{1}{2}$ if n is even and $a_n = \frac{1}{3}$ if n is odd.

Note. The ratio of each term to its predecessor is unbounded; so the ratio test does not work.

(b) $\displaystyle\sum_{n=1}^{\infty} \frac{(-1)^n \ln n}{n}$

(c) $\displaystyle\sum_{n=1}^{\infty} \frac{\ln n}{n^2}$

(d) $\displaystyle\sum_{n=1}^{\infty} \frac{n!}{n^n}$

(e) $\displaystyle\sum_{n=1}^{\infty} \left(1 - \cos\left(\frac{1}{n}\right)\right)$

(f) $\displaystyle\sum_{n=1}^{\infty} \frac{a^n}{b^n + c^n}$,

where a, b and c are positive.

(g) $\displaystyle\sum_{n=1}^{\infty} \frac{(-1)^n}{n} \left(1 + \frac{1}{2} + \frac{1}{3} + \cdots + \frac{1}{n}\right)$.

3. Compute the limit

$$\lim_{n \to \infty} \sum_{k=-n}^{n} \frac{1}{k+i}.$$

4. Let $(a_n)_{n=1}^{\infty}$ and $(b_n)_{n=1}^{\infty}$ be sequences of complex numbers.

   (a) Suppose that the series $\sum_{n=1}^{\infty} |a_n|^2$ and $\sum_{n=1}^{\infty} |b_n|^2$ are convergent. Show that the series $\sum_{n=1}^{\infty} a_n b_n$ is absolutely convergent.
   (b) More generally, let $p > 1$ and $q > 1$ and suppose that $(1/p) + (1/q) = 1$. Suppose that the series $\sum_{n=1}^{\infty} |a_n|^p$ and $\sum_{n=1}^{\infty} |b_n|^q$ are convergent. Show that the series $\sum_{n=1}^{\infty} a_n b_n$ is absolutely convergent.

5. Let $(a_n)_{n=1}^{\infty}$ be a convergent sequence of complex numbers and let $\lim_{n\to\infty} a_n = w$. Show that every rearrangement of the sequence has the same limit. More precisely, if $\phi : \mathbb{N}_+ \to \mathbb{N}_+$ is a bijection then $\lim_{n\to\infty} a_{\phi(n)} = w$.
6. Let $(a_n)_{n=1}^{\infty}$ be a bounded complex sequence and let $\sum_{n=1}^{\infty} b_n$ be an absolutely convergent complex series. Show that the series $\sum_{n=1}^{\infty} a_n b_n$ is absolutely convergent.

10.3   Product of Series

Given two sequences $(a_n)_{n=0}^{\infty}$ and $(b_n)_{n=0}^{\infty}$ we can consider the set of all products of the form $a_n b_m$. The reason for beginning at place number $n = 0$, instead of $n = 1$ as before, is that this topic has important applications to power series, to be considered in Chap. 11.

The set of products $a_n b_m$ do not form a sequence as they stand; they constitute a family of elements indexed by the set of all pairs (n, m) of natural numbers. It is most natural to see this family as an infinite two-dimensional array, as pictured here:

$$a_0 b_0 \quad a_0 b_1 \quad a_0 b_2 \quad a_0 b_3 \; \dots$$

$$a_1 b_0 \quad a_1 b_1 \quad a_1 b_2 \quad a_1 b_3 \; \dots$$

$$a_2 b_0 \quad a_2 b_1 \quad a_2 b_2 \quad a_2 b_3 \; \dots$$

$$a_3 b_0 \quad a_3 b_1 \quad a_3 b_2 \quad a_3 b_3 \; \dots$$

$$\vdots \qquad \vdots \qquad \vdots \qquad \vdots$$

There are many ways to arrange the elements of the array as a sequence indexed as usual by the natural numbers. We can walk through the array taking in each element, a bit like walking through a large shopping mall and visiting every shop. The following diagram shows one way to do this:

$$a_0b_0 \rightarrow a_0b_1 \qquad a_0b_2 \rightarrow a_0b_3 \ \ldots$$
$$\downarrow \qquad\qquad \uparrow \qquad\qquad \downarrow$$
$$a_1b_0 \leftarrow a_1b_1 \qquad a_1b_2 \qquad a_1b_3 \ \ldots$$
$$\downarrow \qquad\qquad\qquad \uparrow \qquad\qquad \downarrow$$
$$a_2b_0 \rightarrow a_2b_1 \rightarrow a_2b_2 \qquad a_2b_3 \ \ldots$$
$$\downarrow$$
$$a_3b_0 \leftarrow a_3b_1 \leftarrow a_3b_2 \leftarrow a_3b_3 \ \ldots$$
$$\downarrow \qquad \vdots \qquad \vdots \qquad \vdots$$

To interpret the diagram you must follow the arrows. The effect is to take in square blocks of the array; a typical block consists of all terms $a_n b_m$ for which $\max(n, m) \leq N$, say, and is completed each time you visit the upper or left-hand edge of the array.

Another way through the array, which turns out to be very important, is shown in the next diagram:

$$a_0b_0 \rightarrow a_0b_1 \qquad a_0b_2 \rightarrow a_0b_3 \ \ldots$$
$$\swarrow \qquad \nearrow \qquad \swarrow \qquad \nearrow$$
$$a_1b_0 \qquad a_1b_1 \qquad a_1b_2 \qquad a_1b_3 \ \ldots$$
$$\downarrow \ \nearrow \qquad \swarrow \qquad \nearrow \qquad \swarrow$$
$$a_2b_0 \qquad a_2b_1 \qquad a_2b_2 \qquad a_2b_3 \ \ldots$$
$$\swarrow \qquad \nearrow \qquad \swarrow \qquad \nearrow$$
$$a_3b_0 \qquad a_3b_1 \qquad a_3b_2 \qquad a_3b_3 \ \ldots$$
$$\downarrow \ \nearrow \ \vdots \ \swarrow \ \vdots \ \nearrow \ \vdots \ \swarrow$$

The procedure indicated is to collect whole diagonals. A diagonal consists of the $n + 1$ terms of the form $a_k b_{n-k}$ for some n. These are assembled in increasing order, starting at $n = 0$, then $n = 1, 2, 3$, etc.

These are just the two most important ways to arrange the elements $a_n b_m$ in a simple sequence. There are clearly infinitely many ways to do it. In the following proposition we make a remarkable claim that applies to any possible arrangement.

Proposition 10.9 *Assume that the series $\sum_{n=0}^{\infty} a_n$ and $\sum_{n=0}^{\infty} b_n$ are absolutely convergent. Set $A = \sum_{n=0}^{\infty} a_n$ and $B = \sum_{n=0}^{\infty} b_n$. Consider an arrangement of the products $a_n b_m$, $(m \in \mathbb{N}, n \in \mathbb{N})$ in a sequence $(d_n)_{n=0}^{\infty}$. Then the series $\sum_{n=0}^{\infty} d_n$ is absolutely convergent and its sum is AB.*

Proof Given the natural number n there exists a natural number N, such that all the products $d_0, d_1, ..., d_n$ appear among the products that arise by multiplying out the expression

$$(a_0 + a_1 + \cdots + a_N)(b_0 + b_1 + \cdots + b_N).$$

But then we have

$$|d_0| + |d_1| + \cdots + |d_n| \leq (|a_0| + |a_1| + \cdots + |a_N|)(|b_0| + |b_1| + \cdots + |b_N|)$$

$$\leq \left(\sum_{k=0}^{\infty} |a_k| \right) \left(\sum_{k=0}^{\infty} |b_k| \right).$$

We conclude that the series $\sum_{n=0}^{\infty} d_n$ is absolutely convergent. Its sum is therefore independent of the order in which the terms are arranged.

One possible arrangement is

$$a_0 b_0 + a_0 b_1 + a_1 b_1 + a_1 b_0 + a_0 b_2 + a_1 b_2 + a_2 b_2 + a_2 b_1 + a_2 b_0$$
$$+ a_0 b_3 + a_1 b_3 + a_2 b_3 + a_3 b_3 + a_3 b_2 + a_3 b_1 + a_3 b_0 + \cdots$$
$$= \overbrace{a_0 b_0}^{e_0} + \overbrace{(a_0 b_1 + a_1 b_1 + a_1 b_0)}^{e_1} + \overbrace{(a_0 b_2 + a_1 b_2 + a_2 b_2 + a_2 b_1 + a_2 b_0)}^{e_2}$$
$$+ \overbrace{(a_0 b_3 + a_1 b_3 + a_2 b_3 + a_3 b_3 + a_3 b_2 + a_3 b_1 + a_3 b_0)}^{e_3} + \cdots \qquad (10.2)$$

Beneath "e_n" appear all products in which one of the factors is a_n or b_n. It corresponds to the collection-by-blocks arrangement pictured in the first diagram above.

Now we have

$$e_0 + e_1 + \cdots + e_n = (a_0 + a_1 + \cdots + a_n)(b_0 + b_1 + \cdots + b_n),$$

which tends to AB as $n \to \infty$. The sequence of partial sums of the series $\sum_{n=0}^{\infty} e_n$ is a subsequence of the sequence of partial sums of the series in the first line of (10.2). Since the latter series is known to be convergent we conclude that its sum is AB. The sum of the arrangement $\sum_{n=0}^{\infty} d_n$ is therefore also AB. $\qquad\square$

10.3.1   Cauchy Product of Series

By far the most important arrangement of the products $a_n b_m$ is by diagonals as shown in the second diagram. This leads to the series:

$$a_0 b_0 + a_0 b_1 + a_1 b_0 + a_0 b_2 + a_1 b_1 + a_2 b_0 + a_0 b_3 + a_1 b_2 + a_2 b_1 + a_3 b_0 + \cdots$$
$$= \overbrace{a_0 b_0}^{c_0} + \overbrace{a_0 b_1 + a_1 b_0}^{c_1} + \overbrace{a_0 b_2 + a_1 b_1 + a_2 b_0}^{c_2}$$
$$+ \overbrace{a_0 b_3 + a_1 b_2 + a_2 b_1 + a_3 b_0}^{c_3} + \cdots$$

Here we have

$$c_n = \sum_{i+j=n} a_i b_j = \sum_{k=0}^{n} a_k b_{n-k}.$$

The series $\sum_{n=0}^{\infty} c_n$ is called the Cauchy product. By Proposition 10.9, if $\sum_{n=0}^{\infty} a_n$ and $\sum_{n=0}^{\infty} b_n$ are both absolutely convergent, we can conclude that

$$\sum_{n=0}^{\infty} c_n = \left(\sum_{n=0}^{\infty} a_n \right) \left(\sum_{n=0}^{\infty} b_n \right).$$

10.3.2 Exercises

1. The series $\sum_{n=0}^{\infty} (-1)^n / \sqrt{n+1}$ converges by Leibniz's test. Let its Cauchy product with itself be the series $\sum_{n=0}^{\infty} c_n$. Show that $|c_n| \geq 1$, so that $\sum_{n=0}^{\infty} c_n$ diverges.
2. There are more ways to collect the pairs (n, m) into a simple sequence than the two covered in the text. For example, we can group together all pairs (n, m) for which n and m have a given product. This is like a Cauchy product, but uses the product of place numbers instead of their sum.
   Let the series $\sum_{n=1}^{\infty} a_n$ and $\sum_{n=1}^{\infty} b_n$ be absolutely convergent, and let their sums be A and B, respectively. For each n we can take all pairs of place numbers $(d, n/d)$ where d ranges over all the divisors of n (including 1 and n). This gives the series $\sum_{n=1}^{\infty} c_n$ where

$$c_n = \sum_{d \mid n} a_d b_{n/d}.$$

   Here, $d \mid n$ means that d divides n, and instructs us to sum over all divisors of n. By Proposition 10.9 the series $\sum_{n=1}^{\infty} c_n$ is convergent with sum AB.
   The Riemann zeta function $\zeta(s)$ is defined for $s > 1$ by

$$\zeta(s) = \sum_{n=1}^{\infty} \frac{1}{n^s}.$$

   Show that

$$\left(\zeta(s) \right)^2 = \sum_{n=1}^{\infty} \frac{\sigma(n)}{n^s}$$

   where $\sigma(n)$ is the *divisor function*, that is, $\sigma(n)$ is the number of positive integral divisors of n, including 1 and n.
3. Prove Mertens' theorem. Let the series $\sum_{n=0}^{\infty} a_n$ and $\sum_{n=0}^{\infty} b_n$ be convergent, let $\sum_{n=0}^{\infty} a_n = A$ and $\sum_{n=0}^{\infty} b_n = B$. Assume that one of the series, let us say the first, is absolutely convergent. Then the Cauchy product is convergent and its sum is AB.
   Hint. Let

$$A_n = \sum_{k=0}^{n} a_k, \quad B_n = \sum_{k=0}^{n} b_k, \quad c_n = \sum_{k=0}^{n} a_k b_{n-k}, \quad C_n = \sum_{k=0}^{n} c_k.$$

Show that $C_n = \sum_{k=0}^{n} a_k B_{n-k}$ and use this to estimate $C_n - B \sum_{k=0}^{n} a_k$ for large n.

4. In several places we have had occasions to sum a series by grouping its terms. This appeared in the proof of Proposition 10.9 and in our study of rearranging the series $\sum_{n=1}^{\infty} (-1)^{n-1}/n$ in Sect. 10.2. It was also used in the treatment of the positive series $\sum_{n=1}^{\infty} n^{-p}$ (see Sect. 3.8).

In this exercise we shall define and study a general version of grouping. Begin with the series $\sum_{n=1}^{\infty} a_n$. We do not assume that it is convergent. Let $(k_n)_{n=1}^{\infty}$ be a strictly increasing sequence of positive integers, beginning with $k_1 = 1$. We replace the series $\sum_{n=1}^{\infty} a_n$ by the grouped series $\sum_{n=1}^{\infty} b_n$, where

$$b_n = \sum_{j=k_n}^{k_{n+1}-1} a_j.$$

(a) Suppose that the series $\sum_{n=1}^{\infty} a_n$ is convergent. Show that the grouped series $\sum_{n=1}^{\infty} b_n$ is convergent and has the same sum.

(b) Suppose that $a_n \to 0$ and that the sequence $k_{n+1} - k_n$ is bounded above (in other words there is a cap on the number of terms that are grouped together). Show that the series $\sum_{n=1}^{\infty} a_n$ is convergent if and only if the grouped series $\sum_{n=1}^{\infty} b_n$ is convergent.

(c) Suppose that $a_n \geq 0$ for each n. Show that the series $\sum_{n=1}^{\infty} a_n$ is convergent if and only if the grouped series $\sum_{n=1}^{\infty} b_n$ is convergent.

(d) Give an example where the series $\sum_{n=1}^{\infty} a_n$ is divergent but the grouped series $\sum_{n=1}^{\infty} b_n$ is convergent.

10.4 ($\Diamond$) Riemann's Rearrangement Theorem

Riemann proved a striking theorem that throws light on our experiments on rearranging the series $\sum_{n=1}^{\infty} (-1)^{n-1}/n$.

Proposition 10.10 *Let the real number series $\sum_{n=1}^{\infty} a_n$ be conditionally convergent and let t be either a real number, or else $+\infty$ or $-\infty$. Then there is a rearrangement $\sum_{n=1}^{\infty} a_{\phi(n)}$ that has the sum t.*

Proof It is important that the terms are real numbers. For each real number x, let $x^+ = \max(x, 0)$ and $x^- = -\min(x, 0)$. Then $x = x^+ - x^-$ and $|x| = x^+ + x^-$. Now the series $\sum |a_n|$ is divergent (because $\sum a_n$ is *conditionally* convergent) and $|a_n| = a_n^+ + a_n^-$. Hence it is impossible for both the series, $\sum_{n=1}^{\infty} a_n^+$ and $\sum_{n=1}^{\infty} a_n^-$, to be convergent. But if one is convergent so is the other, since $\sum_{n=1}^{\infty} (a_n^+ - a_n^-)$ is convergent. We conclude therefore that both the series $\sum_{n=1}^{\infty} a_n^+$ and $\sum_{n=1}^{\infty} a_n^-$ are divergent.

From this point we will assume that no term in the sequence a_n is 0 (we can strike out all the zeros and put them back later if we so wish). We form two increasing sequences of integers, k_n and ℓ_n, which are such that the terms a_{k_n} are all the positive terms of the sequence $(a_n)_{n=1}^{\infty}$ in order of increasing index; and the terms a_{ℓ_n} are all the negative terms of the sequence $(a_n)_{n=1}^{\infty}$, again in order of increasing index. Both these sequences can be defined formally by induction if we so wish.

Now set $c_n = a_{k_n}$ and $d_n = a_{\ell_n}$. It is helpful to imagine these two sequences spread out before us from left to right like cards from a pack:

$$c_1, c_2, c_3, c_4, c_5, c_6, \ldots \qquad d_1, d_2, d_3, d_4, d_5, d_6, \ldots$$

The first sequence comprises all the positive terms, and the second all negative terms, of the sequence $(a_n)_{n=1}^{\infty}$, taken in order of increasing n. We know from the first paragraph that $\sum_{n=1}^{\infty} c_n = \infty$ and $\sum_{n=1}^{\infty} d_n = -\infty$. But also that $\lim_{n\to\infty} c_n = \lim_{n\to\infty} d_n = 0$, as follows from the fact that $\sum_{n=1}^{\infty} a_n$ is convergent.

We now describe a rearrangement that sums to t. We take the case that t is a finite number, leaving the two cases of infinite t to the reader. Starting with the left-hand sequence c_n (of positive terms) we take as many terms from left to right as are needed to make a sum higher than t, stopping as soon as t is surpassed. Note that we want to go higher than t; if we land on t we take an additional term. This can be carried out because $\sum_{n=1}^{\infty} c_n = \infty$, so any number can be surpassed by a sum $\sum_{n=1}^{N} c_n$ if we take N large enough.

Let the last term taken from the left-hand sequence be c_{p_1}. It is clear that the error, that is the amount by which $\sum_{n=1}^{p_1} c_n$ exceeds t, is at most c_{p_1}. We continue the sequence $c_1, \ldots, c_{p_1}$ with as many terms from the right-hand sequence as are needed to make, together with the positive terms already taken, a sum lower than t. Again this is feasible because $\sum_{n=1}^{\infty} d_n = -\infty$. Again we take only as many as are needed stopping as soon as the sum passes below t (if we land on t we take an additional term).

Suppose the last term taken to be d_{q_1}. Then the sum so far,

$$c_1 + \cdots + c_{p_1} + d_1 + \cdots + d_{q_1}$$

differs from t by at most $|d_{q_1}|$. We return to the left-hand sequence from where we left off and take just enough positive terms to surpass t again, say from c_{p_1+1} to c_{p_2}, then just enough negative terms from the right-hand sequence to pass below t again, say from d_{q_1+1} to d_{q_2} and so on. This process will never terminate, because all terms are non-zero and the series $\sum c_n$ and $\sum d_n$ both diverge, the first to $+\infty$ and the second to $-\infty$. So in this fashion we construct a rearrangement $\sum a_{\phi(n)}$ of $\sum a_n$.

The rearrangement sums to t. To see this observe first that if all the terms of a_n satisfy $|a_n| < M$, then the error, the absolute difference between a partial sum of our rearrangement and t, is less than M as soon as we reach the partial sum $\sum_{n=1}^{p_1} c_n$ as described above. It remains less than M from there on.

Now let $\varepsilon > 0$. From some point on, from an index r say, all the terms c_n and d_n with $n \geq r$ have absolute value less than ε. Eventually in our rearrangement we will have exhausted all terms to the left of c_r and d_r. After this all unused terms have absolute value less than ε; so by the argument of the last paragraph, the error will drop below ε after a finite number of terms are added on, and will remain below ε. In other words our rearrangement sums to t. $\square$

10.4.1 Exercises

1. A proof like the one above that contains so much written text can leave a lingering doubt about its correctness. Dispel some of these doubts by writing a nice definition by induction of the sequences k_n, ℓ_n, p_n, q_n and the rearrangement mapping $\phi(n)$. You may wish to use Proposition 2.1.
2. Prove Riemann's theorem for the cases $t = \infty$ and $t = -\infty$.
3. Show that a complex series $\sum_{n=1}^{\infty} z_n$ is absolutely convergent if and only if the real series $\sum_{n=1}^{\infty} \operatorname{Re} z_n$ and $\sum_{n=1}^{\infty} \operatorname{Im} z_n$ are absolutely convergent. Hence show that if a complex series $\sum_{n=1}^{\infty} z_n$ has the property that all its rearrangements are convergent, then the series is absolutely convergent and all its rearrangements have the same sum.

10.4.2 Pointers to Further Study

$\rightarrow$ Series in Banach spaces
$\rightarrow$ Orthogonal series

10.5 ($\Diamond$) Gauss's Test

When the ratio test is used to test a positive series $\sum_{n=1}^{\infty} a_n$ for convergence, it often happens that no conclusion is obtained because $\lim_{n\to\infty} a_{n+1}/a_n = 1$. A more delicate test is required. It is a trade secret that it is best to try Gauss's test.

Proposition 10.11 (Gauss's test) *Let* $\sum_{n=1}^{\infty} a_n$ *be a positive series, such that* $a_n \neq 0$ *for all n. Assume that*

$$\frac{a_{n+1}}{a_n} = 1 - \frac{\mu}{n} + \frac{K_n}{n^{1+r}}$$

where $(K_n)_{n=1}^{\infty}$ *is a bounded sequence,* μ *a constant, and r a strictly positive constant. Then the series* $\sum_{n=1}^{\infty} a_n$ *converges if* $\mu > 1$ *and diverges if* $\mu \leq 1$.

Here are some interesting points about Gauss's test:

(a) There is no indeterminate case. Once μ is found a conclusion is reached.
(b) The equation is often written as

$$\frac{a_{n+1}}{a_n} = 1 - \frac{\mu}{n} + O\left(\frac{1}{n^{1+r}}\right).$$

The sequence c_n is said to be $O(d_n)$ (and we write $c_n = O(d_n)$) if c_n/d_n is bounded as $n \to \infty$. See the nugget "Asymptotic orders of magnitude" for more about this notation.

(c) If there exists a function $g(x)$, such that $g(0) = 1$, g is twice differentiable at $x = 0$, and $a_{n+1}/a_n = g(1/n)$, then $\mu = -g'(0)$.

(d) It is commonly the case that a_{n+1}/a_n is a rational function of n and then μ is easily found by point (c). This explains the great utility of Gauss's test and was the historical context for introducing it.

The proof of Gauss's test is in several steps, in each of which another test is introduced.

Step 1. Kummer's tests.

Let $(D_n)_{n=1}^{\infty}$ be a positive sequence and set

$$\phi(n) = D_n - D_{n+1}\frac{a_{n+1}}{a_n}.$$

Then

(1) *Suppose there exists $h > 0$ and N, such that $\phi(n) > h$ for all $n \geq N$. Then $\sum_{n=1}^{\infty} a_n$ converges.*

(2) *Suppose $\sum_{n=1}^{\infty} D_n^{-1}$ is divergent and $\phi(n) \leq 0$ for all $n \geq N$. Then $\sum_{n=1}^{\infty} a_n$ diverges.*

Proof of Kummer's Tests (1) We have that

$$D_k a_k - D_{k+1}a_{k+1} = a_k\phi(k) \geq ha_k, \quad (k \geq N).$$

Sum from $k = N$ to $k = n$:

$$D_N a_N - D_n a_n \geq h\sum_{k=N}^{n} a_k, \quad (n \geq N).$$

But then $\sum_{k=N}^{n} a_k \leq D_N a_N/h$ and it is bounded above as $n \to \infty$.

(2) We have that $D_k a_k \leq D_{k+1}a_{k+1}$ for $k \geq N$, so that $D_k a_k \geq D_N a_N$ and $a_k \geq D_N a_N D_k^{-1}$. But then $\sum_{n=1}^{\infty} a_n$ diverges since $\sum_{n=1}^{\infty} D_n^{-1}$ diverges. □

Step 2. There are as many Kummer's tests as there are ways to select the numbers D_n. Let us look at some examples:

(a) $D_n = 1$ for all n.
Conclusions. If

$$1 - \frac{a_{n+1}}{a_n} > h > 0$$

for $n \geq N$, then the series converges. If

$$1 - \frac{a_{n+1}}{a_n} \leq 0$$

for $n \geq N$, then the series diverges. This is D'Alembert's test (ratio test) in a slightly more general form.

(b) $D_n = n - 1$.
Conclusions. If

$$n\left(1 - \frac{a_{n+1}}{a_n}\right) > 1 + h$$

for $n \geq N$, then the series converges. If

$$n\left(1 - \frac{a_{n+1}}{a_n}\right) \leq 1$$

for $n \geq N$ then the series diverges. This is called Raabe's test.

Step 3. Completion of the proof of Gauss's test.
The condition that the series satisfies implies that

$$n\left(1 - \frac{a_{n+1}}{a_n}\right) \to \mu.$$

By Raabe's test ((b) of step 2) the series converges if $\mu > 1$ and diverges if $\mu < 1$.

Finally we consider the case $\mu = 1$. So far Gauss's test is just a special case of Raabe's test. It is because Gauss's test resolves the case $\mu = 1$ that it merits its special status. We apply Kummer's test with $D_n = (n-1)\ln(n-1)$. We know that $\sum_{n=3}^{\infty} D_n^{-1}$ diverges (or if not known it can be seen by Cauchy's condensation test, Sect. 3.8 Exercise 12, for example). We have

$$\phi(n) = D_n - D_{n+1}\frac{a_{n+1}}{a_n}$$

$$= (n-1)\ln(n-1) - n(\ln n)\left(1 - \frac{1}{n} + \frac{K_n}{n^{1+r}}\right)$$

$$= (n-1)\ln\left(1 - \frac{1}{n}\right) - \frac{K_n \ln n}{n^r}.$$

Now K_n is bounded and $\lim_{n\to\infty} \ln n/n^r = 0$, so that $\lim_{n\to\infty} K_n \ln n/n^r = 0$. We also have

$$\lim_{n\to\infty} (n-1) \ln \left(1 - \frac{1}{n}\right) = \lim_{t\to 0+} \left(\frac{1}{t} - 1\right) \ln(1-t) = \lim_{t\to 0+} (1-t) \frac{\ln(1-t)}{t} = -1.$$

We find that $\lim_{n\to\infty} \phi(n) = -1$, so that $\phi(n) \le 0$ when n is sufficiently big. We conclude that $\sum_{n=1}^{\infty} a_n$ diverges. □

10.5.1 Exercises

1. Verify point (c). Let $\sum a_n$ be a positive series. Suppose there exists a function $g(x)$, such that $g(0) = 1$, g is twice differentiable at $x = 0$ and $a_{n+1}/a_n = g(1/n)$. Show that Gauss's test can be applied with $\mu = -g'(0)$ and $r = 1$.
2. Use Gauss's test to study the series $\sum_{n=1}^{\infty} n^{-p}$ where p is a real number.
3. Use Gauss's test to study the series

$$\sum_{n=0}^{\infty} \frac{(a)_n (b)_n}{(c)_n (d)_n}$$

where, for a real number t and natural number n, we define

$$(t)_n = \begin{cases} 1 & \text{if } n = 0 \\ t(t+1)...(t+n-1) & \text{if } n \ge 1. \end{cases}$$

We assume that neither c nor d is a non-positive integer in order to avoid zero denominators. Careful! The series terminates for certain values of a and b.

10.5.2 Pointers to Further Study

→ Convergence tests.

Chapter 11
Function Sequences and Function Series

I shall apply all my strength to bring more light into the tremendous obscurity which one unquestionably finds in analysis. It lacks so completely all plan and system that it is peculiar that so many have studied it. The worst of it is, it has never been treated stringently. There are very few theorems in advanced analysis which have been demonstrated in a logically tenable manner.

N. H. Abel

11.1 Problems with Convergence

Consider the following example of a function series:

$$\sum_{n=1}^{\infty} \left(\frac{x}{x+1} \right)^n, \quad (0 \le x \le 1).$$

With a function series, as here, it is important to specify carefully the domain of the functions in the series. It should be the same for all the terms. In this case it is the interval $[0, 1]$. If we fix a value for x within the interval $[0, 1]$, we obtain a number series, in fact, a geometric series which is convergent, having the sum

$$\left(\frac{x}{x+1} \right) \cdot \frac{1}{1 - \left(\frac{x}{x+1} \right)} = x.$$

The function series is convergent for each x in the interval $[0, 1]$ and its sum is the function x.

R. Magnus, *Fundamental Mathematical Analysis*, Springer Undergraduate Mathematics Series, https://doi.org/10.1007/978-3-030-46321-2_11

Consider next the function series:

$$\sum_{n=1}^{\infty} \frac{x}{(x+1)^{n-1}}, \quad (0 \le x \le 1).$$

For each x in $[0, 1]$, such that $x > 0$, this is a geometric series with ratio $1/(1 + x)$, and the first term is x. This geometric series is convergent, since we have $1/(1 + x) < 1$, and its sum is

$$x \cdot \frac{1}{1 - \frac{1}{1+x}} = x \cdot \frac{1+x}{x} = 1 + x.$$

But if we try to put $x = 0$ in the above equalities the first two members make no sense. However, there is an obvious limit as x tends to 0, namely, 1. On the other hand it is also obvious that if we put $x = 0$ in the terms of the series we obtain a series in which every term is 0. The sum is therefore 0. So the sum of the function series is a discontinuous function on the interval $[0, 1]$, in spite of the fact that each term is a continuous function on the same interval.

Of course the convergence of a series depends on the convergence of a sequence of partial sums. So the phenomenon exhibited here should first be studied in the case of function sequences.

Consider then the function sequence

$$f_n(x) = x^n, \quad (0 \le x \le 1), \quad n = 1, 2, 3, \ldots$$

Computing the limit for each x in the domain, we obtain the function $g : [0, 1] \to \mathbb{R}$, where $g(x) = 0$ for $0 \le x < 1$ and $g(1) = 1$. Again we have a discontinuous limit of a sequence of continuous functions.

In order to understand better what is happening here, let us fix x in the domain, such that $x < 1$. Now $\lim_{n \to \infty} x^n = 0$. Let $\varepsilon > 0$. We ask: how large must we choose N in order that $|x^n - g(x)| < \varepsilon$ for all $n \ge N$? Here $g(x) = 0$. The obvious answer is: it suffices if N exceeds $|\ln \varepsilon|/|\ln x|$.

It is interesting how the lowest N found in the previous paragraph depends on x. Let us ask: can we choose N independent of x, such that $|x^n - g(x)| < \varepsilon$ for all $n \ge N$ and x in the domain? Can we use the same N for all x and achieve an error less than ε? The answer is no. We saw that the lowest available N for a given x must exceed $|\ln \varepsilon|/|\ln x|$. But now $\lim_{x \to 1-} |\ln \varepsilon|/|\ln x| = \infty$. Ever larger values of N are required as x approaches 1. On the other hand, when $x = 1$ it suffices to choose $N = 1$.

The phenomenon studied in the last paragraph, and pictured in Fig. 11.1, is particularly sensitive to the choice of domain. On the domain $[0, 1 - \delta]$ (we fix $\delta > 0$) we can find N, that suffices, independent of x, namely, we can take N as the smallest integer that exceeds $|\ln \varepsilon|/|\ln(1 - \delta)|$ and achieve an error less than ε. For $n \ge N$ and for all x in this *restricted domain* we have $|x^n - g(x)| < \varepsilon$.

Fig. 11.1 Non-uniform convergence

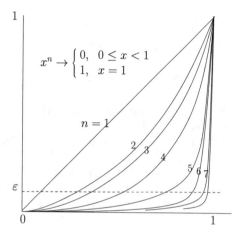

$$x^n \rightarrow \begin{cases} 0, & 0 \le x < 1 \\ 1, & x = 1 \end{cases}$$

Fig. 11.2 Uniform convergence

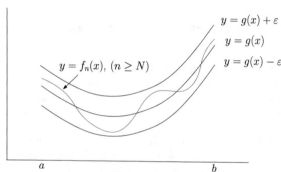

11.2 Pointwise Convergence and Uniform Convergence

Let $(f_n)_{n=1}^{\infty}$ be a sequence of functions defined in a common domain A.

Definition The function sequence $(f_n)_{n=1}^{\infty}$ is said to converge pointwise to a function g in the domain A if, for each x in A, we have $\lim_{n \to \infty} f_n(x) = g(x)$.

Definition The function sequence $(f_n)_{n=1}^{\infty}$ is said to converge uniformly to a function g in the domain A (or sometimes "with respect to A") if the following condition is satisfied: for each $\varepsilon > 0$ there exists a natural number N, such that for all x in A and for all $n \ge N$ we have $|f_n(x) - g(x)| < \varepsilon$.

Obviously if f_n converges uniformly to g, it also converges pointwise to the same limit. Uniform convergence is stronger in that we require that N should be specifiable, for each given ε, independently of x in the domain. Uniform convergence is illustrated in Fig. 11.2.

The difference is apparent in the corresponding statements in quantifier logic, which eliminate all the ambiguity that may reside in everyday language. The order of the quantifiers is the only difference. First, f_n converges to g pointwise:

$$(\forall x \in A)(\forall \varepsilon > 0)(\exists N \in \mathbb{N})(\forall n \in \mathbb{N})(n \geq N \Rightarrow |f_n(x) - g(x)| < \varepsilon).$$

Next, f_n converges to g uniformly:

$$(\forall \varepsilon > 0)(\exists N \in \mathbb{N})(\forall n \in \mathbb{N})(\forall x \in A)(n \geq N \Rightarrow |f_n(x) - g(x)| < \varepsilon).$$

The reader should practise reading aloud these two sentences in a literal translation to ordinary speech.

Changing the domain may make a difference. The function sequence $(x^n)_{n=1}^{\infty}$ converges pointwise in the domain $[0, 1]$. The convergence is not uniform. The same function sequence, but in the domain $[0, 1 - \delta]$, converges uniformly, as we saw in the last section. If there is some ambiguity about the domain in question we may say, "The sequence $(f_n)_{n=1}^{\infty}$ converges to g, uniformly with respect to the domain B".

11.2.1 Cauchy's Principle for Uniform Convergence

Just as Cauchy's principle for real number sequences gives a necessary and sufficient condition for convergence without needing a candidate for the limit, for function sequences there is a Cauchy's principle for uniform convergence, that does not require us to guess the limit in advance.

Proposition 11.1 Let $(f_n)_{n=1}^{\infty}$ be a sequence of functions with common domain A. The following condition, called Cauchy's condition, is necessary and sufficient for uniform convergence of the sequence $(f_n)_{n=1}^{\infty}$: for all $\varepsilon > 0$ there exists N, such that for all $n \geq N$, $m \geq N$ and x in A we have $|f_m(x) - f_n(x)| < \varepsilon$.

Proof Suppose the function sequence is uniformly convergent and let the function g be its limit. Let $\varepsilon > 0$. There exists N, such that for all $n \geq N$ and all x in A we have $|f_n(x) - g(x)| < \varepsilon/2$. Now for all $n \geq N$, $m \geq N$ and x in A we have

$$|f_m(x) - f_n(x)| < |f_m(x) - f(x)| + |f(x) - f_n(x)| < \frac{\varepsilon}{2} + \frac{\varepsilon}{2} = \varepsilon.$$

Conversely, suppose that the function sequence $(f_n)_{n=1}^{\infty}$ satisfies Cauchy's condition. Now for each fixed x the numerical sequence $(f_n(x))_{n=1}^{\infty}$ satisfies Cauchy's condition for a real sequence, and hence is convergent as a sequence of real numbers. Let its limit be the number $g(x)$. This defines a function g with domain A. We shall show that f_n converges uniformly to g.

Let $\varepsilon > 0$. There exists N, such that for all $m \geq N$, $n \geq N$ and x in A we have $|f_m(x) - f_n(x)| < \varepsilon$. We may let m tend to infinity in this inequality and deduce that $|g(x) - f_n(x)| \leq \varepsilon$, and this therefore holds for all $n \geq N$ and all x in A. This shows that f_n converges uniformly to g. $\square$

11.2.2   Uniform Convergence and Continuity

There follows a key reason why uniform convergence is so important. The proof is the *locus classicus* for what is called the $\varepsilon/3$-argument.

Proposition 11.2 *Let $(f_n)_{n=1}^{\infty}$ be a sequence of continuous functions on the domain A. Assume that the sequence converges uniformly on the domain A to a function g. Then g is continuous in the domain A.*

Proof Let $c \in A$. We shall show that g is continuous at c. Let $\varepsilon > 0$. There exists N, such that $|f_n(x) - g(x)| < \varepsilon/3$ for all $n \geq N$ and all $x \in A$. The function f_N is continuous by assumption. Hence there exists $\delta > 0$, such that $|f_N(x) - f_N(c)| < \varepsilon/3$ for all x in A that satisfy $|x - c| < \delta$. We now find, if x is in A and $|x - c| < \delta$, that

$$|g(x) - g(c)| \leq |g(x) - f_N(x)| + |f_N(x) - f_N(c)| + |f_N(c) - g(c)|$$
$$< \frac{\varepsilon}{3} + \frac{\varepsilon}{3} + \frac{\varepsilon}{3} = \varepsilon.$$

$\square$

The proof actually shows that g is continuous at c, given only that, firstly, each function f_n is continuous at c, and secondly, f_n converges uniformly to g in an open interval containing c.

11.2.3   Uniform Convergence of Series

Let $\sum_{n=1}^{\infty} f_n$ be a function series, such that each term has the same domain A.

Definition The series $\sum_{n=1}^{\infty} f_n$ is pointwise convergent on the domain A and its sum is the function g, if, for each x in A, we have $\lim_{n \to \infty} \sum_{k=1}^{n} f_k(x) = g(x)$.

Definition The series $\sum_{n=1}^{\infty} f_n$ is uniformly convergent on the domain A and its sum is the function g, if $\lim_{n \to \infty} \sum_{k=1}^{n} f_k(x) = g(x)$ uniformly with respect to x in A.

We may be able to infer about a series $\sum_{n=0}^{\infty} f_n(x)$, that there exists a function g, such that the series converges uniformly to g, and yet we may not know anything about g. So to eliminate mentioning g at all, we simply say "The series $\sum_{k=1}^{\infty} f_k(x)$ is uniformly convergent".

The importance of uniform convergence of a function series is obvious: if each term is a continuous function, so is the sum function. For this to be useful we need a convenient test for uniform convergence of a series. Fortunately we have one.

Proposition 11.3 (Weierstrass M-test) *Let $\sum_{n=1}^{\infty} f_n$ be a function series where the terms have a common domain A. Assume that there exists a sequence of positive* *numbers* $(M_n)_{n=1}^\infty$, *such that the series* $\sum_{n=1}^\infty M_n$ *is convergent, and such that for all* x *in A and for all n we have* $|f_n(x)| \le M_n$. *Then the series* $\sum_{n=1}^\infty f_n$ *is uniformly convergent.*

Proof The numerical series $\sum_{n=1}^\infty f_n(x)$ is convergent (indeed absolutely convergent) for each x in A by the comparison test. The sum is then a function g with domain A.

So far, the convergence to g is only pointwise. Let $\varepsilon > 0$. Since the series $\sum_{n=1}^\infty M_n$ is convergent, there exists N, such that $\sum_{k=N}^\infty M_k < \varepsilon$. It follows, for all $n \ge N$ and all $x \in A$, that

$$\left| g(x) - \sum_{k=1}^n f_k(x) \right| = \left| \sum_{k=n+1}^\infty f_k(x) \right| \le \sum_{k=n+1}^\infty |f_k(x)| \le \sum_{k=n+1}^\infty M_k \le \varepsilon.$$

The convergence is therefore uniform.   □

11.2.4 Cauchy's Principle for Uniform Convergence of Function Series

We can apply Cauchy's principle for uniform convergence of function sequences to the study of the function series $\sum_{n=1}^\infty f_n(x)$, with common domain A. We find that the series is uniformly convergent if and only if it satisfies the condition: for all $\varepsilon > 0$ there exists N, such that for all m and n that satisfy $n \ge m \ge N$, and all x in A, we have

$$\left| \sum_{k=n+1}^m f_k(x) \right| < \varepsilon.$$

Weierstrass's M-test supposes that the terms $f_n(x)$ are bounded in modulus by constants $M_n > 0$, such that $\sum_{n=1}^\infty M_n < \infty$. Then we have, on choosing N so that $\sum_{k=N}^\infty M_k < \varepsilon$, that

$$\left| \sum_{k=n+1}^m f_k(x) \right| < \sum_{k=n+1}^m |f_k(x)| < \sum_{k=n+1}^m M_k < \varepsilon,$$

for all x in A, provided only that $m \ge n \ge N$. This is another proof of Weierstrass's test, appealing to Cauchy's principle, but the perspicacious reader will see that it is really the same as the first one.

The Weierstrass M-test can only prove uniform convergence of a function series if, for each x, it is an absolutely convergent numerical series. To prove that a function series, that is conditionally convergent for certain values of x, is uniformly convergent can be trickier, but Cauchy's principle can be a valuable tool as we shall see.

11.2.5 Integration and Uniform Convergence

Let $(f_n)_{n=1}^{\infty}$ be a sequence of functions integrable on the interval $[a, b]$. If f_n converges pointwise to a function g on $[a, b]$, we cannot in general infer that

$$\lim_{n \to \infty} \int_a^b f_n = \int_a^b \lim_{n \to \infty} f_n = \int_a^b g \,.$$

The first equality, involving the interchange of limit and integral, may be inadmissible. In fact the function g may even fail to be integrable. And even if it is integrable it is not guaranteed that equality holds.

We ask: when is it permissible to interchange the operations of limit and integral? We have an extremely useful sufficient condition.

Proposition 11.4 *Let* $(f_n)_{n=1}^{\infty}$ *be a sequence of continuous functions on the interval* $[a, b]$ *and assume that* $\lim_{n \to \infty} f_n = g$ *uniformly on* $[a, b]$. *Then* g *is integrable and*

$$\lim_{n \to \infty} \int_a^b f_n = \int_a^b g.$$

Proof In the first place g is continuous and therefore integrable on $[a, b]$. Let $\varepsilon > 0$. We choose N, such that $|f_n(x) - g(x)| < \varepsilon/(b - a)$ for all $n \geq N$ and all x in $[a, b]$. For such n we find

$$\left| \int_a^b f_n - \int_a^b g \right| \leq \int_a^b |f_n - g| < (b - a) \frac{\varepsilon}{b - a} = \varepsilon.$$

That is, $\lim_{n \to \infty} \int_a^b f_n = \int_a^b g$. $\qquad\qquad\square$

And its counterpart for series, for which the proof should be obvious.

Proposition 11.5 *Let* $\sum_{n=1}^{\infty} f_n$ *be a function series, where each term is a continuous function on the domain* $[a, b]$. *Assume that the series* $\sum_{n=1}^{\infty} f_n$ *is uniformly convergent on the domain* $[a, b]$ *and let its sum be the function* g. *Then*

$$\int_a^b g = \sum_{n=1}^{\infty} \int_a^b f_n.$$

The proposition is sometimes loosely described in the following way: it is permissible to integrate a function series term-by-term when the series in question is uniformly convergent. It is surprising how often one wants to integrate a function series term-by-term, so we are very glad to have this proposition.

It also introduces a seminal theme. Often it happens, just when we wish to integrate term-by-term, that uniform convergence is wanting. More flexible criteria for allowing this do exist for the Riemann integral, but it is preferable to adopt a more advanced integration theory, to wit, the Lebesgue integral, that allows for term-by-term integration under much weaker conditions.

11.2.6   Differentiation and Uniform Convergence of Series

After term-by-term integration we turn to term-by-term differentiation. First we look at the interchange of differentiation and limit for function sequences. This is a trifle more complicated than for integration.

Proposition 11.6 *Let* $(f_n)_{n=1}^\infty$ *be a sequence of differentiable functions on a common open interval A such that the derivatives* f_n' *are all continuous. Suppose that there exist functions g and h with domain A, such that for each x in A we have*

$$\lim_{n\to\infty} f_n(x) = g(x), \quad and \quad \lim_{n\to\infty} f_n'(x) = h(x).$$

Assume that the second limit is uniform with respect to A. Then g is differentiable in A and $g' = h$.

Proof The function h is continuous by Proposition 11.2. Fix a in A. By the previous proposition and the fundamental theorem we find, for all $x \in A$, that

$$\int_a^x h = \lim_{n\to\infty} \int_a^x f_n' = \lim_{n\to\infty} \left(f_n(x) - f_n(a) \right) = g(x) - g(a).$$

But then $g' = h$ by the fundamental theorem.   □

Now for series we have the following:

Proposition 11.7 *Let* $\sum_{n=1}^\infty f_n$ *be a function series on the open interval A, such that each function* f_n *is differentiable and the derivative* f_n' *is continuous. Suppose that the series* $\sum_{n=1}^\infty f_n(x)$ *and* $\sum_{n=1}^\infty f_n'(x)$ *are convergent for each* $x \in A$ *and set*

$$g(x) = \sum_{n=1}^\infty f_n(x), \quad h(x) = \sum_{n=1}^\infty f_n'(x), \quad (x \in A).$$

Assume that the second series is uniformly convergent. Then g is differentiable in A and $g' = h$.

Proof Fix $a \in A$. For all $x \in A$ we have

$$g(x) - g(a) = \sum_{n=1}^\infty \left(f_n(x) - f_n(a) \right) = \sum_{n=1}^\infty \int_a^x f_n' = \int_a^x h$$

and the conclusion follows by the fundamental theorem.   □

11.2.7 Exercises

In problems where the uniform convergence of a function sequence $(f_n)_{n=1}^{\infty}$ is to be assessed on an interval A, it may be straightforward to determine the pointwise limit g. It can then be useful, in order to decide about uniform convergence, to determine the maximum of $|f_n - g|$ in A. Sketching the graph of f_n and g can also be helpful to orient one's thinking.

1. Determine whether the following limits exist, and are attained uniformly or merely pointwise, on the stated intervals A:

   (a) $\displaystyle\lim_{n\to\infty} \frac{1}{nx+1}$, $A = [0, 1]$

   (b) $\displaystyle\lim_{n\to\infty} \frac{1}{nx+1}$, $A = [0, 1]$

   (c) $\displaystyle\lim_{n\to\infty} \frac{1}{nx+1}$, $A = [\delta, 1]$, where $0 < \delta < 1$.

   (d) $\displaystyle\lim_{n\to\infty} \frac{x}{nx+1}$, $A = [0, 1]$.

2. Determine whether the following limits exist, and are attained uniformly or merely pointwise, on the stated intervals A:

   (a) $\displaystyle\lim_{n\to\infty} nx(1-x)^n$, $A = [0, 1]$

   (b) $\displaystyle\lim_{n\to\infty} x^n(1-x^n)$, $A = [0, 1]$

   (c) $\displaystyle\lim_{n\to\infty} e^{-nx^2}$, $A = \mathbb{R}$

   (d) $\displaystyle\lim_{n\to\infty} \frac{1}{n}e^{-nx^2}$, $A = \mathbb{R}$

   (e) $\displaystyle\lim_{n\to\infty} \frac{x}{nx+1}$, $A = [0, 1]$

   (f) $\displaystyle\lim_{n\to\infty} xe^{-nx}$, $A = [0, \infty[$

   (g) $\displaystyle\lim_{n\to\infty} xe^{-nx}$, $A = [\delta, \infty[$, where $\delta > 0$.

   (h) $\displaystyle\lim_{n\to\infty} \sqrt{x^2 + \frac{1}{n^2}}$, $A = \mathbb{R}$.

3. The period of a pendulum of length ℓ swinging in a uniform gravitational field of strength (acceleration) g is given by

$$T(k) = 4\sqrt{\frac{\ell}{g}} \int_0^{\pi/2} \frac{1}{\sqrt{1 - k^2 \sin^2\phi}} \, d\phi$$

where $k = \sin(\theta_0/2)$ and θ_0 is the angular amplitude of the swing. Prove that

$$\lim_{k\to 0} T(k) = 2\pi\sqrt{\frac{\ell}{g}}.$$

Note. The limit is called the period of small oscillations and is the familiar approximate formula for the period of swing given in books on elementary mechanics. For large values of k, the approximation may be too inaccurate and then one would want to calculate the integral. We have here another of Legendre's standard elliptic integrals. For a surprising method to approximate it see Exercise 5 below.

4. Let $(a_n)_{n=1}^\infty$ and $(b_n)_{n=1}^\infty$ be positive sequences and suppose that

$$\lim_{n\to\infty} a_n = \lim_{n\to\infty} b_n = 1.$$

Show that

$$\lim_{n\to\infty} \frac{1}{\sqrt{a_n \sin^2 t + b_n \cos^2 t}} = 1$$

and that the limit is attained uniformly for $0 \le t \le 2\pi$.
Deduce that

$$\lim_{n\to\infty} \int_0^{2\pi} \frac{1}{\sqrt{a_n \sin^2 t + b_n \cos^2 t}}\, dt = 2\pi.$$

5. The integral in the preceding exercise (actually another so-called elliptic integral) can be computed numerically by using the arithmetic-geometric mean (Sect. 3.4 Exercise 10 and Sect. 5.12 Exercise 6). This discovery is due to Gauss, who wrote down the change of variables in item (a) without giving the rather lengthy calculations needed, merely writing that if they are done correctly this is the result. Perseverance and a cool head are needed to do them correctly, but the payoff in item (b) is worth the effort.

   (a) Let a and b be distinct positive numbers and set

$$I(a,b) = \int_0^{\pi/2} \frac{1}{\sqrt{a^2 \cos^2 \theta + b^2 \sin^2 \theta}}\, d\theta.$$

   Carry out a change of variables, from θ to ϕ, where

$$\sin\theta = \frac{2a \sin\phi}{a + b + (a-b)\sin^2\phi}.$$

   Show that the change of variables leads to the conclusion

$$I(a,b) = I(a_1, b_1)$$

   where $a_1 = \frac{1}{2}(a+b)$ and $b_1 = \sqrt{ab}$.
   Hint. It helps to take it in steps, by verifying the following formulas:

   (i) $\cos\theta = \dfrac{2\cos\phi \left(a_1^2 \cos^2\phi + b_1^2 \sin^2\phi\right)^{1/2}}{a + b + (a-b)\sin^2\phi}$

(ii) $(a^2 \cos^2 \phi + b^2 \sin^2 \phi)^{1/2} = a \dfrac{a + b - (a - b) \sin^2 \phi}{a + b + (a - b) \sin^2 \phi}$

(iii) $\cos \theta \, d\theta = \dfrac{(2a \cos \phi)\left(a + b - (a - b) \sin^2 \phi\right)}{\left(a + b + (a - b) \sin^2 \phi\right)^2} \, d\phi.$

(b) Applying the results of Exercise 4 together with Sect. 3.4 Exercise 10, show that

$$I(a, b) = \frac{\pi}{2M(a, b)}$$

where $M(a, b)$ is the arithmetic geometric mean of a and b.

Note. This gives an efficient way to calculate the integral. To see why consult Sect. 5.12 Exercise 6.

(c) Show that the period of swing of the pendulum in Exercise 3 is given by

$$T(k) = \frac{2\pi}{M(1, \sqrt{1 - k^2})} \sqrt{\frac{\ell}{g}} = \frac{2\pi}{M(1, \cos(\theta_0/2))} \sqrt{\frac{\ell}{g}}.$$

6. Using Weierstrass's test on a function series $\sum_{n=1}^{\infty} f_n(x)$ is all about finding those constants M_n. A useful approach can be to set M_n equal to the maximum of $|f_n|$ in the interval A, if it exists and can be found.

   Determine whether the following series converge uniformly or merely pointwise, on the stated intervals A:

   (a) $\displaystyle\sum_{n=1}^{\infty} \frac{1}{n^2} \sin nx, \quad A = \mathbb{R}$

   (b) $\displaystyle\sum_{n=1}^{\infty} 2^n \sin\left(\frac{1}{3^n x}\right), \quad A =]0, \infty[$

   (c) $\displaystyle\sum_{n=1}^{\infty} 2^n \sin\left(\frac{1}{3^n x}\right), \quad A = [\delta, \infty[, \text{ where } \delta > 0.$

   (d) $\displaystyle\sum_{n=1}^{\infty} x^n e^{-nx^2}, \quad A = \mathbb{R}$

   (e) $\displaystyle\sum_{n=1}^{\infty} \frac{x}{n^p (1 + nx^2)}, \text{ where } p > \frac{1}{2}, \text{ and } A = \mathbb{R}.$

   Note. If $0 < p \le \frac{1}{2}$ the question of uniformity is trickier to resolve. The simplest way to examine this is by comparing the sum for a given x to an improper integral, a topic studied in Chap. 12.

   (f) $\displaystyle\sum_{n=1}^{\infty} \frac{x}{n^p (1 + nx^2)}, \text{ where } p > 0, \text{ and } A = [\delta, \infty[, \text{ where } \delta > 0.$

7. Express the notions of pointwise convergence and uniform convergence of a function sequence using set theory. More precisely let $(f_n)_{n=1}^{\infty}$ be a sequence

of functions whose common domain is a set A of real numbers, and let g be a function with domain A. For each pair (n, k) of positive integers we define a subset of A by the specification

$$B_{n,k} = \left\{ x \in A : |f_j(x) - g(x)| < 1/k \text{ for all } j \geq n \right\}.$$

(a) Show that for each k, the sequence of sets $(B_{n,k})_{n=1}^{\infty}$ is increasing, and for each n, the sequence of sets $(B_{n,k})_{k=1}^{\infty}$ is decreasing. In other words show that $B_{n,k} \subset B_{n+1,k}$ and $B_{n,k+1} \subset B_{n,k}$.
(b) Show that $f_n \to g$ pointwise if and only if for each k we have

$$\bigcup_{n=1}^{\infty} B_{n,k} = A.$$

(c) Show that $f_n \to g$ uniformly if and only if, for each k, there exists n, such that

$$B_{n,k} = A.$$

Note. Here, at last, we get to use a union of infinitely many sets. Precisely, the union of a sequence of sets $\bigcup_{n=1}^{\infty} C_n$ is the set of all x, such that there exists n, such that $x \in C_n$. The formulation of convergence of a function sequence described in this problem is the key to some important propositions; we can mention Dini's theorem and Egorov's theorem.

8. The assumptions of Proposition 11.6 can be weakened. Assume as in the proposition that $f_n'(x) \to h(x)$ uniformly with respect to A, but regarding the convergence of $f_n(x)$ assume only that there exists a point x_0 in A such that the numerical sequence $(f_n(x_0))_{n=1}^{\infty}$ converges. Prove that there exists a function g, such that $f_n(x) \to g(x)$ pointwise in A and $g'(x) = h(x)$ for all x in A.
   Hint. Show that for each x the sequence $(f_n(x) - f_n(x_0))_{n=1}^{\infty}$ satisfies Cauchy's condition.

11.3 Power Series

A power series is a function series of the form $\sum_{n=0}^{\infty} a_n x^n$. The numbers a_n are constants, called the coefficients of the power series. The terms $a_n x^n$ are meaningful if a_n and x are complex numbers. So in our initial study of power series we shall assume that this is the case, and write it as $\sum_{n=0}^{\infty} a_n z^n$, where z is a complex variable and the numbers a_n are complex coefficients.

The initial term $a_0 z^0$ is always interpreted as the constant function a_0 (this obviates the need to interpret 0^0). The translated series $\sum_{n=0}^{\infty} a_n (z - c)^n$ is called a power series with midpoint c.

11.3.1 Radius of Convergence

Given a power series the question of interest is: for which complex numbers z is it convergent?

Proposition 11.8 *If the power series $\sum_{n=0}^{\infty} a_n z^n$ is convergent for a given value $z = z_0 \neq 0$, then it is absolutely convergent for all complex z that satisfy $|z| < |z_0|$.*

Proof Since the terms of a convergent series are bounded, there exists $K > 0$, such that $|a_n z_0^n| < K$ for all n. Let $|z| < |z_0|$ and set $r = |z|/|z_0|$. Then $r < 1$ and

$$|a_n z^n| = |a_n z_0^n| \left| \frac{z}{z_0} \right|^n < K r^n.$$

The geometric series $\sum_{n=0}^{\infty} K r^n$ is convergent, and by the comparison test the series $\sum_{n=0}^{\infty} a_n z^n$ is also convergent, in fact absolutely convergent. $\quad\square$

Proposition 11.9 *For a given power series $\sum_{n=0}^{\infty} a_n z^n$, exactly one of the following is true:*

(1) $\displaystyle\sum_{n=0}^{\infty} a_n z^n$ *is convergent only if $z = 0$.*

(2) $\displaystyle\sum_{n=0}^{\infty} a_n z^n$ *is absolutely convergent for all z.*

(3) *There exists a real number $R > 0$, such that $\sum_{n=0}^{\infty} a_n z^n$ is absolutely convergent for all z that satisfy $|z| < R$, and divergent for all z that satisfy $|z| > R$.*

Proof If $\sum_{n=0}^{\infty} a_n z^n$ is convergent for all z then it is absolutely convergent for all z by Proposition 11.8. If it is convergent for some $z = z_0 \neq 0$ (which excludes case 1), and divergent for some $z = z_1$ (which excludes case 2), then we set

$$R = \sup \left\{ |z| : \sum_{n=0}^{\infty} a_n z^n \text{ is convergent} \right\}.$$

If $|z| < R$ then there exists z_0, such that $|z| < |z_0| < R$ and $\sum_{n=0}^{\infty} a_n z_0^n$ is convergent. But then $\sum_{n=0}^{\infty} a_n z^n$ is absolutely convergent. If $|z| > R$ then $\sum_{n=0}^{\infty} a_n z^n$ is divergent. $\quad\square$

The number R is called the *radius of convergence* of the power series. We extend the notion of radius of convergence to cases 1 and 2 as follows. If $\sum_{n=0}^{\infty} a_n z^n$ is convergent only for $z = 0$ then the radius of convergence is 0. If $\sum_{n=0}^{\infty} a_n z^n$ is convergent for all z then the radius of convergence is infinity, or symbolically $R = \infty$.

If z is on the circle of convergence $|z| = R$, there is no general conclusion about convergence. The series could be convergent for all z on the circle, divergent for all z on the circle, or convergent at some points and divergent at others.

An example of the last kind is the series $\sum_{n=1}^{\infty} \left((-1)^{n-1}/n\right) z^n$. This has radius of convergence 1. It is easy to see that it is convergent for $z = 1$ but divergent for $z = -1$. Actually, as we shall see later, it is convergent for all z on the unit circle, except for $z = -1$, a conclusion that requires a fairly delicate test (Dirichlet's test).

11.3.2 Determining the Radius of Convergence by the Ratio Test

In most practical cases, the radius of convergence can be found using the ratio test. We shall look at three examples, that adequately convey the method. In addition the conclusions stated will prove useful.

(a) The series $\displaystyle\sum_{n=0}^{\infty} nz^n$.

Here we have

$$\lim_{n\to\infty} \frac{|(n+1)z^{n+1}|}{|nz^n|} = \lim_{n\to\infty} \frac{n+1}{n}|z| = |z|.$$

We conclude that the series is convergent for $|z| < 1$ and divergent for $|z| > 1$. The radius of convergence is therefore 1.

(b) The series $\displaystyle\sum_{n=0}^{\infty} \frac{z^n}{n!}$.

We have

$$\lim_{n\to\infty} \left| \frac{z^{n+1}/(n+1)!}{z^n/n!} \right| = \lim_{n\to\infty} \frac{|z|}{n+1} = 0.$$

The series is convergent for every z. The radius of convergence is therefore infinity.

(c) The series $\displaystyle\sum_{n=0}^{\infty} n!\,z^n$.

In this case

$$\lim_{n\to\infty} \left| \frac{(n+1)!z^{n+1}}{n!z^n} \right| = \lim_{n\to\infty} (n+1)|z| = \infty$$

provided $z \neq 0$. The series is convergent only when $z = 0$. The radius of convergence is 0.

11.3.3 Uniform Convergence of Power Series

Now we restrict our study to real power series $\sum_{n=0}^{\infty} a_n x^n$, with real coefficients a_n and real variable x. We can still speak of the radius of convergence R. It is the radius of convergence of the complex series $\sum_{n=0}^{\infty} a_n z^n$.

Most of the material of this and subsequent sections carry over to the case of a complex variable, but require considerations of continuity and differentiability with respect to a complex variable that would carry us beyond the planned confines of this text.

Proposition 11.10 *Let R be the radius of convergence of the power series $\sum_{n=0}^{\infty} a_n x^n$. Let $[c_1, c_2]$ be a bounded and closed interval, such that $-R < c_1 < c_2 < R$. Then the function series $\sum_{n=0}^{\infty} a_n x^n$ is uniformly convergent on the domain $[c_1, c_2]$.*

Proof If R is finite we choose K so that $-R < -K < c_1 < c_2 < K < R$. If $R = \infty$ we choose K so that $-K < c_1 < c_2 < K$. For $x \in [c_1, c_2]$ we have $|x| < K$, and therefore

$$|a_n x^n| < |a_n| K^n.$$

Now the series $\sum_{n=0}^{\infty} |a_n| K^n$ is convergent since $K < R$. We conclude by the Weierstrass M-test that the series $\sum_{n=0}^{\infty} a_n x^n$ is uniformly convergent on the domain $[c_1, c_2]$. $\qquad \square$

Proposition 11.11 *Let $\sum_{n=0}^{\infty} a_n x^n$ be a power series with real coefficients and real variable x. Let the radius of convergence satisfy $R > 0$ (includes $R = \infty$). Let*

$$f(x) = \sum_{n=0}^{\infty} a_n x^n, \quad -R < x < R.$$

Then the function f is continuous in the interval $]-R, R[$.

Proof Let $-R < c_1 < c_2 < R$. The series is uniformly convergent on the domain $[c_1, c_2]$. We conclude that f is continuous in the interval $[c_1, c_2]$. But then f must be continuous in $]-R, R[$, since every x in $]-R, R[$ lies in some interval of the form $[c_1, c_2]$. $\qquad \square$

11.3.4 The Exponential Series

The series $\sum_{n=0}^{\infty} z^n/n!$ is called the exponential series. We saw that it has infinite radius of convergence. Let $f(z) = \sum_{n=0}^{\infty} z^n/n!$ for each $z \in \mathbb{C}$. This defines a function of the complex variable z that has remarkable properties.

Proposition 11.12 *The function f of the complex variable z defined by*

$$f(z) = \sum_{n=0}^{\infty} \frac{z^n}{n!}, \quad (z \in \mathbb{C})$$

satisfies the functional equation

$$f(z + w) = f(z)f(w)$$

for all complex z and w.

Proof Let z and w be complex numbers. We calculate

$$
\begin{aligned}
f(z)f(w) &= \sum_{n=0}^{\infty} \frac{z^n}{n!} \sum_{n=0}^{\infty} \frac{w^n}{n!} \\
&= \sum_{n=0}^{\infty} \left(\sum_{k=0}^{n} \frac{z^k}{k!} \frac{w^{n-k}}{(n-k)!} \right) \quad \text{(Cauchy product, Sect. 10.3)} \\
&= \sum_{n=0}^{\infty} \frac{1}{n!} \left(\sum_{k=0}^{n} \frac{n! z^k w^{n-k}}{k!(n-k)!} \right) \\
&= \sum_{n=0}^{\infty} \frac{1}{n!} \left(\sum_{k=0}^{n} \binom{n}{k} z^k w^{n-k} \right) \\
&= \sum_{n=0}^{\infty} \frac{1}{n!} (z+w)^n \quad \text{(by the binomial rule)} \\
&= f(z+w).
\end{aligned}
$$

$\square$

The foregoing proof is a beautiful example of how power series can be used to obtain algebraic properties of functions. Note how Proposition 10.9 provides the justification for the second equality sign.

Proposition 11.13 *For all real x we have*

$$e^x = \sum_{n=0}^{\infty} \frac{x^n}{n!}.$$

Proof Let $f(x) = \sum_{n=0}^{\infty} x^n/n!$ for all real x. We have $f(x+h) = f(x)f(h)$ for all x and h. Therefore

$$\frac{f(x+h) - f(x)}{h} = f(x) \frac{f(h) - 1}{h}.$$

For $h \neq 0$ we have

$$\frac{f(h) - 1}{h} = 1 + \frac{h}{2!} + \frac{h^2}{3!} + \cdots,$$

which is a continuous function of h, and hence has the limit 1 at $h = 0$. We conclude, by taking the limit of the difference quotient as $h \to 0$, that f is differentiable at x and the derivative satisfies $f'(x) = f(x)$, that is, f is a solution of the differential equation $y' = y$. Hence, by Proposition 7.13, we must have $f(x) = Ce^x$ for some constant C, and it is easily seen, for example by putting $x = 0$, that $C = 1$. $\qquad \square$

The famous formula of this proposition motivates the extension of the exponential function to complex variables. For each complex z we define

$$\exp z = \sum_{n=0}^{\infty} \frac{z^n}{n!}.$$

Then $\exp z$ extends to $\mathbb{C}$ the exponential function e^x of the real variable x.

11.3.5 The Number e

Now we have

$$e = \exp(1) = \sum_{n=0}^{\infty} \frac{1}{n!} = 1 + \frac{1}{1!} + \frac{1}{2!} + \frac{1}{3!} + \cdots$$

This provides a practical method (already studied in Sect. 7.2 Exercise 6) to calculate e since the series converges fast. If we take terms up to and including $1/N!$ then the error is

$$\frac{1}{(N+1)!} + \frac{1}{(N+2)!} + \frac{1}{(N+3)!} + \cdots$$
$$< \frac{1}{(N+1)!} \left(1 + \frac{1}{N+2} + \frac{1}{(N+2)^2} + \cdots\right) = \frac{N+2}{(N+1)(N+1)!}$$

For $N = 10$ this upper bound for the error is $0.0000000273...$, whilst the sum to $N = 10$ is $2.718281801...$ giving seven correct decimal digits of e.

The calculation of π is more difficult. In Sect. 8.1 Exercise 1 the value $1979/630$ was obtained which is short of π by around $3/10000$. Archimedes gave the bounds

$$\frac{223}{71} < \pi < \frac{22}{7}.$$

The difference between these bounds is around $1/500$. In fact the lower bound is accurate to around $7/10000$. Soon we shall exhibit a series that can be used to calculate π with arbitrary accuracy.

11.3.6 Differentiating a Power Series

The series $\sum_{n=0}^{\infty}(n+1)a_{n+1}z^n$, or equivalently $\sum_{n=1}^{\infty}na_nz^{n-1}$, is obtained formally by differentiating the series $\sum_{n=0}^{\infty}a_nz^n$ term-by-term with respect to z. We shall call it the derived series of the series $\sum_{n=0}^{\infty}a_nz^n$. Whatever may be the relationship between the sums of the two series, we can study the derived series in its own right.

Proposition 11.14 *The derived series $\sum_{n=0}^{\infty}(n+1)a_{n+1}z^n$ has the same radius of convergence as the original series $\sum_{n=0}^{\infty}a_nz^n$.*

Proof Let R be the radius of convergence of the original series $\sum_{n=0}^{\infty}a_nz^n$, and R' the radius of convergence of the derived series $\sum_{n=0}^{\infty}(n+1)a_{n+1}z^n$.

Suppose first that $0 < |z| < R$ and choose z_0, so that $|z| < |z_0| < R$. Then the series $\sum_{n=0}^{\infty}a_nz_0^n$ is convergent, so that there exists K, such that $|a_nz_0^n| < K$ for all n. Now we find

$$\left|(n+1)a_{n+1}z^n\right| = \frac{(n+1)|a_{n+1}z_0^{n+1}|}{|z|}\left|\frac{z}{z_0}\right|^{n+1} < \frac{K}{|z|}(n+1)\left|\frac{z}{z_0}\right|^{n+1}.$$

Since $|z/z_0| < 1$, we can refer to example (a) under "Determining the radius of convergence by the ratio test" and conclude that the series

$$\sum_{n=0}^{\infty}(n+1)\left|\frac{z}{z_0}\right|^{n+1}$$

is convergent. Hence, by the comparison test, the series

$$\sum_{n=0}^{\infty}(n+1)a_{n+1}z^n$$

is convergent if $|z| < R$. But this means that $R' \geq R$. Note that if $R = 0$ the above arguments are invalid but then it is trivial that $R' \geq R$.

Next we suppose that $|z| < R'$. The series $\sum_{n=0}^{\infty}|(n+1)a_{n+1}z^n|$ is now convergent and for $n \geq 1$ we have

$$|a_nz^n| \leq |z| |na_nz^{n-1}|,$$

so that the series $\sum_{n=0}^{\infty}a_nz^n$ is also convergent. This means that $R \geq R'$. Again if $R' = 0$ the deduction is invalid but the inequality is then trivial.

Putting the inequalities together gives $R = R'$. $\square$

Proposition 11.15 *Let the power series $\sum_{n=0}^{\infty}a_nx^n$ have real coefficients and let R be its radius of convergence. For all x in the interval $]-R, R[$ we define the function $f(x) = \sum_{n=0}^{\infty}a_nx^n$. We have the following conclusions:*

(1) The derivatives $f^{(k)}$ of all orders exist for every x in $]-R, R[$.
(2) The formula

$$f^{(k)}(x) = \sum_{n=k}^{\infty} n(n-1)...(n-k+1)a_n x^{n-k} = \sum_{n=k}^{\infty} \frac{n!}{(n-k)!} a_n x^{n-k}$$

holds for every x in $]-R, R[$. The equality is formally obtained by differentiating the series $\sum_{n=0}^{\infty} a_n x^n$ term-by-term, k times.
(3) For each n we have $a_n = f^{(n)}(0)/n!$.

Proof However often we differentiate formally term-by-term we obtain a power series with the same radius of convergence R. It therefore suffices to show that f' exists and equals $\sum_{n=1}^{\infty} na_n x^{n-1}$ for $|x| < R$. The rest is just repetition.

Let $0 < r < R$. The series $\sum_{n=1}^{\infty} na_n x^{n-1}$ is uniformly convergent on the domain $[-r, r]$. We conclude by Proposition 11.7 that f is differentiable for $-r < x < r$ and $f'(x) = \sum_{n=1}^{\infty} na_n x^{n-1}$. This holds for all r that satisfy $0 < r < R$. This means that the equation $f'(x) = \sum_{n=1}^{\infty} na_n x^{n-1}$ must hold for $-R < x < R$.

By repeating this argument we find

$$f^{(k)}(x) = \sum_{n=k}^{\infty} n(n-1)...(n-k+1)a_n x^{n-k} = \sum_{n=k}^{\infty} \frac{n!}{(n-k)!} a_n x^{n-k}$$

for all x in $]-R, R[$. Finally, setting $x = 0$ gives $f^{(k)}(0) = k!\, a_k$. □

If a function f has derivatives of all orders at a point at $x = 0$, we may form the power series (which merits a display because of its great importance):

$$\sum_{n=0}^{\infty} \frac{1}{n!} f^{(n)}(0)x^n.$$

Whether or not this series has a positive radius of converges (and if it does it need not converge to $f(x)$ in its interval of convergence), we call it the *Maclaurin series* of f. Proposition 11.15 shows that if a power series $\sum_{n=0}^{\infty} a_n x^n$ has a positive radius of convergence, then it is the Maclaurin series of its sum function.

The permissibility of differentiating power series term-by-term makes them into a powerful tool for investigating transcendental functions. As function series go it is quite a luxury; one is apt to forget that term-by-term differentiation is not generally permissible for function series, and not even for commonly used function series such as Fourier series, without some extra conditions. In the next section we shall use power series to study the elementary transcendental functions.

11.3.7 Exercises

1. Determine the radius of convergence of the following power series:

   (a) $\displaystyle\sum_{n=0}^{\infty} \frac{1}{n+1} z^n$

   (b) $\displaystyle\sum_{n=0}^{\infty} \frac{2^n}{n+1} z^n$

   (c) $\displaystyle\sum_{m=0}^{\infty} \frac{(-1)^m}{m!} z^{2m}$

   Note. This type of formulation, where not all powers are displayed, is common. Here only the even powers are displayed; the odd powers are understood to have the coefficient 0. In other cases only odd powers may be displayed, or powers of the form z^{3k+1} where k is an integer. The variations on this are many.

   (d) $\displaystyle\sum_{n=0}^{\infty} \frac{(2+n)^n}{n+1} z^n$

   (e) $\displaystyle\sum_{n=0}^{\infty} \frac{(2+n)^n}{n!} z^n$

   (f) $\displaystyle\sum_{n=0}^{\infty} \frac{(-1)^{n-1}(2n)!}{(n!)^2(2n-1)2^{2n}} z^n$

2. ($\Diamond$) Prove the following formula for the radius of convergence R of a power series $\sum_{n=0}^{\infty} a_n x^n$:

$$R = \frac{1}{\limsup_{n\to\infty} |a_n|^{\frac{1}{n}}}.$$

   The formula is interpreted as $R = 0$ when the denominator is ∞ and as $R = \infty$ when the denominator is 0.
   Hint. Consult Sect. 10.2 under "Extended forms of the ratio and Cauchy's test". Refer to Sect. 3.11 for an account of limit superior.

3. ($\Diamond$) Use the formula obtained in the previous exercise to give a short proof that the derived series $\sum_{n=0}^{\infty}(n+1)a_{n+1}x^n$ has the same radius of convergence as the original series $\sum_{n=0}^{\infty} a_n x^n$.
   Hint. Prove and use the rule

$$\limsup_{n\to\infty} c_n d_n = (\lim_{n\to\infty} c_n)(\limsup_{n\to\infty} d_n),$$

   valid on the assumption that c_n converges to a positive limit.

11.4 The Power Series of Common Elementary Functions

The power series expansions of $\sin x$ and $\cos x$ were among the spectacular results of the early decades of calculus. They led to the biggest advance in the practical calculation of the circular functions since Ptolemy of Alexandria. The connection between the exponential function and the circular functions followed and was seen to justify fully the controversial introduction of the imaginary unit i.

The binomial series was studied by Newton, who conjectured its sum by extrapolating from the case of a positive integer exponent. He probably found further support for the conjecture by using the series to calculate some square roots. A reasonably satisfactory proof was lacking until Abel gave one.

11.4.1 Unification of Exponential and Circular Functions

The power series defining the exponential of a complex number

$$\exp z = \sum_{n=0}^{\infty} \frac{1}{n!} z^n$$

is convergent for all complex z. We have seen that the function $\exp z$ satisfies the functional equation $\exp(z + w) = \exp z \exp w$ for all complex z and w. We also know that for real x we have $\exp x = e^x$. This motivates the notion of e raised to the power z, defined by

$$e^z := \exp z,$$

and justifies the use of the name exponential.

Now we consider the function e^{ix} of the real variable x, the restriction of e^z to the imaginary axis. We know that $i^2 = -1$, $i^3 = -i$ and $i^4 = 1$. In general $i^{2m} = (-1)^m$, $i^{2m+1} = (-1)^m i$. We therefore have

$$e^{ix} = \sum_{n=0}^{\infty} \frac{1}{n!} (ix)^n$$

$$= \sum_{m=0}^{\infty} \frac{1}{(2m)!} i^{2m} x^{2m} + \sum_{m=0}^{\infty} \frac{1}{(2m+1)!} i^{2m+1} x^{2m+1}$$

$$= \sum_{m=0}^{\infty} \frac{(-1)^m}{(2m)!} x^{2m} + i \sum_{m=0}^{\infty} \frac{(-1)^m}{(2m+1)!} x^{2m+1}.$$

This expresses the real and imaginary parts of e^{ix} as power series. Let us write $u(x) = \operatorname{Re} e^{ix}$, $v(x) = \operatorname{Im} e^{ix}$. For all real x we have

$$u(x) = \sum_{m=0}^{\infty} \frac{(-1)^m}{(2m)!} x^{2m} = 1 - \frac{x^2}{2!} + \frac{x^4}{4!} - \frac{x^6}{6!} + \frac{x^8}{8!} - \frac{x^{10}}{10!} + \cdots$$

$$v(x) = \sum_{m=0}^{\infty} \frac{(-1)^m}{(2m+1)!} x^{2m+1} = \frac{x}{1!} - \frac{x^3}{3!} + \frac{x^5}{5!} - \frac{x^7}{7!} + \frac{x^9}{9!} - \frac{x^{11}}{11!} + \cdots$$

We can compute the derivatives of u and v by differentiating the series term-by-term:

$$u'(x) = -\frac{x}{1!} + \frac{x^3}{3!} - \frac{x^5}{5!} + \frac{x^7}{7!} - \frac{x^9}{9!} + \frac{x^{11}}{11!} - \cdots = -v(x)$$

$$v'(x) = 1 - \frac{x^2}{2!} + \frac{x^4}{4!} - \frac{x^6}{6!} + \frac{x^8}{8!} - \frac{x^{10}}{10!} + \cdots = u(x)$$

and from this we conclude

$$u''(x) = -u(x), \quad v''(x) = -v(x).$$

We see that both u and v satisfy the differential equation $y'' + y = 0$. Hence $u(x)$ and $v(x)$ are each of the form $A \cos x + B \sin x$ where A and B are constants (see Proposition 7.2). But from the power series we see that $u(0) = 1$ and $u'(0) = 0$, giving $u(x) = \cos x$, and that $v(0) = 0$ and $v'(0) = 1$, giving $v(x) = \sin x$. To summarise, we have the following conclusions.

Proposition 11.16 *For all real x the following formulas are valid:*

$$(1) \quad \cos x = \sum_{m=0}^{\infty} \frac{(-1)^m}{(2m)!} x^{2m} = 1 - \frac{x^2}{2!} + \frac{x^4}{4!} - \frac{x^6}{6!} + \frac{x^8}{8!} - \frac{x^{10}}{10!} + \cdots$$

$$(2) \quad \sin x = \sum_{m=0}^{\infty} \frac{(-1)^m}{(2m+1)!} x^{2m+1} = \frac{x}{1!} - \frac{x^3}{3!} + \frac{x^5}{5!} - \frac{x^7}{7!} + \frac{x^9}{9!} - \frac{x^{11}}{11!} + \cdots$$

$(3) \quad e^{ix} = \cos x + i \sin x.$

The third formula in the proposition (sometimes called Euler's formula) tells us that e^{ix} parametrises the unit circle. Setting $x = \pi$ we derive the most beautiful formula in mathematics,

$$e^{i\pi} + 1 = 0.$$

The five most important numbers of analysis, 0, 1, i, π and e are here quite unexpectedly, and one might even say poetically, intertwined.

11.4.2 The Binomial Series

The series in question is

$$\sum_{n=0}^{\infty} \frac{a(a-1)...(a-n+1)}{n!} z^n.$$

The number a is a constant, which could be complex. The term for $n = 0$ is always interpreted as 1. Compare the binomial rule for a positive integral exponent m:

$$(1+x)^m = \sum_{n=0}^{m} \frac{m(m-1)...(m-n+1)}{n!} x^n = \sum_{n=0}^{m} \binom{m}{n} x^n$$

and we see why it is a plausible conjecture that for real a the binomial series sums to $(1+x)^a$.

If a is not a natural number then the radius of convergence of the binomial series is 1. We only have to apply the ratio test:

$$\frac{a(a-1)...(a-n)}{(n+1)!} z^{n+1} \Big/ \frac{a(a-1)...(a-n+1)}{n!} z^n = \frac{a-n}{n+1} z \to -z.$$

Hence the binomial series converges for $|z| < 1$ and diverges for $|z| > 1$.

Proposition 11.17 *Let a be a real number. For all x in the open interval* $-1 < x < 1$ *we have*

$$(1+x)^a = \sum_{n=0}^{\infty} \frac{a(a-1)...(a-n+1)}{n!} x^n.$$

Proof Let $f(x)$ be the sum of the series for those x that satisfy $|x| < 1$. We shall show that $(x+1)f'(x) - af(x) = 0$.

Set

$$c_n = \frac{a(a-1)...(a-n+1)}{n!}, \quad n = 1, 2, 3, ... \quad c_0 = 1,$$

so that $f(x) = \sum_{n=0}^{\infty} c_n x^n$, for $-1 < x < 1$. We saw, in the calculation of the radius of convergence, that

$$\frac{c_{n+1}}{c_n} = \frac{a-n}{n+1},$$

that is

$$(n+1)c_{n+1} = (a-n)c_n.$$

Differentiating the series term-by-term, and using this, we find

$$f'(x) = \sum_{n=1}^{\infty} n c_n x^{n-1}$$

$$= \sum_{n=0}^{\infty} (n+1)c_{n+1} x^n$$

$$= \sum_{n=0}^{\infty} (a - n)c_n x^n$$

$$= a \sum_{n=0}^{\infty} c_n x^n - x \sum_{n=1}^{\infty} n c_n x^{n-1}$$

$$= af(x) - xf'(x).$$

We conclude that $(x + 1)f'(x) - af(x) = 0$, that is, f is a solution to the differential equation

$$(x + 1)y' - ay = 0.$$

To see why this implies that $f(x) = (1 + x)^a$ we observe that

$$\frac{d}{dx}\left((1 + x)^{-a} f(x)\right) = (1 + x)^{-a} f'(x) - a(1 + x)^{-a-1} f(x)$$

$$= (1 + x)^{-a-1}\left((1 + x)f'(x) - af(x)\right)$$

$$= 0,$$

and conclude that $(1 + x)^{-a} f(x)$ is a constant C for $-1 < x < 1$. By setting $x = 0$ we find that $C = 1$. $\qquad\square$

The product $a(a - 1)...(a - n + 1)$ that appears in the binomial series can often be tidied up. Of course one way is to write it as $\Pi_{k=1}^{n}(a - k + 1)$. There are others that drastically alter its appearance, but are quite common. Consider the series

$$(1 + x)^{-1/2} = \sum_{n=0}^{\infty} \frac{\left(-\frac{1}{2}\right)\left(-\frac{3}{2}\right)...\left(\frac{1}{2} - n\right)}{n!} x^n.$$

We rewrite the product in the numerator, noting that there are n factors:

$$\left(-\frac{1}{2}\right)\left(-\frac{3}{2}\right)...\left(\frac{1}{2} - n\right) = (-1)^n \frac{1.3....(2n - 1)}{2^n}.$$

Here we have a new product of n factors, increasing in steps of 2. We insert the even numbers, above and below the line, to obtain

$$(-1)^n \frac{1.2.3....(2n - 1)(2n)}{2^n \ 2.4...(2n)} = (-1)^n \frac{(2n)!}{2^{2n} n!}.$$

This produces the series

$$(1 + x)^{-1/2} = \sum_{n=0}^{\infty} (-1)^n \frac{(2n)!}{2^{2n}(n!)^2} x^n,$$

not easily recognisable as a binomial series.

The binomial series makes sense when a is a complex number. This provides the opportunity to take up again the story of the power function x^a, prematurely pronounced concluded in Sect. 7.2, and extend it to complex powers a. In what follows x will be a positive real variable but a may be complex. We define

$$x^a := \exp(a \ln x), \quad (x > 0).$$

Exercise Show that complex powers obey the laws of exponents:

$$x^{a+b} = x^a x^b, \quad (x^a)^b = x^{ab}$$

where $x > 0$ and a and b are complex numbers.

Now we can calculate

$$
\begin{aligned}
\frac{d}{dx} x^a &= \frac{d}{dx} \exp(a \ln x) \\
&= \frac{a}{x} \exp(a \ln x) \\
&= a \exp(-\ln x) \exp(a \ln x) \\
&= a \exp\big((a-1) \ln x\big) \\
&= a x^{a-1}.
\end{aligned}
$$

Exercise Justify the second equality sign by differentiating the real and imaginary parts of the function.

Although we took a to be a real exponent in the proof of Proposition 11.17, there is nothing in the proof that does not work if a is complex. All we need is the formula giving the derivative of x^a as ax^{a-1}, which we obtained just now for a complex exponent a and a real variable x. In this text we do not go as far as considering differentiation with respect to a complex variable.

Exercise Verify that term-by-term differentiation is valid for a power series $\sum_{n=0}^{\infty} a_n x^n$ with complex coefficients within its interval of convergence $-R < x < R$.

11.4.3 Series for Arctangent

The case $a = -1$ of the binomial series gives us the formula

$$\frac{1}{1+x^2} = \sum_{n=0}^{\infty} (-1)^n x^{2n}, \quad -1 < x < 1.$$

If $0 < r < 1$ the series converges uniformly for $-r \leq x \leq r$. We may therefore integrate term-by-term from 0 to x and find

$$\arctan x = \sum_{n=0}^{\infty} \int_0^x (-1)^n t^{2n}\, dt = \sum_{n=0}^{\infty} \frac{(-1)^n}{2n+1} x^{2n+1}, \quad -1 < x < 1.$$

It should be clear that this holds for all x in the interval $]-1, 1[$ since we can move r as near to 1 as we wish. But we should also notice that the series is convergent for $x = 1$ (by Leibniz's test), although the series for $1/(1 + x^2)$ is divergent when $x = 1$.

It is natural to ask whether $\arctan 1$ is equal to $\sum_{n=0}^{\infty} (-1)^n/(2n+1)$. The arguments given above do not settle this. If the answer is yes, then we obtain a series for π, namely,

$$\frac{\pi}{4} = \sum_{n=0}^{\infty} \frac{(-1)^n}{2n+1}.$$

11.4.4   Abel's Lemma and Dirichlet's Test

The answer to the question posed at the end of the last section will soon be obtained by the use of *Abel's theorem on power series*. We postpone this topic to the next section and turn instead to *Abel's lemma*. On the face of it, this is a diversion, referring as it does to numerical series (rather than function series); so it might have fitted better into the previous chapter. However, its main use is in proving Dirichlet's test, an important application of which is to study the convergence of power series on the circle of convergence. And that is exactly where we have arrived in the discussion of power series.

Proposition 11.18 (Abel's lemma) *Let $(a_n)_{n=1}^{\infty}$ be a complex sequence and let $(b_n)_{n=1}^{\infty}$ be a decreasing sequence of positive, real numbers. Set $s_n = \sum_{k=1}^{n} a_k$, and assume that there exists $M > 0$, such that $|s_n| \leq M$ for all n. Then*

$$\left| \sum_{k=1}^{n} a_k b_k \right| \leq M b_1.$$

Proof We have (as an exercise, the reader might try writing the calculation using $\sum$-notation throughout)

$$\sum_{k=1}^{n} a_k b_k = s_1 b_1 + (s_2 - s_1) b_2 + (s_3 - s_2) b_3 + \cdots + (s_n - s_{n-1}) b_n$$

$$= s_1(b_1 - b_2) + s_2(b_2 - b_3) + \cdots + s_{n-1}(b_{n-1} - b_n) + s_n b_n.$$

Since $b_k - b_{k+1} \geq 0$ for all k, we find

$$\left| \sum_{k=1}^{n} a_k b_k \right| \leq M\Big((b_1 - b_2) + (b_2 - b_3) + \cdots + (b_{n-1} - b_n) + b_n\Big) = M b_1.$$

$\square$

Abel's lemma leads to a convergence test for complex series that does not test for absolute convergence. It is sometimes useful for proving that a power series converges at a given point on the circle of convergence, where, naturally, absolute convergence may fail.

Proposition 11.19 (Dirichlet's test) *Let $(a_n)_{n=1}^{\infty}$ be a complex sequence, let $(b_n)_{n=1}^{\infty}$ be a decreasing sequence of positive real numbers, and assume that $\lim_{n \to \infty} b_n = 0$. Set $s_n = \sum_{k=1}^{n} a_k$ and assume that there exists M, such that $|s_n| \leq M$ for all n. Then the series $\sum_{k=1}^{\infty} a_k b_k$ is convergent.*

Proof For integers m and ℓ such that $m < \ell$ we have

$$\left| \sum_{k=m}^{\ell} a_k \right| \leq \left| \sum_{k=1}^{\ell} a_k \right| + \left| \sum_{k=1}^{m-1} a_k \right| \leq 2M.$$

By Abel's lemma we now find, for $m < n$, that

$$\left| \sum_{k=m}^{n} a_k b_k \right| \leq \left(\sup_{\ell \geq m} \left| \sum_{k=m}^{\ell} a_k \right| \right) b_m \leq 2M b_m.$$

Since $\lim_{m \to \infty} b_m = 0$ the series $\sum_{k=1}^{\infty} a_k b_k$ is convergent by Cauchy's criterion (Proposition 10.4).

$\square$

A simple case of Dirichlet's test is to take $a_n = (-1)^{n-1}$, since then s_n is alternately 1 and 0. We obtain Leibniz's test, that $\sum_{k=1}^{\infty} (-1)^k b_k$ is convergent if b_k is decreasing and tends to 0.

A popular choice that gives new conclusions is to take $a_n = \eta^n$, where η is a complex number such that $|\eta| = 1$, but $\eta \neq 1$. Then we have

$$\left| \sum_{k=0}^{n} a_k \right| = \left| \sum_{k=0}^{n} \eta^k \right| = \left| \frac{\eta^{n+1} - 1}{\eta - 1} \right| \leq \frac{2}{|\eta - 1|}.$$

Now suppose that the power series $\sum_{n=0}^{\infty} b_n z^n$ has real coefficients and radius of convergence $R < \infty$, and assume that the sequence $(b_n R^n)_{n=0}^{\infty}$ is decreasing and tends to 0. If z is on the circle of convergence then $z = R\eta$ where $|\eta| = 1$, and we can write

$$\sum_{n=0}^{\infty} b_n z^n = \sum_{n=0}^{\infty} (b_n R^n) \eta^n.$$

We conclude that the series converges if $\eta \neq 1$, that is, the power series $\sum_{n=0}^{\infty} b_n z^n$ converges on its circle of convergence except possibly at $z = R$.

11.4.5 Abel's Theorem on Power Series

We come to the result of Abel postponed from a previous section. It is also commonly known as Abel's limit theorem.

Proposition 11.20 (Abel's theorem) *Let the power series $\sum_{n=0}^{\infty} a_n x^n$ have real coefficients and let its radius of convergence R be finite. Set $f(x) = \sum_{n=0}^{\infty} a_n x^n$ for $-R < x < R$. Assume that the series $\sum_{n=0}^{\infty} a_n R^n$ is convergent and let L be its sum. Then $\lim_{x \to R-} f(x) = L$.*

Proof We first treat the special case when $R = 1$ and $\sum_{n=0}^{\infty} a_n = 0$. We must show that $\lim_{x \to 1-} f(x) = 0$.

Let $s_n = \sum_{k=0}^{n} a_k$, so that $\lim_{n \to \infty} s_n = 0$. For $0 < x < 1$ we have

$$f(x) = (1-x)(1-x)^{-1} \sum_{n=0}^{\infty} a_n x^n$$

$$= (1-x)\left(\sum_{n=0}^{\infty} x^n\right)\left(\sum_{n=0}^{\infty} a_n x^n\right)$$

$$= (1-x) \sum_{n=0}^{\infty} s_n x^n$$

where the product in the second line is computed as the Cauchy product.

Let $\varepsilon > 0$. Choose N, such that $|s_n| < \varepsilon$ for all $n \geq N$. Then for $0 < x < 1$ we have

$$|f(x)| \leq (1-x)\left|\sum_{n=0}^{N-1} s_n x^n\right| + \varepsilon(1-x)\sum_{n=N}^{\infty} x^n = (1-x)\left|\sum_{n=0}^{N-1} s_n x^n\right| + \varepsilon x^N.$$

The right-hand member of the equality has the limit ε when $x \to 1-$. Hence there exists $\delta > 0$, such that it is below 2ε for all x that satisfy $1 - \delta < x < 1$. For such x we have $|f(x)| < 2\varepsilon$. This shows that $\lim_{x \to 1-} f(x) = 0$.

Finally the general case. Let $s = \sum_{n=0}^{\infty} a_n R^n$. We note that the power series

$$(a_0 - s) + \sum_{n=1}^{\infty} (a_n R^n) x^n$$

satisfies the conditions of the special case treated first. We conclude that

$$\lim_{x \to 1-} \left((a_0 - s) + \sum_{n=1}^{\infty} a_n R^n x^n \right) = (a_0 - s) + \sum_{n=1}^{\infty} a_n R^n$$

which is equivalent to

$$\lim_{x \to R-} \sum_{n=0}^{\infty} a_n x^n = \sum_{n=0}^{\infty} a_n R^n.$$

□

We can give some nice applications of Abel's theorem.

(a) We saw that

$$\arctan x = \sum_{n=0}^{\infty} \frac{(-1)^n}{2n+1} x^{2n+1}, \quad -1 < x < 1$$

and that the series converges for $x = 1$ by Leibniz's test. We conclude that

$$\frac{\pi}{4} = \arctan 1 = \sum_{n=0}^{\infty} \frac{(-1)^n}{2n+1}.$$

This is sometimes called Gregory's series for π.

(b) We know that $(1+x)^{-1} = \sum_{n=0}^{\infty} (-1)^n x^n$ for $-1 < x < 1$. Integrating we find

$$\ln(1+x) = \sum_{n=1}^{\infty} \frac{(-1)^{n-1}}{n} x^n, \quad -1 < x < 1.$$

The series is convergent for $x = 1$. We conclude that

$$\ln 2 = \sum_{n=1}^{\infty} \frac{(-1)^{n-1}}{n}.$$

These are beautiful results. But neither series is very good for practical computation since they converge so slowly. This is not surprising as we are operating on the circle of convergence.

11.4.6 Exercises

1. Show that $\overline{e^z} = e^{\bar{z}}$ for all complex z.
2. Show that $|e^z| = e^{\operatorname{Re} z}$ for all complex z.
3. Show that if θ is real then

$$|e^{i\theta} - 1| = 2 \left| \sin \frac{\theta}{2} \right|.$$

4. Show that the set of all solutions in $\mathbb{C}$ of the equation $e^z = 1$ consists of the numbers $2\pi n i$ where n is an integer, positive, negative or 0, in short the integer multiples of $2\pi i$.

5. Show that $\exp z$ is a periodic function of the complex variable z with basic period $2\pi i$. In other words show that $\exp z = \exp(z + 2\pi i)$ for all z, and that if $\exp z = \exp(z + T)$ for all z, then T is an integer multiple of $2\pi i$.

6. Show that the range of exp is the whole complex plane with the exclusion of the point 0.

7. Recall the principal logarithm of a complex variable, $\mathrm{Log}\, z$, defined in 9.1 for all z that are not negative real numbers or 0. The excluded set is the interval $]-\infty, 0]$ considered as a subset of $\mathbb{C}$. Show that Log and exp are inverses to each other if we restrict the domain of exp suitably.
   More precisely, let S be the strip $\{z \in C : -\pi i < \mathrm{Im}\, z < \pi i\}$. Show that if $z \in S$ and $w \in \mathbb{C}\backslash]-\infty, 0]$, then $w = \exp z$ if and only if $z = \mathrm{Log}\, w$.

8. Prove that if θ is not an integer multiple of 2π then

$$\sum_{k=0}^{n} e^{ki\theta} = \frac{e^{(n+1)i\theta} - 1}{e^{i\theta} - 1}.$$

   What is the correct formula if θ is an integer multiple of 2π?

9. Let

$$C_n(\theta) = \sum_{k=0}^{n} \cos k\theta, \qquad S_n(\theta) = \sum_{k=0}^{n} \sin k\theta$$

   for all real θ and positive integers n.

   (a) Prove that if θ is not an integer multiple of 2π then

$$C_n(\theta) = \frac{\cos \frac{1}{2}n\theta \, \sin \frac{1}{2}(n+1)\theta}{\sin \frac{1}{2}\theta}, \qquad S_n(\theta) = \frac{\sin \frac{1}{2}n\theta \, \sin \frac{1}{2}(n+1)\theta}{\sin \frac{1}{2}\theta}.$$

   What are the correct formulas if θ is an integer multiple of 2π?

   (b) Show that for all $\delta > 0$ there exists $K > 0$, such that for all n, and for all θ in the interval $\delta < \theta < 2\pi - \delta$, we have

$$|C_n(\theta)| \leq K \quad \text{and} \quad |S_n(\theta)| \leq K.$$

   (c) In the previous item K depends on δ and we cannot replace δ by 0. Show that

$$\lim_{n\to\infty} \sup_{0<\theta<2\pi} C_n(\theta) = \lim_{n\to\infty} \sup_{0<\theta<2\pi} S_n(\theta) = \infty.$$

10. (a) Derive a power series for the function $\dfrac{1}{2} \ln\left(\dfrac{1+x}{1-x}\right)$ from the power series for $\ln(1+x)$.

(b) Set $x = \frac{1}{3}$ in the power series obtained in the previous item and compute $\ln 2$. Use enough terms to get three correct decimal digits and for this purpose estimate the tail of the series. You should only need a very few terms. Compare with the series $\sum_{n=1}^{\infty} (-1)^{n-1}/n = \ln 2$. How many terms are needed of the latter series to get $\ln 2$ to 3 decimal places?

(c) Set $x = \frac{1}{9}$ and compute $\ln \frac{5}{4}$.

(d) Find a nice approximation to $\ln 10$ (and recall Sect. 7.2 Exercise 1).

11. Show that the series $\sum_{n=0}^{\infty} (1/(n+1))z^n$, which has radius of convergence 1, converges for all complex z that satisfy $|z| = 1$, except for $z = 1$, where it diverges.

12. Using the binomial series, obtain series for the following functions, and tidy them up after the fashion of the text.

(a) $(1+x)^{1/2}$
(b) $(1+x^2)^{-1/2}$
(c) $(1-x^2)^{-1/2}$.

13. ($\lozenge$) Let a be a real number, but not a natural number. You are asked to give an exhaustive description of the convergence, or otherwise, of the binomial series

$$\sum_{k=0}^{m} \frac{a(a-1)...(a-k+1)}{k!} x^k$$

at the endpoints of its interval of convergence. For which values of a (excluding here the natural numbers) is the following true of the series?

(a) It is absolutely convergent for $x = 1$.
(b) It is absolutely convergent for $x = -1$.
(c) It is conditionally convergent for $x = 1$.
(d) It is conditionally convergent for $x = -1$.
(e) It is divergent for $x = 1$.
(f) It is divergent for $x = -1$.

Hint. Let c_n be the binomial coefficient. A good place to start is from the ratio c_{n+1}/c_n. For absolute convergence Gauss's test is useful. For conditional convergence a useful first question is: for what a is the sequence of absolute values $|c_n|$ decreasing for sufficiently large n? And for what a does $|c_n|$ tend to 0?

14. Calculate some terms of the Maclaurin series (see Sect. 11.3 under "Differentiating a power series" for the definition of Maclaurin series) for $\tan x$, for example up to the term a_3x^3.

15. In the previous exercise the going gets tougher when higher powers are needed. Obtain a simple recurrence formula for the coefficients in the Maclaurin series for $\tan x$, by showing that $\tan x$ satisfies the differential equation $y' = 1 + y^2$, and using Leibniz's formula for the nth derivative of a product.

16. Using only the series for $\cos x$ and $\sin x$ and not relying on knowledge about these functions, prove that $\cos x$ has a lowest positive zero, and that it lies between $\sqrt{2}$ and $\sqrt{3}$.

    Note. In texts that use the power series of $\sin x$ and $\cos x$ as definitions of the circular functions, the result of this exercise is used to define the number π as twice the lowest positive zero of $\cos x$.

17. Suppose that the function $f(x)$ has derivatives of all orders at $x = 0$, and that $f(0) \neq 0$. Show that $1/f$ has derivatives of all orders at $x = 0$.

18. Suppose that f has the Maclaurin series $\sum_{n=0}^{\infty} a_n x^n$, with $a_0 \neq 0$, and that $1/f$ has the Maclaurin series $\sum_{n=0}^{\infty} b_n x^n$. Show that the coefficients b_n can be calculated from the recurrence relations

$$ b_n = -\frac{1}{a_0} \sum_{k=1}^{n} a_k b_{n-k} \quad (n \geq 1), \quad b_0 = \frac{1}{a_0}. $$

This holds irrespectively of whether either of the two Maclaurin series has a positive radius of convergence.

19. Suppose that f has the Maclaurin series $\sum_{n=0}^{\infty} a_n x^n$, with $a_0 \neq 0$, and that $1/f$ has the Maclaurin series $\sum_{n=0}^{\infty} b_n x^n$. Suppose further that the series $\sum_{n=0}^{\infty} a_n x^n$ converges to $f(x)$ in an interval $]-R, R[$ centred at 0. It is natural to ask whether the series $\sum_{n=0}^{\infty} b_n x^n$ converges to $1/f(x)$ in some interval centred at 0. This question can be answered most satisfactorily by means of complex analysis. However, using only methods of this text, one can produce an interval in which $\sum_{n=0}^{\infty} b_n x^n$ converges to $1/f(x)$, though it may fall far short of the largest one. Assume that $0 < r < R$ and set $M = \max_{k \geq 1} |a_k| r^k$.

    (a) Show that

    $$ |b_n| r^n \leq \frac{M}{|a_0|} \sum_{k=0}^{n-1} |b_k| r^k, \quad (k \geq 1). $$

    (b) Deduce that

    $$ |b_n| r^n \leq \frac{M}{|a_0|^2} \left(1 + \frac{M}{|a_0|} \right)^{n-1}, \quad (k \geq 1). $$

    (c) Deduce that the series $\sum_{n=0}^{\infty} b_n x^n$ converges if

    $$ |x| \leq \frac{|a_0| r}{M + |a_0|} $$

    and has the sum $1/f(x)$.

    (d) Obtain some lower estimates for the interval in which $\tan x$ can be represented by its Maclaurin series.

20. (a) Obtain the power series representation

$$\arcsin x = \sum_{n=1}^{\infty} \frac{(2n)!}{2^{2n}(n!)^2} \frac{x^{2n+1}}{2n+1}$$

for $-1 < x < 1$.

(b) ($\Diamond$) Show that the representation of $\arcsin x$ obtained in the previous item remains valid at $x = \pm 1$. This gives two further series for π.

21. Let the series $\sum_{k=1}^{\infty} a_k$ be convergent. Without using Abel's theorem on power series, show that the power series $\sum_{k=1}^{\infty} a_k x^k$ is uniformly convergent with respect to x in the closed interval $[0, 1]$.

    Hint. Study the tail $\sum_{k=n}^{\infty} a_k x^k$, using a treatment similar to that in the proof of Abel's theorem.

    Note. This gives another proof of Abel's theorem.

22. Show that the Maclaurin series for $f(x) := (1 - x)^{1/2}$ converges uniformly to $f(x)$ on the closed interval $[0, 1]$.

    Hint. By the previous exercise it is enough to show that the Maclaurin series converges at $x = 1$. If you haven't read the nugget on Gauss's test you might try to prove, since the terms plainly alternate in sign, that the coefficients tend to 0.

    Note. The result shows that the function $\sqrt{x}$ can be approximated uniformly by polynomials in the interval $[0, 1]$. This is the first step in proving the Stone–Weierstrass theorem, a very general result of metric space theory that includes as a special case the Weierstrass approximation theorem, which tells us that a continuous function can be approximated uniformly in a bounded interval by polynomials with arbitrary accuracy.

23. The convergence of the series for $\arctan x$ and $\ln(1 + x)$ at $x = 1$ can be obtained, without using Abel's theorem, by keeping a close eye on the remainder terms in the series being integrated. More precisely, write

$$\frac{1}{1+x} = \sum_{k=0}^{n-1}(-1)^k x^k + \frac{(-1)^n x^n}{1+x}$$

and derive

$$\ln(1 + x) = \sum_{k=0}^{n-1} \frac{(-1)^k x^{k+1}}{k+1} + \int_0^x \frac{(-1)^n t^n}{1+t}\, dt,$$

valid for all $x > -1$. Show that for $x = 1$ the remainder tends to 0 as $n \to \infty$. Carry out a similar analysis for $\arctan x$.

24. Gregory's series for π converges too slowly to be practical as a way of computing π. Some games with the addition formula for arctangent give better series. Prove the following formulas, and use the second to compute π.

(a) $\quad \dfrac{\pi}{4} = \arctan \dfrac{1}{2} + \arctan \dfrac{1}{3}$

(b) $\quad \dfrac{\pi}{4} = 4 \arctan \dfrac{1}{5} - \arctan \dfrac{1}{239}.$

In the remaining exercises in this section, we look at some consequences of Abel's lemma and Dirichlet's test. One benefit is a test that can decide uniform convergence of a function series that is not absolutely convergent.

25. Show that the series $\sum_{n=1}^{\infty} \sin(nx)/n$ is convergent for every x.

26. Prove the following useful variant of Dirichlet's test which imposes stronger conditions on a_n but weaker ones on b_n. Suppose that the series $\sum_{k=1}^{\infty} a_k$ (which may be complex) is convergent and that $(b_k)_{k=1}^{\infty}$ is a monotonic and bounded sequence of real numbers. Then the series $\sum_{k=1}^{\infty} a_k b_k$ is convergent.

27. Let $(a_n)_{n=1}^{\infty}$ and $(b_n)_{n=1}^{\infty}$ be complex sequences. Let $s_n = \sum_{k=1}^{n} a_k$ for each $n \geq 1$. Rewrite the calculation in the proof of Abel's lemma to give the formula:

$$\sum_{k=1}^{n} a_k b_k = s_n b_{n+1} - \sum_{k=1}^{n} s_k (b_{k+1} - b_k).$$

Deduce from this another variant of Dirichlet's test: if $\sum_{k=1}^{\infty} a_k$ is convergent and $\sum_{k=1}^{\infty} (b_{k+1} - b_k)$ absolutely convergent, then $\sum_{k=1}^{n} a_k b_k$ is convergent.

Note. This includes the result of the previous exercise as a special case, but is stronger; for example, b_n can be complex. For an example of its use see Exercise 32.

28. Dirichlet's test is the basis of a test for uniform convergence, which is sometimes useful for function series $\sum_{n=1}^{\infty} f_n(x)$, that are not absolutely convergent for some values of x. For such series the Weierstrass M-test cannot succeed. Prove the following proposition. The proof is the same as for Dirichlet's test but functions replace numbers.

Let $(u_n)_{n=1}^{\infty}$ and $(v_n)_{n=1}^{\infty}$ be function sequences on the same domain A. Suppose that there exists M, such that for all n and for all x in A we have $|\sum_{k=1}^{n} v_k(x)| < M$. Suppose further that the functions u_n are real valued, that $u_n(x)$ decreases with increasing n for each x in A, and that u_n tends uniformly to 0 on A. Then the series $\sum_{n=1}^{\infty} u_n(x)v_n(x)$ converges uniformly with respect to x in A.

There is a variant of this, similar to Exercise 26:

Let $(u_n)_{n=1}^{\infty}$ and $(v_n)_{n=1}^{\infty}$ be function sequences on the same domain A. Assume that the function series $\sum_{k=1}^{\infty} v_k(x)$ is uniformly convergent for x in A. Suppose further that the functions u_n are real valued, that $u_n(x)$ decreases with increasing n for each x in A, and that there exists $K > 0$, such that $|u_n(x)| \leq K$ for all n and for all x in A. Then the function series $\sum_{n=1}^{\infty} u_n(x)v_n(x)$ converges uniformly with respect to x in A.

29. Let $f(x) = \sum_{n=1}^{\infty} \sin(nx)/n$. Show that for all $\delta > 0$ the series is uniformly convergent with respect to x in the interval $[\delta, 2\pi - \delta]$. Conclude that f is continuous in the open interval $]0, 2\pi[$.

    Hint. Use Exercise 9 and the previous exercise.

    Note. The function f is actually discontinuous at multiples of 2π and exhibits a "saw tooth" pattern. The series is a basic example of a Fourier series representation of a discontinuous function.

30. Let s be a complex number. Show that the series $\sum_{n=1}^{\infty} n^{-s}$ is absolutely convergent if $\operatorname{Re} s > 1$..

    Note. The function $\zeta(s) = \sum_{n=1}^{\infty} n^{-s}$ is the Riemann zeta function and was referred to in Sect. 10.3 Exercise 2. It plays a really important role in number theory. In this exercise we have defined it for $\operatorname{Re} s > 1$ but it can be extended to the left of the line $\operatorname{Re} s = 1$ in an essentially unique way.

31. Series of the form $\sum_{n=1}^{\infty} a_n n^{-s}$, with complex coefficients a_n, are known as Dirichlet series. They generalise the series of the previous exercise. Normally they are studied with complex s, but in this exercise we restrict ourselves to real s.

    (a) Suppose that the series $\sum_{n=1}^{\infty} a_n$ is absolutely convergent. Show that for every $s \geq 0$ the series $\sum_{n=1}^{\infty} a_n n^{-s}$ is absolutely convergent.
    (b) Suppose that the sequence of partial sums $\sum_{k=1}^{n} a_k$ is bounded (in particular, this is the case if the series $\sum_{n=1}^{\infty} a_n$ is convergent). Show that for every $s > 0$ the series $\sum_{n=1}^{\infty} a_n n^{-s}$ is convergent.

32. Extend the conclusions of the previous exercise to complex s. The main difference lies in the test needed for item (b).

    (a) Suppose that the series $\sum_{n=1}^{\infty} a_n$ is absolutely convergent. Show that for every complex s such that $\operatorname{Re} s \geq 0$ the series $\sum_{n=1}^{\infty} a_n n^{-s}$ is absolutely convergent.
    (b) Suppose that the sequence of partial sums $\sum_{k=1}^{n} a_k$ is bounded. Show that for every complex s such that $\operatorname{Re} s > 0$ the series $\sum_{n=1}^{\infty} a_n n^{-s}$ is convergent.
    Hint. Use the fact that if $\operatorname{Re} t > 0$ the series $\sum_{n=1}^{\infty} \left(n^{-t} - (n+1)^{-t} \right)$ is absolutely convergent. This can be seen by computing the limit

$$\lim_{x \to \infty} \frac{x^{-t} - (x+1)^{-t}}{x^{-t-1}},$$

    where t is complex. Then use the version of Dirichlet's test given in Exercise 27.

11.5   (◊) Summability Theories

There is an old puzzle about the light switch that is turned on at one minute to midnight, off at half-a-minute to midnight, on again at a quarter-of-a-minute to midnight, and so on, always halving the time between successive switchings. We ask: is the light on or off at midnight?

It seems we are asking whether the infinite sum

$$\sum_{n=1}^{\infty}(-1)^{n-1} = 1 - 1 + 1 - 1 + 1 - 1 + \cdots$$

is 0 or 1. Of course the series is divergent; having a correct definition of limit seems to dispense with the question as being meaningless. However, mathematicians are not so ready to give up, and have invented summability theories to shed light on this question.

A summability theory is a procedure for assigning values to infinite sums $\sum_{n=1}^{\infty} a_n$, that assigns the correct value to convergent series, assigns a value to some divergent series, and satisfies certain natural rules. We want the sum to be a linear operation, that is,

$$\sum_{n=1}^{\infty}(\alpha a_n + \beta b_n) = \alpha \sum_{n=1}^{\infty} a_n + \beta \sum_{n=1}^{\infty} b_n,$$

and we want it to behave naturally with regard to tacking on extra terms at the front,

$$\alpha + \sum_{n=1}^{\infty} a_n = \sum_{n=1}^{\infty} b_n,$$

where $b_1 = \alpha$ and $b_{n+1} = a_n$ for $n = 1, 2, 3, \ldots$

These rules alone fix the value of $\sum_{n=1}^{\infty}(-1)^{n-1}$ in any summability theory, for calling the sum s, we have

$$1 - s = 1 - \sum_{n=1}^{\infty}(-1)^{n-1} = \sum_{n=1}^{\infty}(-1)^{n-1} = s$$

so that $s = \frac{1}{2}$. Quite a sensible conclusion which suggests that the light is equally likely to be on or off.

There are two principal summability theories that are commonly seen.

Cesaro summation

Also known as (C,1)-summation or summation by arithmetic means. We let

$$s_n = \sum_{k=1}^{n} a_k, \quad \text{and} \quad \sigma_n = \frac{1}{n}\sum_{k=1}^{n} s_k.$$

The Cesaro sum of the series $\sum_{k=1}^{\infty} a_k$ is the limit $\lim_{n \to \infty} \sigma_n$, provided the limit exists. By Proposition 3.23 this assigns the correct value to convergent series.

One can take this further by considering the sequence

$$\sigma_n' = \frac{1}{n} \sum_{k=1}^{n} \sigma_k,$$

of arithmetic means of the sequence $(\sigma_n)_{n=1}^{\infty}$ of arithmetic means. The limit $\lim_{n \to \infty} \sigma_n'$, if it exists, is the (C,2)-sum of $\sum_{k=1}^{\infty} a_k$. In this way we can produce a whole scale of summability methods, (C,m)-summability, for $m = 1, 2, 3, \ldots$

Abel summation

We consider the power series $\sum_{n=0}^{\infty} a_n x^n$. If its radius of convergence is greater than or equal to 1 we set $F(x) = \sum_{n=0}^{\infty} a_n x^n$ for $|x| < 1$. The limit, $\lim_{x \to 1^-} F(x)$, if it exists, is called the Abel sum of the series $\sum_{n=0}^{\infty} a_n$. By Abel's theorem on power series this assigns the correct value to convergent series.

11.5.1 Abelian Theorems and Tauberian Theorems

It may be that a series, whose normal convergence status is unknown, can be shown to be Abel summable, or Cesaro summable. We cannot of course deduce that the series in question is convergent in the normal sense from either of these facts, as the case of $\sum_{n=1}^{\infty} (-1)^{n-1}$ shows. But an additional condition imposed on the sequence $(a_n)_{n=1}^{\infty}$ may suffice to make the deduction. Many such conditions are now known. The following result is their prototype.

Proposition 11.21 (Tauber's theorem) *Suppose that the series $\sum_{n=0}^{\infty} a_n$ is Abel summable with sum s, assumed to be a finite number. Assume that $\lim_{n \to \infty} n a_n = 0$. Then the series is convergent in the normal sense, with sum s.*

Proof Let $F(x) = \sum_{k=0}^{\infty} a_k x^k$ for $0 < x < 1$. The assumption of Abel summability means that $\lim_{x \to 1^-} F(x) = s$. Let $s_n = \sum_{k=0}^{n} a_k$. For each n and each x in the interval $]0, 1[$ we have

$$s_n - s = F(x) - s + \sum_{k=0}^{n} a_k (1 - x^k) + \sum_{k=n+1}^{\infty} a_k x^k.$$

Since $0 < x < 1$ we have

$$1 - x^k = (1 - x)(1 + x + x^2 + \cdots + x^{k-1}) < k(1 - x).$$

And for $k \geq n + 1$ we have $1 < k/(n + 1)$. Therefore

$$|s_n - s| \le |F(x) - s| + (1 - x) \sum_{k=0}^{n} k|a_k| + \frac{1}{n+1} \sum_{k=n+1}^{\infty} k|a_k|x^k.$$

Let $\varepsilon > 0$. Since $\lim_{n \to \infty} na_n = 0$ there exists N, such that

$$n|a_n| < \varepsilon \quad \text{and} \quad \frac{1}{n+1} \sum_{k=0}^{n} k|a_k| < \varepsilon$$

for all $n \ge N$. Then for $0 < x < 1$ and $n \ge N$ we have

$$|s_n - s| \le |F(x) - s| + (1 - x)(n + 1)\varepsilon + \frac{\varepsilon}{(n+1)(1-x)}.$$

In particular, putting $x = x_n = 1 - 1/(n + 1)$, we find for $n \ge N$ that

$$|s_n - s| \le |F(x_n) - s| + 2\varepsilon.$$

Letting $n \to \infty$ we obtain

$$\limsup_{n \to \infty} |s_n - s| \le 2\varepsilon$$

and since ε is arbitrary we conclude $\lim_{n \to \infty} s_n = s$ as required. $\qquad\square$

In general a Tauberian theorem enables one to conclude that a series that is summable by a given summability method is also convergent in the usual sense, given some additional condition (a Tauberian condition) imposed on the terms. An Abelian theorem enables one to conclude that a prospective summability theory assigns the correct value to series convergent in the normal sense.

11.5.2 Exercises

1. Show that both the Abel sum and the (C,1)-sum of $\sum_{n=1}^{\infty}(-1)^{n-1}$ is $\frac{1}{2}$.
2. Show that a series that is (C,1)-summable is also Abel summable, and has the same sum by both methods.
   Hint. Copy the proof of Abel's theorem but using $(1 - x)^{-2}$ instead of $(1 - x)^{-1}$. This might suggest the more general result that (C,m)-summability implies Abel-summability, and a strategy for proving it.
3. Prove the following Tauberian theorem. Let $a_n \ge 0$ for all n and suppose that the series $\sum_{n=0}^{\infty} a_n$ is Abel-summable. Then the series is convergent in the normal sense.
4. Suppose that the series $\sum_{n=0}^{\infty} a_n$ and $\sum_{n=0}^{\infty} b_n$ are convergent, with sums s and t respectively. Show that their Cauchy product is (C,1)-summable, with sum st.

This implies a theorem of Abel: if the Cauchy product is convergent then its sum is st.

11.5.3 Pointers to Further Study

→ Abelian and Tauberian theorems.
→ Fejer's theorem.

11.6 (◊) The Irrationality of e and π

Euler published a proof that e is irrational in 1740. We present here a simple proof due to Fourier (early nineteenth century).

We begin with

$$e = \sum_{k=0}^{\infty} \frac{1}{k!} = \sum_{k=0}^{m} \frac{1}{k!} + \sum_{k=m+1}^{\infty} \frac{1}{k!}.$$

We can estimate the tail as follows:

$$\sum_{k=m+1}^{\infty} \frac{1}{k!}$$

$$= \frac{1}{(m+1)!}\left(1 + \frac{1}{m+2} + \frac{1}{(m+2)(m+3)} + \frac{1}{(m+2)(m+3)(m+4)} + \cdots\right)$$

$$< \frac{1}{(m+1)!}\left(1 + \frac{1}{m+2} + \frac{1}{(m+2)^2} + \frac{1}{(m+2)^3} + \cdots\right)$$

$$= \frac{1}{(m+1)!}\frac{m+2}{m+1}$$

since the series after the inequality sign is a geometric series. We obtain an estimate for the error

$$0 < e - \sum_{k=0}^{m} \frac{1}{k!} < \frac{1}{(m+1)!}\frac{m+2}{m+1}$$

which leads to

$$0 < m!\left(e - \sum_{k=0}^{m} \frac{1}{k!}\right) < \frac{m+2}{(m+1)^2}. \tag{11.1}$$

If $e = a/b$, where a and b are integers, then the central quantity in the inequalities (11.1) is an integer when $m \geq b$. On the other hand the right-hand member

$(m + 2)/(m + 1)^2$ tends to 0 as $m \to \infty$. It is clearly impossible that an integer can lie strictly between 0 and $(m + 2)/(m + 1)^2$ for sufficiently large m, in fact not even for $m \geq 1$. This proves the irrationality of e.

The first proof of the irrationality of π was published by Lambert in 1761. Another proof, due to Hermite (second half of nineteenth century) is widely available in the books.

Here is a proof, possibly due to Mary Cartwright (early twentieth century), who was mistress of Girton College, Cambridge. In any case she set it as an examination problem. The proof is in several steps, which Cartwright would have called exercises. To encourage the reader to adopt the right spirit, we give the solutions in a separate section.

For each real α let

$$I_n(\alpha) = \int_{-1}^{1} \left(1 - x^2\right)^n \cos \alpha x \, dx, \quad n = 0, 1, 2, \ldots$$

Step 1. Derive the reduction formula

$$\alpha^2 I_n(\alpha) = 2n(2n - 1)I_{n-1}(\alpha) - 4n(n - 1)I_{n-2}(\alpha), \quad n = 2, 3, 4, \ldots$$

Step 2. Show that there exist polynomials $P_n(\alpha)$ and $Q_n(\alpha)$, with degree less than or equal to n, and with integer coefficients, such that

$$\alpha^{2n+1} I_n(\alpha) = n! \left(P_n(\alpha) \sin \alpha + Q_n(\alpha) \cos \alpha\right), \quad n = 0, 1, 2, \ldots$$

Step 3. Prove the following: if $\pi/2 = b/a$, where a and b are integers, then

$$\frac{b^{2n+1}}{n!} I_n\left(\frac{b}{a}\right)$$

is an integer.

Step 4. Show that the assumption that $\pi/2 = b/a$, where a and b are integers, leads to a contradiction. In other words: π is irrational.

11.6.1   Solutions

Step 1. Let $n \geq 2$. Integrate twice by parts and note that $(1 - x^2)^n$ and $(1 - x^2)^{n-1}$ are both 0 for $x = \pm 1$:

$$I_n(\alpha) = \frac{2n}{\alpha} \int_{-1}^{1} (1-x^2)^{n-1} x \sin \alpha x \, dx$$

$$= \frac{2n}{\alpha^2} \int_{-1}^{1} \left((1-x^2)^{n-1} - 2x^2(1-x^2)^{n-2} \right) \cos \alpha x \, dx$$

$$= \frac{2n}{\alpha^2} \int_{-1}^{1} \Big((1-x^2)^{n-1} + 2(n-1)(1-x^2)^{n-1}$$

$$- 2(n-1)(1-x^2)^{n-2} \Big) \cos \alpha x \, dx$$

$$= \frac{2n}{\alpha^2} \int_{-1}^{1} \left((2n-1)(1-x^2)^{n-1} - 2(n-1)(1-x^2)^{n-2} \right) \cos \alpha x \, dx$$

$$= \frac{2n(2n-1)}{\alpha^2} I_{n-1}(\alpha) - \frac{4n(n-1)}{\alpha^2} I_{n-2}(\alpha)$$

and the reduction formula is proved.

Step 2. We use induction. We assume that the claim is true for place numbers up to $n-1$, that is, we assume that P_k and Q_k exist for $k \leq n-1$ and satisfy

$$\alpha^{2k+1} I_k(\alpha) = k! \left(P_k(\alpha) \sin \alpha + Q_k(\alpha) \cos \alpha \right).$$

By step 1 we then have

$$\alpha^{2n+1} I_n = 2n(2n-1)\alpha^{2n-1} I_{n-1} - 4n(n-1)\alpha^{2n-1} I_{n-2}$$

$$= 2n(2n-1)(n-1)! \left(P_{n-1} \sin \alpha x + Q_{n-1} \cos \alpha x \right)$$

$$- 4n(n-1)(n-2)!\alpha^2 \left(P_{n-2} \sin \alpha x + Q_{n-2} \cos \alpha x \right)$$

and we obtain recurrence relations for P_n and Q_n:

$$P_n = 2(2n-1)P_{n-1} - 4\alpha^2 P_{n-2}$$

$$Q_n = 2(2n-1)Q_{n-1} - 4\alpha^2 Q_{n-2}$$

valid for all $n \geq 2$.

From these relations we conclude that if P_{n-1} and Q_{n-1} are polynomials with degree less than or equal to $n-1$ and with integer coefficients, and if P_{n-2} and Q_{n-2} are polynomials with degree less than or equal to $n-2$ and with integer coefficients, then P_n and Q_n are polynomials with degree less than or equal to n and have integer coefficients.

To complete the induction we must examine the initial values. They are

$$P_0 = 2, \quad P_1 = 4, \quad Q_0 = 0, \quad Q_1 = -4\alpha,$$

so that P_0, P_1, Q_0 and Q_1 are polynomials with integer coefficients, and their degrees satisfy the required bounds.

Step 3. Suppose that $\pi/2 = b/a$. Since $\cos(\pi/2) = 0$ and $\sin(\pi/2) = 1$ the result of step 2 with $\alpha = b/a$ implies

$$\frac{b^{2n+1}}{a^{2n+1}} I_n\left(\frac{b}{a}\right) = n!\, P_n\left(\frac{b}{a}\right).$$

Now the degree of P_n is less than $2n + 1$, and so $a^{2n+1} P_n(b/a)$ is an integer. Hence $(b^{2n+1}/n!) I_n(b/a)$ is an integer. This completes step 3.

Step 4. We show that $(b^{2n+1}/n!) I_n(b/a)$ cannot be an integer when n is sufficiently high. In fact we have $\lim_{n\to\infty} b^{2n+1}/n! = 0$, but also $I_n(\alpha) \le 2$ (an obvious bound; actually $I_n(\alpha)$ tends to 0 for each α). From this we find that $(b^{2n+1}/n!) I_n(b/a)$ tends to 0, but it is manifestly never equal to 0, for $\cos(\pi x/2) > 0$ on the interval $]-1, 1[$ and so the integral is strictly positive. This is impossible because $(b^{2n+1}/n!) I_n(b/a)$ is an integer for all n.

11.6.2   Pointers to Further Study

→ Irrationality theory
→ Diophantine approximation
→ Transcendence of e and π

11.7   Taylor Series

Let $f(x)$ be a function with domain A, an open interval, and suppose that f has derivatives of all orders. Fix a point c in A. The power series

$$\sum_{n=0}^{\infty} \frac{f^{(n)}(c)}{n!}(x - c)^n$$

is called the Taylor series of f with centre c. The Taylor series with centre 0 (formable if 0 is in A) is, as we mentioned in Sect. 11.3, called the Maclaurin series of f.
    We list some important points:

(a) The Taylor series of f can have radius of convergence 0. In fact, according to a theorem of E. Borel, any sequence of real numbers is the coefficient sequence of the Maclaurin series of some function. So, for example, there is a function having derivatives of all orders, whose Maclaurin series is $\sum_{n=0}^{\infty} n!\, x^n$.
(b) Even if the radius of convergence R is not 0, the sum of the Taylor series of f, for $c - R < x < c + R$, need not be equal to $f(x)$.
(c) The partial sums of the Taylor series of f are the Taylor polynomials with centre c.

(d) According to the theorem on the differentiation of power series (Proposition 11.15), if a power series $\sum_{n=0}^{\infty} a_n (x - c)^n$ is convergent, with sum $g(x)$, for $c - R < x < c + R$, then the series is the Taylor series of g with centre c.

(e) A given power series can be the Taylor series of more than one function.

A key example for understanding the pitfalls of Taylor series is the function f with domain $\mathbb{R}$ given by

$$f(x) = \begin{cases} e^{-\frac{1}{x}}, & \text{if } x > 0 \\ 0, & \text{if } x \leq 0. \end{cases}$$

Obviously f has derivatives of all orders for $x \neq 0$; but what about $x = 0$? By induction one may show that for $x > 0$ we can write

$$f^{(n)}(x) = P_n\left(\frac{1}{x}\right) e^{-\frac{1}{x}},$$

where P_n is a polynomial of degree $2n$. Since the exponential function overwhelms any polynomial we see that $\lim_{x \to 0+} f^{(n)}(x) = 0$ for all n. A further induction (see Sect. 5.6 Exercises 6 and 7) now shows that all derivatives $f^{(n)}(0)$ exist and are 0.

But there is a further, immediate and surprising conclusion. All Taylor polynomials with centre 0 are 0. The Maclaurin series has every coefficient a_n equal to 0, and its sum is 0 for every x.

11.7.1 Taylor's Theorem with Lagrange's Remainder

Proposition 11.22 *Assume that* $f : A \to \mathbb{R}$ *has derivatives of order up to n, let $c \in A$, and let $x \in A$. Then there exists ξ in the interval $]c, x[$ (if $x > c$) or in the interval $]x, c[$ (if $x < c$), such that*

$$f(x) = \sum_{k=0}^{n-1} \frac{f^{(k)}(c)}{k!} (x - c)^k + \frac{f^{(n)}(\xi)}{n!} (x - c)^n.$$

Before proving this we shall discuss its meaning. The first expression on the right is the Taylor polynomial of degree $n - 1$ centred at c (as a polynomial it may have degree less than $n - 1$), and the second is the remainder, or error, that is incurred when the Taylor polynomial is used to approximate $f(x)$. The phrasing implies that $x \neq c$; there can be no ξ in $]x, c[$ if $x = c$ as the interval is empty.

The exclusion of $x = c$ is of no real consequence. There is another form of the same expression that is sometimes easier to handle. Set $x - c = h$ and set $\xi = c + \theta h$ where $0 < \theta < 1$. Then

$$f(c+h) = \sum_{k=0}^{n-1} \frac{f^{(k)}(c)}{k!} h^k + \frac{f^{(n)}(c+\theta h)}{n!} h^n.$$

When using this formula one must remember that h can be negative, just as in the first version x can be below c. Similarly $|h|$ could be big. The only restriction on h is that $c + h$ should lie in the interval A, which could be unbounded. In this version there is no problem with having $h = 0$.

Both forms of the formula fail to indicate the important point that ξ and θ depend on x, c and n.

Taylor's theorem takes as its starting point the formula

$$f(x) = P_{n-1}(x,c) + R_n(x,c)$$

involving the Taylor polynomial and the remainder, and the whole content is an assertion that the remainder may be expressed variously as

$$R_n(x,c) = \frac{f^{(n)}(\xi)}{n!}(x-c)^n$$

where ξ is between c and x, or as

$$R_n(c+h,c) = \frac{f^{(n)}(c+\theta h)}{n!} h^n,$$

where $0 < \theta < 1$. Both of these expressions are called Lagrange's form of the remainder.

First proof of Taylor's theorem with Lagrange's remainder We prove the proposition in the second version, the one in which x is replaced by $c + h$. Suppose that $h \neq 0$. For $0 \le t \le h$ or $h \le t \le 0$ (depending on the sign of h), we set

$$g(t) = f(c+t) - \sum_{k=0}^{n-1} \frac{f^{(k)}(c)}{k!} t^k - \frac{t^n}{n!} B \tag{11.2}$$

where the constant B is chosen so that $g(h) = 0$, that is, B is determined by

$$f(c+h) = \sum_{k=0}^{n-1} \frac{f^{(k)}(c)}{k!} h^k + \frac{h^n}{n!} B.$$

We only need to show that B is of the form

$$B = f^{(n)}(c+\theta h)$$

for some θ between 0 and 1.

Differentiating (11.2) repeatedly with respect to t we obtain

$$g^{(j)}(t) = f^{(j)}(c+t) - \sum_{k=j}^{n-1} \frac{f^{(k)}(c)}{(k-j)!} t^{k-j} - \frac{t^{n-j}}{(n-j)!} B, \quad (j = 0, 1, \ldots n - 1),$$

and therefore

$$g^{(j)}(0) = 0, \quad (j = 0, 1, \ldots n - 1)$$
$$g^{(n)}(t) = f^{(n)}(c+t) - B. \tag{11.3}$$

We apply the mean value theorem (actually Rolle's theorem) repeatedly and exploit the first equation of (11.3). Recall that B was chosen so that $g(h) = 0$, but that we also have $g(0) = 0$. Hence, there exists h_1 between 0 and h, such that $g'(h_1) = 0$; then, if $n \geq 2$ we have $g'(0) = 0$, and so there exists h_2 between 0 and h_1, such that $g''(h_2) = 0$; then, if $n \geq 3$ we have $g''(0) = 0$, and so there exists h_3 between 0 and h_2, such that $g'''(h_3) = 0$, and so on, terminating in h_n between 0 and h_{n-1}, such that $g^{(n)}(h_n) = 0$. But then h_n is between 0 and h and by the second equation of (11.3) satisfies

$$f^{(n)}(c + h_n) - B = 0.$$

Set $h_n = \theta h$. We have $0 < \theta < 1$ and $B = f^{(n)}(c + \theta h)$. $\qquad \square$

Second proof of Taylor's theorem with Lagrange's remainder For $0 \leq t \leq h$ or $h \leq t \leq 0$ (depending on the sign of h), we set

$$g(t) = f(c+t) - \sum_{k=0}^{n-1} \frac{f^{(k)}(c)}{k!} t^k.$$

Differentiating repeatedly we obtain

$$g^{(j)}(t) = f^{(j)}(c+t) - \sum_{k=j}^{n-1} \frac{f^{(k)}(c)}{(k-j)!} t^{k-j}, \quad (j = 0, 1, \ldots n - 1),$$

and therefore

$$g^{(j)}(0) = 0, \quad (j = 0, 1, \ldots n - 1),$$
$$g^{(n)}(t) = f^{(n)}(c+t). \tag{11.4}$$

We apply Cauchy's mean value theorem repeatedly to the quotient $g(t)/t^n$ and exploit the first equation of (11.4). We obtain the string of equalities:

$$\frac{g(h)}{h^n} = \frac{g'(h_1)}{nh_1^{n-1}} = \frac{g''(h_2)}{n(n-1)h_2^{n-2}} = \cdots = \frac{g^{(n)}(h_n)}{n!}$$

where h_1 lies between 0 and h, and in general h_{j+1} lies between 0 and h_j. Putting $h_n = \theta h$ we see that $0 < \theta < 1$ and the extreme members of the string of equalities together with the second equation of (11.4) give

$$f(c+h) - \sum_{k=0}^{n-1} \frac{f^{(k)}(c)}{k!} h^k = g(h) = \frac{f^{(n)}(c+\theta h)h^n}{n!}.$$

□

11.7.2 Error Estimates

When an approximation is used it can be useful to estimate the error.   The main force of Taylor's theorem resides in the information it gives about the remainder term. This can help us to estimate the error when a function is approximated by one of its Taylor polynomials in a part of its domain of definition.

Let us look at some elementary functions, taking the point c to be 0, and write down Lagrange's remainder. For all $x \neq 0$ and m, there exists, in each of the following cases, ξ between 0 and x (not necessarily the same for each case), such that

(a) $\cos x = \displaystyle\sum_{k=0}^{m} \frac{(-1)^k}{(2k)!} x^{2k} + (-1)^{m+1}(\cos \xi)\frac{x^{2m+2}}{(2m+2)!}$

(b) $\sin x = \displaystyle\sum_{k=0}^{m} \frac{(-1)^k}{(2k+1)!} x^{2k+1} + (-1)^{m+1}(\cos \xi)\frac{x^{2m+3}}{(2m+3)!}$

(c) $e^x = \displaystyle\sum_{k=0}^{m} \frac{x^k}{k!} + e^\xi \frac{x^{m+1}}{(m+1)!}.$

The sums appearing here are in each case partial sums of the Maclaurin series of the corresponding functions. The Maclaurin series for $\cos x$ has only even powers of x, that is, the coefficients of all odd powers are 0. In case (a) we can think of the series as extending up to the unwritten term $0.x^{2m+1}$, and is followed by the remainder that includes the power x^{2m+2}. A similar remark holds for $\sin x$, for which the even powers are missing.

If the remainder tends to 0 as $m \to \infty$ we get a proof that the function in question is the sum of its Taylor series. The problem is that we know nothing about ξ except that it is confined between 0 and x. However we can make the following estimates:

$$|\cos \xi| < 1, \quad 0 < e^\xi < \max(1, e^x).$$

Even without further knowledge of ξ we can conclude in all three cases that the error tends to 0 as $m \to \infty$, by exploiting the very useful limit $\lim_{n\to\infty} x^n/n! = 0$, itself a consequence of the convergence of the exponential series. Taylor's theorem yields nice new proofs of the power series representations of the circular functions and the exponential function.

However, the error estimates may seem disappointing. The problem is that we know nothing about that number ξ. For the circular functions we can only say that the error is bounded by the next unused term. For e^x the Lagrange remainder does not look very useful for estimating the error when $x > 0$, as the only thing one can say about e^ξ is that it is less than e^x.

The truth is that these are very simple Maclaurin series for which it is easy to estimate the tail directly, giving results very similar to estimating the remainder term. But suppose one wants to approximate the function $\sin x$ on the interval $[\pi/6, \pi/4]$ using its fourth-degree Taylor polynomial centred at $\pi/6$? The remainder term here is a valuable source of information (see the exercises).

A computer calculates transcendental functions by using polynomial approximations tailored (pun intended!) to different parts of its domain of definition. An estimate of the error is essential so that we can be confident of delivering a minimum number of correct decimal digits. The remainder term offered by Taylor's theorem was the first important tool for accomplishing this, although more sophisticated ones have since been developed.

11.7.3 Error Estimates for $\ln(1 + x)$

We shall consider the Lagrange remainder for $\ln(1 + x)$, again taking $c = 0$. For each x in the interval $-1 < x < \infty$ there exists ξ between 0 and x, such that

$$\ln(1 + x) = \sum_{n=1}^{m} \frac{(-1)^{n-1}}{n} x^n + \frac{(-1)^m}{(1 + \xi)^{m+1}} \frac{x^{m+1}}{m + 1}.$$

Exercise Obtain this formula from Taylor's theorem.

The first expression on the right, if extended indefinitely, is a power series with radius of convergence 1. Nevertheless the above formula holds for all x and not just those for which the power series converges. If $x > 1$ the series diverges and it is impossible that the remainder should tend to 0 as $m \to \infty$.

Analysis of the remainder term is a little tricky; certain difficulties will become apparent. We have

$$\left| \frac{(-1)^m}{(1 + \xi)^{m+1}} \frac{x^{m+1}}{m + 1} \right| = \left| \frac{x}{1 + \xi} \right|^{m+1} \frac{1}{m + 1}.$$

In order to prove that the remainder tends to 0 it seems that we must show that $|x/(1 + \xi)| \leq 1$, whilst knowing nothing about ξ except that it lies between 0 and x. To make further progress we must consider some special cases:

(a) If $0 < x \le 1$ then $0 < \xi$, $1 + \xi > 1$ and so

$$\left| \frac{x}{1+\xi} \right| < x \le 1.$$

The remainder term is bounded by $1/(m+1)$ and so tends to 0.

(b) If $-\frac{1}{2} < x < 0$ then $-\frac{1}{2} < x < \xi < 0$, $1 + \xi > \frac{1}{2} > |x|$ and so

$$\left| \frac{x}{1+\xi} \right| < \frac{1}{2}.$$

Again the remainder term tends to 0. In fact it is bounded by $2^{-m-1}/(m+1)$ so we have reasonably fast convergence.

(c) What if $-1 < x \le -\frac{1}{2}$? There seems to be no way to ensure that $|x/(1+\xi)| \le 1$, which was the key to showing that the remainder tended to 0.

We seem to have a worse result on the representation of $\ln(1+x)$ as a power series than the one given previously and obtained by integrating the series for $(1+x)^{-1}$. There is though a small bonus. We obtain convergence at $x = 1$ and hence another proof of the formula $\ln 2 = \sum_{n=1}^{\infty} (-1)^{n-1}/n$.

11.7.4 Error Estimates for $(1+x)^a$

We consider the Lagrange remainder for $(1+x)^a$ again taking $c = 0$. Since we wish to admit any real power a, we must assume that $x > -1$. We have

$$\frac{d^k}{dx^k}(1+x)^a = a(a-1)...(a-k+1)(1+x)^{a-k}$$

$$(1+x)^a = \sum_{k=0}^{m} \frac{a(a-1)...(a-k+1)}{k!} x^k + R_{m+1}(x).$$

The series, when extended indefinitely, has radius of convergence 1. Lagrange's remainder gives

$$R_{m+1}(x) = \frac{a(a-1)...(a-m)}{(m+1)!}(1+\xi)^{a-m-1}x^{m+1}, \quad (\xi \text{ between } 0 \text{ and } x)$$

and so

$$|R_{m+1}(x)| = \left| \frac{a(a-1)...(a-m)}{(m+1)!}(1+\xi)^{a-m-1}|x|^{m+1}. \right.$$

Note that $1 + \xi > 0$ because $x > -1$.

The remainder is again tricky to analyse. In order to prove that the remainder tends to 0 we could exploit the limit

$$\lim_{m \to \infty} \frac{a(a-1)...(a-m)}{(m+1)!} x^{m+1} = 0,$$

which is valid if $|x| < 1$ (a detail for the reader to check). We would therefore seek to prove that the factor $(1 + \xi)^{a-m-1}$ is bounded as $m \to \infty$. For this it is enough if $\xi > 0$ for each m (remember that ξ actually depends on m).

We therefore have two cases. For $0 < x < 1$ it is obvious that $\xi > 0$ and the remainder tends to 0. But for $-1 < x < 0$? There seems to be no way to ensure that $(1 + \xi)^{a-m-1}$ is bounded.

11.7.5 Taylor's Theorem with Cauchy's Remainder

The deficiencies of Lagrange's form of the remainder are partially remedied by a different version: Cauchy's remainder. The downside is that the formula is hard to remember, not to mention the proof.

Proposition 11.23 *Assume that $f : A \to \mathbb{R}$ has derivatives of all orders up to n, let $c \in A$, $x \in A$. Then there exists ξ, in the interval $]c, x[$ if $x > c$, and in the interval $]x, c[$ if $x < c$, such that*

$$f(x) = \sum_{k=0}^{n-1} \frac{f^{(k)}(c)}{k!}(x-c)^k + \frac{f^{(n)}(\xi)}{(n-1)!}(x-\xi)^{n-1}(x-c).$$

Again we rewrite this by letting $x - c = h$ and $\xi = c + \theta h$, where $0 < \theta < 1$. Then we have the slightly easier to remember formula:

$$f(c+h) = \sum_{k=0}^{n-1} \frac{f^{(k)}(c)}{k!}h^k + \frac{f^{(n)}(c+\theta h)}{(n-1)!}(1-\theta)^{n-1}h^n.$$

Proof of Taylor's theorem with Cauchy's remainder For all t, such that $c + t$ is in A, we set

$$F(t) = f(c+h) - \sum_{k=0}^{n-1} \frac{(h-t)^k}{k!} f^{(k)}(c+t)$$

and differentiate (for this proof a single differentiation suffices), as follows:

$$F'(t) = \sum_{k=1}^{n-1} \frac{(h-t)^{k-1}}{(k-1)!} f^{(k)}(c+t) - \sum_{k=0}^{n-1} \frac{(h-t)^k}{k!} f^{(k+1)}(c+t)$$

$$= \sum_{k=0}^{n-2} \frac{(h-t)^k}{k!} f^{(k+1)}(c+t) - \sum_{k=0}^{n-1} \frac{(h-t)^k}{k!} f^{(k+1)}(c+t)$$

$$= -\frac{(h-t)^{n-1}}{(n-1)!} f^{(n)}(c+t).$$

We apply the mean value theorem. There exists θ in the interval $]0, 1[$, such that $F(0) - F(h) = -hF'(\theta h)$. Noting that $F(h) = 0$ we obtain

$$f(c+h) - \sum_{k=0}^{n-1} \frac{h^k}{k!} f^{(k)}(c) = \frac{(1-\theta)^{n-1} h^n}{(n-1)!} f^{(n)}(c+\theta h)$$

as was to be proved.   □

Let us return to the problem of estimating the remainder for the function $\ln(1+x)$. We have

$$\ln(1+x) = \sum_{n=1}^{m} \frac{(-1)^{n-1}}{n} x^n + R_{m+1}(x),$$

where Cauchy's remainder gives

$$R_{m+1}(x) = \frac{(-1)^m (x-\xi)^m x}{(1+\xi)^{m+1}}$$

for some ξ between 0 and x.

Previously we were unable to control the error when $-1 < x < -\frac{1}{2}$. Now suppose that $-1 < x < 0$. Then $-1 < x < \xi < 0$ and the reader should check that this implies that

$$0 < \frac{\xi - x}{1+\xi} < -x.$$

We therefore have

$$|R_{m+1}(x)| = \left| \frac{(x-\xi)^m x}{(1+\xi)^{m+1}} \right| = \left(\frac{\xi - x}{1+\xi} \right)^m \frac{|x|}{1+\xi} < \frac{|x|^{m+1}}{1+\xi} < \frac{|x|^{m+1}}{1+x}$$

which tends to 0 as $m \to \infty$.

Finally we return to the function $(1+x)^a$ where a is a real number. We recall that we were unable to control Lagrange's remainder when $-1 < x < 0$. Cauchy's remainder is

$$R_{m+1}(x) = \frac{a(a-1)...(a-m)}{m!}(1+\xi)^{a-m-1}(x-\xi)^m x,$$

(for some ξ between 0 and x)

$$= \frac{a(a-1)...(a-m)}{m!}x^{m+1}(1+\xi)^{a-1}\left(\frac{x-\xi}{x(1+\xi)}\right)^m.$$

We would like to exploit the known limit

$$\lim_{m\to\infty} \frac{a(a-1)...(a-m)}{m!}x^{m+1} = 0,$$

a consequence of the convergence of the binomial series, but this requires gaining control of the remaining factors.

Let $-1 < x < 0$. Then $-1 < x < \xi < 0$, and this gives

$$0 < 1+x < 1+\xi < 1$$

and

$$(1+\xi)^{a-1} \le \max(1, (1+x)^{a-1}).$$

Moreover

$$0 < \frac{x-\xi}{x(1+\xi)} < 1.$$

Again the reader should check these three claims, bearing in mind for the second claim that $a-1$ can be negative.

Now we have

$$|R_{m+1}(x)| \le \left|\frac{a(a-1)...(a-m)}{m!}x^{m+1}\right| \max\left(1, (1+x)^{a-1}\right),$$

and this tends to 0 as $m \to \infty$.

11.7.6 Taylor's Theorem with the Integral Form of the Remainder

Lagrange's remainder and Cauchy's remainder can be maddening because we do not know that crucial number ξ. It is therefore a relief to have a form of the remainder in which everything is known. There is a small price; the premises are slightly stronger, though not such that it matters much in practice.

Proposition 11.24 *Assume that $f : A \to \mathbb{R}$ has continuous derivatives up to order n, let $c \in A$, $x \in A$. Then*

$$f(x) = \sum_{k=0}^{n-1} \frac{f^{(k)}(c)}{k!}(x-c)^k + \frac{1}{(n-1)!} \int_c^x (x-u)^{n-1} f^{(n)}(u)\, du.$$

Proof Write

$$R_n(x) = \frac{1}{(n-1)!} \int_c^x (x-u)^{n-1} f^{(n)}(u)\, du.$$

Integrating by parts we obtain a reduction formula

$$R_n(x) = \frac{1}{(n-1)!} f^{(n-1)}(u)(x-u)^{n-1}\Big|_{u=c}^x + \frac{1}{(n-2)!} \int_c^x (x-u)^{n-2} f^{(n-1)}(u)\, du$$

$$= -\frac{1}{(n-1)!} f^{(n-1)}(c)(x-c)^{n-1} + R_{n-1}(x)$$

provided $n > 1$, and repeated applications lead to

$$R_n(x) = -\frac{1}{(n-1)!} f^{(n-1)}(c)(x-c)^{n-1} - \frac{1}{(n-2)!} f^{(n-2)}(c)(x-c)^{n-2}$$

$$- \cdots - f'(c)(x-c) + R_1(x)$$

$$= -\sum_{k=1}^{n-1} \frac{f^{(k)}(c)}{k!}(x-c)^k + R_1(x).$$

But $R_1(x) = f(x) - f(c)$ and we are done.   □

As in previous versions of Taylor's theorem we rewrite the formula. Set $u = c + s$, $x = c + h$, and write the integral with respect to s instead of u. This gives

$$R_n = \frac{1}{(n-1)!} \int_0^h (h-s)^{n-1} f^{(n)}(c+s)\, ds.$$

Finally let $s = ht$ and write the integral with respect to t. We find

$$R_n = \frac{h^n}{(n-1)!} \int_0^1 (1-t)^{n-1} f^{(n)}(c+th)\, dt.$$

This form of the remainder is often useful. For example, a variety of different remainders, of which Lagrange's and Cauchy's are mere special cases, can be derived from it. Let $1 \le p \le n$. We first write

$$R_n = \frac{h^n}{(n-1)!} \int_0^1 (1-t)^{n-p} f^{(n)}(c+th)(1-t)^{p-1}\, dt.$$

Using the mean value theorem for integrals (Proposition 6.15) we deduce that there exists θ in the interval $]0, 1[$, such that

$$R_n = \frac{h^n(1-\theta)^{n-p} f^{(n)}(c+\theta h)}{(n-1)!} \int_0^1 (1-t)^{p-1} dt$$

$$= \frac{h^n(1-\theta)^{n-p} f^{(n)}(c+\theta h)}{p(n-1)!}.$$

This version of the remainder is known under several names, most frequently Schlömilch's remainder or Roche's remainder. The case $p = 1$ is Lagrange's remainder; the case $p = n$ Cauchy's.

11.7.7 Exercises

1. Calculate $\sqrt[3]{28}$ to four correct places of decimals by using an appropriate Taylor polynomial of $\sqrt[3]{x}$ centred at $x = 27$.
2. It is proposed to approximate $\sqrt{x}$ in the interval $64 < x < 70$ by

   (a) Its first-degree Taylor polynomial centred at $x = 64$.
   (b) Its second-degree Taylor polynomial centred at $x = 64$.
   (c) Its third-degree Taylor polynomial centred at $x = 64$.

   In each case estimate the maximum error.
3. It is proposed to approximate $\sin x$ on the interval $\pi/6 < x < \pi/4$ by its fourth-degree Taylor polynomial centred at $x = \pi/6$. Determine the polynomial and estimate the maximum error.
4. Suppose we want to use the exponential series truncated after the term $x^m/m!$ to calculate e^x for $0 < x < 1$. There are several ways to estimate the error, from using a form of the remainder provided by Taylor's theorem, to estimating the tail of the series directly.

   (a) Obtain the upper bound $ex^{m+1}/(m+1)!$ for the error using Lagrange's form of the remainder.
   (b) Estimate the tail directly. Show that

   $$\sum_{k=m+1}^{\infty} \frac{x^k}{k!} < \frac{(m+2)x^{m+1}}{(m+2-x)(m+1)!}.$$

   Which estimate of the error is lower?

   Note. An algorithm that calculates e^x would only need to calculate it in the first place for $0 \le x < 1$. Values for other inputs can then be found by a further finite number of multiplications or divisions by e.
5. The error function is defined for all real x by

   $$\operatorname{erf}(x) := \frac{2}{\sqrt{\pi}} \int_0^x e^{-t^2} dt.$$

Obtain the Maclaurin series of $\mathrm{erf}(x)$ and show that it converges to $\mathrm{erf}(x)$ for all x.

Note. The error function is a non-elementary transcendental function. It is important in probability and statistics. The factor $2/\sqrt{\pi}$ ensures that $\lim_{x\to\infty}\mathrm{erf}(x)=1$. See Sect. 12.2 Exercise 2.

6. Show that if f is a polynomial of degree m then for every c we have

$$f(x) = \sum_{k=0}^{m} \frac{f^{(k)}(c)}{k!}(x-c)^k.$$

7. In the exercises in Sect. 5.8 some so-called numerical approximations to derivatives were studied. Using Taylor's theorem one can estimate the errors. In each item we assume that a is a point in the interval of definition of f, that $|h|$ is sufficiently small that $a+h$ falls within A, and that f has enough derivatives in A so that the statement makes sense. We begin with the usual difference quotient so that a comparison can be made. The intermediate value property of derivatives (Sect. 5.6 Exercise 8) can be useful.

(a) Show that for each $h \neq 0$ there exists ξ in A, such that

$$\frac{f(a+h)-f(a)}{h} - f'(a) = \frac{1}{2}hf^{(2)}(\xi).$$

(b) Show that for each $h \neq 0$ there exists ξ in A, such that

$$\frac{f(a+h)-f(a-h)}{2h} - f'(a) = \frac{1}{6}h^2 f^{(3)}(\xi).$$

This suggests that $(f(a+h)-f(a-h))/2h$ could be a superior approximation to $f'(a)$ than $(f(a+h)-f(a))/h$.

(c) Show that for each $h \neq 0$ there exists ξ in A, such that

$$\frac{f(a+h)-2f(a)+f(a-h)}{h^2} - f^{(2)}(a) = \frac{1}{12}h^2 f^{(4)}(\xi).$$

8. A popular method to interpolate between known values of a function is to use the chord joining the two points on the graph. Thus if we have points $(a, f(a))$ and $(b, f(b))$ on the graph $y=f(x)$ with $a<b$, the suggestion is to use

$$\frac{(b-x)f(a)+(x-a)f(b)}{b-a}$$

as an approximation to $f(x)$ at points x between a and b. This is known as the method of proportional parts and has a long history.

Assuming that f is twice differentiable, estimate the error by showing that for each x in the interval $]a, b[$, there exists ξ between a and b, such that

$$f(x) - \frac{(b-x)f(a) + (x-a)f(b)}{b-a} = \frac{1}{2}(x-a)(b-x)f^{(2)}(\xi).$$

Deduce that if $|f^{(2)}(x)| < M$ in the interval $[a, b]$, then

$$\left| f(x) - \frac{(b-x)f(a) + (x-a)f(b)}{b-a} \right| < \frac{1}{8}(b-a)^2 M.$$

Hint. Let $h = x - a$ and $k = b - x$. Expand $f(a+h)$ and $f(b-k)$ by Taylor's theorem and eliminate the term involving $f'(a)$.

9. Let the function f be defined on all of $\mathbb{R}$ and have derivatives up to order 2. Suppose that $|f(x)| \le A$ and $|f''(x)| \le B$ for all x. Prove that

$$|f'(x)| \le 2\sqrt{AB}$$

for all x.

Hint. Use Taylor's theorem in the form

$$f(x+h) = f(x) + hf'(x) + \frac{1}{2}h^2 f''(x + \theta h)$$

and show that

$$|f'(x)| \le \frac{2}{h}A + \frac{h}{2}B$$

for all h.

10. Prove the following theorem of S. N. Bernstein:

Let f have derivatives of all orders in an open interval A. Let c and d be points in A such that $c < d$, and assume that $f^{(n)}(x) \ge 0$ for all x in $[c, d]$ and for all n. Then the Taylor series of f with centre c converges to f in the interval $[c, d]$.

You might use the following steps to prove the theorem. Let $r = d - c$, and for each x in $[c, d]$ let $h = x - c$ and view the remainder at x as a function $R_n(h)$ of h for $0 \le h \le r$.

(a) Show that

$$0 \le R_n(h) \le \frac{h^n}{r^n} R_n(r).$$

Hint. Use the integral form of the remainder.

(b) Show that $R_n(r) \le f(d)$.

(c) Deduce that $\lim_{n \to \infty} R_n(h) = 0$ for $0 \le h \le r$.

11. Show that the Maclaurin series of $\tan x$ converges to $\tan x$ throughout the interval $]-\pi/2, \pi/2[$ and deduce that its radius of convergence is $\pi/2$.

12. Let g be an odd function on the interval $]-r, r[$, with derivatives up to the fifth order. Show that for each x in $]-r, r[$ one can write

$$g(x) = \frac{x}{3}\left(g'(x) + 2g'(0)\right) - \frac{x^5}{180}g^{(5)}(\xi)$$

for some ξ between 0 and x.

Hint. Apply Cauchy's mean value theorem to the quotient

$$\frac{g(x) - \frac{x}{3}g'(x) - \frac{2x}{3}g'(0)}{x^5}$$

and remember that g being odd, $g^{(n)}(0) = 0$ for all even n.

13. The argument used in the proof of Proposition 11.23 to obtain Cauchy's form of the remainder can be adapted to obtain Schlömilch's, or Roche's form, under the same conditions. Define the function $F(t)$ as in the proof of Proposition 11.23 and let $\phi(t) = (h - t)^p$ for some exponent $p \geq 1$. Apply Cauchy's mean value theorem to the quotient

$$\frac{F(h) - F(0)}{\phi(h) - \phi(0)}$$

and obtain the remainder in the form

$$R_n = \frac{f^{(n)}(c + \theta h)\, h^n (1 - \theta)^{n-p}}{p\,(n - 1)!}.$$

Note. This was previously obtained from the integral form of the remainder under slightly stronger conditions (the nth derivative of f was required to be continuous).

14. An example of a function whose Maclaurin series has radius of convergence 0. Define the function

$$f(x) = \sum_{n=0}^{\infty} e^{-n}\cos(n^2 x), \quad (x \in \mathbb{R}).$$

(a) Show that f has derivatives of all orders.

(b) Show that

$$f^{(2k)}(0) = (-1)^k \sum_{n=0}^{\infty} e^{-n} n^{4k}$$

for all k.

(c) Estimate $|f^{(2k)}(0)|$ from below and deduce that the radius of convergence of the Maclaurin series of f is 0.

15. In this exercise you are asked to show that the Riemann zeta function $\zeta(s) = \sum_{n=1}^{\infty} n^{-s}$ has derivatives of all orders in the domain $s > 1$. Moreover its derivative can be obtained by differentiating the series term-by-term. We assume that s is real, but the result (as well as the proof as suggested below) holds equally for complex s with $\operatorname{Re} s > 1$ in the context of complex derivatives.

(a) Let k be a positive integer. Show that the series $\sum_{n=1}^{\infty} n^{-s}(\ln n)^k$ is convergent for every $s > 1$, and that for all $\delta > 0$ it is uniformly convergent with respect to $s \geq 1 + \delta$.

(b) Deduce that $\zeta(s)$ has derivatives of all orders and

$$\zeta^{(k)}(s) = \sum_{n=1}^{\infty} n^{-s}(-\ln n)^k.$$

11.8 ($\Diamond$) Bernoulli Numbers

In this nugget we shall meet the Bernoulli numbers, a sequence of numbers that seem to crop up unexpectedly in all areas of pure and applied mathematics. Interest in them was such, that they were the subject of the first computer program (Ada Lovelace, 1843). The pointers to further study given at the end could easily read "all of mathematics".

The rule

$$\sum_{k=1}^{n} k^2 = \frac{1}{6}n(n+1)(2n+1)$$

is an easy consequence of the fact that if $g(x) = \frac{1}{6}x(x+1)(2x+1)$ then

$$x^2 = g(x) - g(x-1).$$

To obtain a comparable rule for the sum $\sum_{k=1}^{n} k^p$, or more generally for a sum $\sum_{k=1}^{n} f(k)$, where $f(x)$ is a polynomial, it would be enough to find another polynomial $g(x)$, such that

$$f(x) = g(x) - g(x-1).$$

Then we would have

$$\sum_{k=1}^{n} f(k) = g(n) - g(0).$$

Actually it is a little more convenient to seek $g(x)$, such that

$$f(x) = g(x+1) - g(x). \qquad (11.5)$$

The conclusion is then

$$\sum_{k=1}^{n} f(k) = g(n+1) - g(1).$$

The involvement of the Bernoulli numbers in the solution of the functional equation (11.5) is one of the surprises of mathematics.

Since $g(x)$ is supposed to be a polynomial, we have, by Taylor's theorem

$$g(x+1) = \sum_{k=0}^{\infty} \frac{1}{k!} g^{(k)}(x).$$

The sum is actually finite since all derivatives $g^{(k)}(x)$ are 0 for k higher than the degree of $g(x)$. Let D denote the operator of differentiation with respect to x, that is, for a given function $u(x)$ we now write Du for u'. The advantage of this notation is that multiple differentiations are written as a power of D acting on u. Formally we would like to write

$$g(x+1) = \sum_{k=0}^{\infty} \frac{1}{k!} D^k g(x) = \Big(\sum_{k=0}^{\infty} \frac{1}{k!} D^k\Big) g(x) = e^D g(x).$$

The operator e^D transforms $g(x)$ to $g(x+1)$. But of course e^D has no properly defined meaning yet. It is a piece of useful nonsense such as has often been the source of important discoveries in mathematics.

Let us continue calculating, using e^D as if it was a well defined thing. Given the polynomial $f(x)$ we seek a polynomial $g(x)$, that satisfies

$$e^D g(x) - g(x) = f(x).$$

Write this as

$$(e^D - 1)g(x) = f(x).$$

Shouldn't the solution be somehow

$$g(x) = \frac{1}{e^D - 1} f(x)?$$

We can even guess how to work this out. We continue without regard for rigour. We write

$$\frac{1}{e^t - 1} = \frac{1}{t} \frac{t}{e^t - 1}.$$

Now $t/(e^t - 1)$ extends to a function on $\mathbb{R}$ that has derivatives of all orders; we only have to define its value as 1 for $t = 0$. It has a Maclaurin series

$$\sum_{k=0}^{\infty} \frac{B_k}{k!} t^k$$

where the coefficients B_k are known as the Bernoulli numbers. They are all rational, as we shall see.

A possible interpretation of $\dfrac{1}{e^D - 1} f(x)$ might begin with

$$\frac{1}{e^D - 1} f(x) = \frac{1}{D} \frac{D}{e^D - 1} f(x) = \frac{1}{D} \Big(\sum_{k=0}^{\infty} \frac{B_k}{k!} D^k \Big) f(x) = \frac{1}{D} \sum_{k=0}^{\infty} \frac{B_k}{k!} f^{(k)}(x).$$

Let m be the degree of the polynomial f. The infinite sum reduces to a finite one and we have the formula

$$\frac{1}{e^D - 1} f(x) = \frac{1}{D} \sum_{k=0}^{m} \frac{B_k}{k!} f^{(k)}(x)$$

$$= \int \sum_{k=0}^{m} \frac{B_k}{k!} f^{(k)}(x)\, dx$$

$$= B_0 \int f(x)\, dx + \sum_{k=1}^{m} \frac{B_k}{k!} f^{(k-1)}(x),$$

where we have naturally interpreted $1/D$ as the instruction to find an antiderivative.

We proceed to justify this calculation. Basic to this is the following proposition, that enables us to calculate the Maclaurin series of $1/f$ from that of f by algebra alone.

Proposition 11.25 *Let f be a function, with domain $]-r, r[$, and derivatives of all orders. Assume that $f(x) \neq 0$ for all x in $]-r, r[$. Then*

(1) The reciprocal $1/f$ has derivatives of all orders in $]-r, r[$.
(2) Let $\sum_{n=0}^{\infty} a_n x^n$ be the Maclaurin series of f and let $\sum_{n=0}^{\infty} b_n x^n$ be the Maclaurin series of $1/f$. Then for each m there is a polynomial $Q(x)$ (depending on m), such that

$$\Big(\sum_{n=0}^{m} a_n x^n \Big) \Big(\sum_{n=0}^{m} b_n x^n \Big) = 1 + x^{m+1} Q(x).$$

Proof Conclusion 1 is a consequence of the formula $(1/f)' = -f'/f^2$. We can clearly differentiate repeatedly.

As for conclusion 2, we know, by Proposition 5.14, that

$$\sum_{n=0}^{m} a_n x^n = f(x) + x^m g(x), \qquad \sum_{n=0}^{m} b_n x^n = \frac{1}{f(x)} + x^m h(x)$$

where g and h are continuous in $]-r, r[$ and $g(0) = h(0) = 0$. Therefore

$$\left(\sum_{n=0}^{m} a_n x^n\right)\left(\sum_{n=0}^{m} b_n x^n\right) = \left(f(x) + x^m g(x)\right)\left(\frac{1}{f(x)} + x^m h(x)\right) = 1 + x^m R(x)$$

where

$$R(x) = \frac{g(x)}{f(x)} + f(x)h(x) + x^m g(x)h(x).$$

The last two equations show that $R(x)$ is of the form $x^{-m} S(x)$, where $S(x)$ is a polynomial, but also that $R(x)$ is continuous in $]-r, r[$. This can only hold if the polynomial $S(x)$ is divisible by x^m, so that in fact $R(x)$ is itself a polynomial. More than that, $R(0) = 0$, so that $R(x)$ is divisible by x. Then $R(x) = xQ(x)$ and $Q(x)$ is a polynomial, as required. $\qquad\square$

It is noteworthy that the proposition in no way requires convergence of the Maclaurin series of f and $1/f$; they may diverge, or even converge but to functions different from f and $1/f$.

Specialising to the case in hand, we have

$$\left(\sum_{k=0}^{m} \frac{1}{(k+1)!} x^k\right)\left(\sum_{k=0}^{m} \frac{B_k}{k!} x^k\right) = 1 + x^{m+1} Q(x),$$

where $Q(x)$ is a polynomial (which depends on m). All the functions in this equation are now polynomials; so we can replace x by the operator D and assert the following relation involving differential operators:

$$\left(\sum_{k=0}^{m} \frac{1}{(k+1)!} D^k\right)\left(\sum_{k=0}^{m} \frac{B_k}{k!} D^k\right) = 1 + D^{m+1} Q(D).$$

This we multiply by D to yield

$$\left(\sum_{k=0}^{m} \frac{1}{(k+1)!} D^{k+1}\right)\left(\sum_{k=0}^{m} \frac{B_k}{k!} D^k\right) = D + D^{m+2} Q(D).$$

Let $f(x)$ now be a polynomial of degree m. We seek $g(x)$ that satisfies

$$g(x+1) - g(x) = f(x).$$

Let $F(x)$ be an antiderivative for $f(x)$. Then $F(x)$ has degree $m+1$ and so

$$\left(\sum_{k=0}^{m} \frac{1}{(k+1)!} D^{k+1}\right)\left(\sum_{k=0}^{m} \frac{B_k}{k!} D^k\right) F(x) = DF(x) + D^{m+2} Q(D) F(x)$$

$$= f(x).$$

The solution to the problem $g(x+1) - g(x) = f(x)$ is therefore

$$g(x) = \left(\sum_{k=0}^{m} \frac{B_k}{k!} D^k\right) F(x) = B_0 F(x) + \sum_{k=1}^{m} \frac{B_k}{k!} D^{k-1} f(x),$$

precisely the formula that we had guessed.

Now we can write down the desired sum formula, the original motivation for all these calculations:

$$\sum_{k=1}^{n} f(k) = B_0 F(n+1) + \sum_{k=1}^{m} \frac{B_k}{k!} D^{k-1} f(n+1) - B_0 F(1) - \sum_{k=1}^{m} \frac{B_k}{k!} D^{k-1} f(1)$$

$$= B_0 \int_{1}^{n+1} f + \sum_{k=1}^{m} \frac{B_k}{k!} \left(D^{k-1} f(n+1) - D^{k-1} f(1)\right).$$

It is interesting to replace n by $n-1$ (on both sides of course), use the facts that $B_0 = 1$ and $B_1 = -\frac{1}{2}$, and add $f(n)$ to both sides. The result can be written as

$$\sum_{k=1}^{n} f(k) = \int_{1}^{n} f + \frac{f(1) + f(n)}{2} + \sum_{k=2}^{m} \frac{B_k}{k!} \left(D^{k-1} f(n) - D^{k-1} f(1)\right).$$

This should be compared to the result of Sect. 8.1 Exercise 2. It is again an instance of the Euler–Maclaurin summation formula.

11.8.1 Computing the Bernoulli Numbers

One can compute the Maclaurin series of $1/f$ from the Maclaurin series of f by purely algebraic means, without requiring convergence of the series (compare Sect. 11.4 Exercise 18). We start from the equation

$$\left(\sum_{k=0}^{m} \frac{1}{(k+1)!} x^k\right) \left(\sum_{k=0}^{m} \frac{B_k}{k!} x^k\right) = 1 + x^{m+1} Q(x).$$

Setting $x = 0$ we deduce that $B_0 = 1$. We can multiply the polynomial factors on the left-hand side, obtain the coefficient of x^n in the product for $n = 1, 2, ..., m$, and equate it to 0. This gives

$$\sum_{k=0}^{n} \frac{B_k}{k!(n-k+1)!} = 0$$

and there is clearly no cap on n since we may raise m as much as we like. It is convenient to multiply by $(n+1)!$ and use binomial coefficients. We obtain the recurrence relations for the Bernoulli numbers

$$\sum_{k=0}^{n} \binom{n+1}{k} B_k = 0, \quad n = 1, 2, 3, \ldots.$$

Some samples, after $B_0 = 1$, are the relations:

$$B_0 + 2B_1 = 0 \quad \Rightarrow \quad B_1 = -\frac{1}{2}$$

$$B_0 + 3B_1 + 3B_2 = 0 \quad \Rightarrow \quad B_2 = \frac{1}{6}$$

$$B_0 + 4B_1 + 6B_2 + 4B_3 = 0 \quad \Rightarrow \quad B_3 = 0$$

$$B_0 + 5B_1 + 10B_2 + 10B_3 + 5B_4 = 0 \quad \Rightarrow \quad B_4 = -\frac{1}{30}$$

$$B_0 + 6B_1 + 15B_2 + 20B_3 + 15B_4 + 6B_5 = 0 \quad \Rightarrow \quad B_5 = 0$$

$$B_0 + 7B_1 + 21B_2 + 35B_3 + 35B_4 + 21B_5 + 7B_6 = 0 \quad \Rightarrow \quad B_6 = \frac{1}{42}$$

and so on.

11.8.2 Exercises

1. Prove that all the Bernoulli numbers are rational.
2. Prove that $B_n = 0$ if n is an odd number higher than 1.
3. Obtain a formula for $\sum_{k=1}^{n} k^p$ as a polynomial $P(n)$ of degree $p + 1$.
4. Prove the following formulas, in each case giving a lower estimate for the number r in the range of validity (not necessarily the same in both formulas):

   (a) $\cot x = \dfrac{1}{x} + \displaystyle\sum_{k=1}^{\infty} \dfrac{(-1)^k 2^{2k} B_{2k}}{(2k)!} x^{2k-1}, \quad (0 < |x| < r).$

   (b) $\tan x = \displaystyle\sum_{k=1}^{\infty} \dfrac{(-1)^{k-1} 2^{2k}(2^{2k} - 1) B_{2k}}{(2k)!} x^{2k-1}, \quad (-r < x < r).$

11.8.3 Pointers to Further Study

→ Euler–Maclaurin summation
→ Faulhaber's formula
→ Number theory
→ Combinatorics

11.9   (◊) **Asymptotic Orders of Magnitude**

Consider the function $f(x) = x^5 + 2x^4 + 3x^2 + 4x + 5$. Suppose that we want to give some idea of $f(x)$ when x is large. Of course the question is rather vague. What meaning should we attach to "some idea"? What meaning to "large"?

It all depends on context and purpose. It might be adequate to point out that $\lim_{x \to \infty} f(x) = \infty$; $f(x)$ is simply really big when x is really big. We may want to give some idea of how big $f(x)$ is when x is really big. Then we might point out that $f(x)$ grows at about the same rate as x^5, vanishingly slowly compared to x^6, but much faster than x^4.

To express such statements concisely, a notation has been devised and was popularised largely by the mathematician Edmund Landau. In the above example we had a function $f(x)$ and we wished to compare it to powers of x for large x. More generally let $h(x)$ be a function, positive on an interval $]K, \infty[$, which will be used for comparison. We write

$$f(x) = O(h(x)), \quad (x \to \infty)$$

to mean that $f(x)/h(x)$ is bounded on some interval $[L, \infty[$. We write

$$f(x) = o(h(x)), \quad (x \to \infty)$$

to mean that $\lim_{x \to \infty} f(x)/h(x) = 0$.

The notation can be used for sequences. Let $(a_n)_{n=1}^{\infty}$ be a sequence and let $(h_n)_{n=1}^{\infty}$ be a positive sequence, intended for comparison. We write

$$a_n = O(h_n), \quad (n \to \infty)$$

to mean that a_n/h_n is bounded. We write

$$a_n = o(h_n), \quad (n \to \infty)$$

to mean that $\lim_{n \to \infty} a_n/h_n = 0$.
Thus

$$x^5 + 2x^4 + 3x^2 + 4x + 5 = O(x^5), \quad (x \to \infty)$$

$$x^5 + 2x^4 + 3x^2 + 4x + 5 = o(x^6), \quad (x \to \infty).$$

It should be understood that these are statements about the functions named here; the variable x is in fact a bound variable. This can be seen by spelling out the meaning in full, which requires the quantifier "for all $x > L$".

Now it has to be said that there is something a little strange about the use of an equality sign here. The apparent equation expresses that f has a certain property; for example $f(x) = O(1)$ says that f is bounded in some interval $[K, \infty[$. And we certainly cannot write $O(1) = f(x)$ to express the same thing. Nevertheless, certain properties of equality do obtain here; for example if $f(x) = O(h(x))$ then it may be possible, if care is taken, to substitute $O(h)$ for f in a relation that contains f. We shall see examples of this.

The notation extends naturally to other destinations than ∞. For example, the statements

$$f(x) = O(h(x)), \quad (x \to a)$$

and

$$f(x) = o(h(x)), \quad (x \to a)$$

mean, respectively, that there exists $\delta > 0$, such that $f(x)/h(x)$ is bounded for $0 < |x - a| < \delta$, and $\lim_{x \to a} f(x)/h(x) = 0$.

What makes the notation useful is a certain flexibility. For example an expression $O(h)$ can appear in algebraic combinations. Some examples follow with elucidations of the meaning where appropriate. The proofs are left as exercises.

(a) $O(x^n) + O(x^m) = O(x^{\min(m,n)}), \quad (x \to 0)$.

Meaning. If $f(x) = O(x^n)$ and $g(x) = O(x^m)$ then $f(x) + g(x) = O(x^{\min(m,n)})$. Here m and n are integers, positive or negative. Real m and n could be admitted if x approaches 0 from above.

Exercise   Write the corresponding statement in the case $x \to \infty$.

(b) $x^m O(x^n) = O(x^{m+n}), \quad (x \to 0)$.

Meaning. If $f(x) = O(x^n)$ then $x^m f(x) = O(x^{m+n})$.

(c) $\dfrac{1}{1+x} = 1 - x + O(x^2), \quad (x \to 0)$.

Meaning. $\dfrac{1}{1+x} - 1 + x = O(x^2)$.

(d) $\dfrac{1}{1+x+O(x^2)} = 1 - x + O(x^2), \quad (x \to 0)$.

Meaning. If $f(x) = O(x^2)$ then

$$\frac{1}{1+x+f(x)} = 1 - x + O(x^2).$$

Formula (d) can be thought of as the result of substituting $x + O(x^2)$ for x in formula (c), thus exploiting the true formula $x = x + O(x^2)$. It is particularly useful for obtaining asymptotic information about quotients from a few basic Maclaurin series.

Here are some further examples. The derivations are left as exercises.

(e) $\dfrac{1}{x} - \dfrac{1}{\sin x} = -\dfrac{x}{6} + O(x^3), \quad (x \to 0)$.

One can obtain this by using $\sin x = x - (x^3/6) + O(x^5)$. We deduce from (e) that

$$\lim_{x \to 0} \frac{1}{x^2} - \frac{1}{x \sin x} = -\frac{1}{6}.$$

The calculation of this limit using L'Hopital's rule is much longer.

(f) $\quad \dfrac{1}{\sin x} = \dfrac{1}{x} + \dfrac{x}{6} + \dfrac{7x^3}{360} + O(x^5), \quad (x \to 0).$

We deduce from this that

$$\lim_{x \to 0} \frac{1}{x^3} \left(\frac{1}{\sin x} - \frac{1}{x} - \frac{x}{6} \right) = \frac{7}{360}.$$

Obtaining this by L'Hopital's rule is a long haul; compare the limits in Sect. 5.8 Exercise 2.

(g) $\quad \dfrac{n^5 - n^3 + 1}{n^5 + n^3 + 1} = 1 - \dfrac{2}{n^2} + \dfrac{2}{n^4} + O\left(\dfrac{1}{n^6}\right), \quad (n \to \infty).$

11.9.1 Asymptotic Expansions

We recall Proposition 5.14, sometimes called Peano's form of Taylor's theorem. This may be expressed using Landau's notation. Suppose that f has derivatives of all orders at the point a. Then, for each n we have

$$f(x) = \sum_{k=0}^{n} \frac{f^{(k)}(a)}{k!}(x - a)^k + o\big((x - a)^n\big), \quad (x \to a).$$

This is an example of an asymptotic expansion. Note that the convergence, or otherwise, of the Taylor series

$$\sum_{k=0}^{\infty} \frac{f^{(k)}(a)}{k!}(x - a)^k$$

is irrelevant. It could even be divergent for all $x \neq a$.

The sequence of functions $(x - a)^n$, $(n = 1, 2, 3, ...)$, form what is called an asymptotic scale as $x \to a$. More generally an asymptotic scale (as $x \to a$) is a sequence of functions $h_n(x)$, $(n = 1, 2, 3, ...)$, such that for each n we have

$$h_{n+1}(x) = o\big(h_n(x)\big), \quad (x \to a).$$

An asymptotic expansion of a function f relative to the asymptotic scale $(h_n)_{n=1}^{\infty}$ is then of the form

$$f(x) = \sum_{k=1}^{n} c_k h_k(x) + o(h_n(x)), \quad (x \to a)$$

where the coefficients c_k are real numbers, which can be continued up to some index $k = N$, or else indefinitely.

In spite of the ubiquity of Taylor series it seems unusual for them to be divergent in practical applications. The most common type of asymptotic expansion is one that gives successively better approximations as $x \to \infty$.

We shall explore one simple example of this. For $x > 0$ we define

$$E(x) = \int_0^1 \frac{1}{t} e^{-x/t}\, dt.$$

There is no problem with the integral at $t = 0$ since the integrand tends to 0.

Successive integration by parts leads to a reversed reduction formula. It is simplest to describe it by letting

$$I_n = \frac{(-1)^n n!}{x^n} \int_0^1 t^{n-1} e^{-x/t}\, dt, \quad n = 0, 1, 2, \ldots$$

Then integration by parts gives

$$I_n = \frac{(-1)^n n! e^{-x}}{x^{n+1}} + I_{n+1}.$$

Exercise Derive this formula.

Now we obtain, starting at I_0,

$$E(x) = e^{-x} \sum_{k=0}^{n} \frac{(-1)^k k!}{x^{k+1}} + I_{n+1}.$$

We shall see that this is an asymptotic expansion as $x \to \infty$ relative to the asymptotic scale

$$h_n(x) = e^{-x} \frac{n!}{x^{n+1}}, \quad (n = 1, 2, 3, \ldots).$$

Exercise Check that $h_{n+1}(x) = o(h_n(x))$, $(x \to \infty)$.

Next one has to check that $I_{n+1}(x) = o(h_n(x))$. This reduces to showing that

$$\lim_{x \to \infty} e^x \int_0^1 t^n e^{-x/t}\, dt = \lim_{x \to \infty} \int_0^1 t^n e^{-x(\frac{1}{t} - 1)}\, dt = 0.$$

Exercise Prove this. One way is to show that for each $\delta > 0$ the integral from 0 to $1 - \delta$ tends to 0 as $x \to \infty$, by noting that the integrand converges to 0 as $x \to \infty$, uniformly with respect to t in the interval $[0, 1 - \delta]$. The remaining part of the integral, from $1 - \delta$ to 1, is bounded by δ.

In this example the series $\sum_{k=0}^{n} (-1)^k k!/x^{k+1}$ is divergent for all $x > 0$. Nevertheless its terms initially decrease; in fact they do so as long as $k < x$. Looking ahead to the next nugget, on Stirling's approximation to $k!$, we can see that with increasing k, the term $k!/x^{k+1}$ reaches a minimum at around $k = x$ of around $\sqrt{2\pi k}\, e^{-k}$. For $k = 10$ this is about 10^{-4} so we might expect the sum of ten terms to give a nice approximation to $E(10)$. This is the case.

Exercise Show that $|I_{n+1}| \leq e^{-x} n!/x^{n+1}$, and therefore the error is less than the last term used of the expansion.

11.9.2 Pointers to Further Study

$\rightarrow$ Special functions
$\rightarrow$ Asymptotic expansions

11.10 (◊) Stirling's Approximation

The factorial function $n!$ grows rapidly with increasing n, and it quickly becomes impossible to compute it by multiplying together all integers from 1 to n. In the simplest instance Stirling's approximation compares $n!$ to n^n, which is easier to compute.

A good clue as to how to approximate $n!$ comes from writing

$$\ln n! = \sum_{k=2}^{n} \ln k.$$

This suggests that we could approximate $\ln n!$, or equivalently the sum $\sum_{k=2}^{n} \ln k$, by comparing it to the integral $\int_1^n \ln x\, dx$, which is equal to $n \ln n - n + 1$. So we should compare $n!$ to $n^n e^{-n+1}$. We have arrived very near to Stirling's approximation.

Proposition 11.26 (Stirling's approximation)

$$\lim_{n \to \infty} \frac{n!}{n^{n+\frac{1}{2}} e^{-n}} = \sqrt{2\pi}.$$

According to this, we can use $n^{n+\frac{1}{2}} e^{-n} \sqrt{2\pi}$ as an approximation to $n!$ when n is big. This is an example of an asymptotic approximation. For example $10! = 3628800$

and Stirling's approximation is 3598696. The error is within 1%. For larger n we can expect the percentage error to decrease, though the actual difference error may increase. A stronger result yields upper and lower bounds:

$$n^{n+\frac{1}{2}}e^{-n}\sqrt{2\pi} < n! < n^{n+\frac{1}{2}}e^{-n+\frac{1}{12n}}\sqrt{2\pi}.$$

The proof of Stirling's approximation is in several steps and builds strongly on the strict concavity of the function $\ln x$. The reader might like to treat these steps as exercises; so we give the proofs in a separate section.

Step 1. If r is a natural number and $r \geq 2$ then

$$\int_{r-\frac{1}{2}}^{r+\frac{1}{2}} \ln x \, dx < \ln r, \qquad \frac{1}{2}\big(\ln(r-1) + \ln r\big) < \int_{r-1}^{r} \ln x \, dx.$$

Step 2.

$$\int_{\frac{3}{2}}^{n} \ln x \, dx < \ln(n!) - \frac{1}{2}\ln n < \int_{1}^{n} \ln x \, dx.$$

Step 3. Let $u_n = \ln(n!) - (n + \frac{1}{2})\ln n + n$. Then

$$\frac{3}{2}\left(1 - \ln\left(\frac{3}{2}\right)\right) < u_n < 1.$$

Step 4. $(u_n)_{n=1}^{\infty}$ is a decreasing sequence.

Step 5. The limit

$$\lim_{n\to\infty} \frac{n!}{n^{n+\frac{1}{2}}e^{-n}} = A$$

exists and

$$\left(\frac{2e}{3}\right)^{\frac{3}{2}} \leq A < e.$$

Step 6. $A = \sqrt{2\pi}$.

11.10.1 Proofs

Proof of step 1 For the first inequality we use the strict concavity of the function $\ln x$. Its graph lies beneath its tangent at the point $(r, \ln r)$. The slope of the tangent is $1/r$. Hence we find

$$\ln x < \ln r + \frac{x - r}{r}, \qquad (x \neq r)$$

Integrate from $r - \frac{1}{2}$ to $r + \frac{1}{2}$. We find

$$\int_{r-\frac{1}{2}}^{r+\frac{1}{2}} \ln x \, dx < \ln r.$$

For the second inequality we again use the strict concavity of $\ln x$. The chord joining $(r - 1, \ln(r - 1))$ and $(r, \ln r)$ lies beneath the graph. That is,

$$(x - r + 1) \ln r + (r - x) \ln(r - 1) < \ln x, \quad (r - 1 < x < r).$$

Integrate from $r - 1$ to r. We find

$$\frac{1}{2} \ln r + \frac{1}{2} \ln(r - 1) < \int_{r-1}^{r} \ln x \, dx.$$

□

Proof of step 2 The first inequality from step 1, together with the fact that $\ln x$ is increasing, gives

$$\int_{\frac{3}{2}}^{n} \ln x \, dx = \sum_{r=2}^{n} \int_{r-\frac{1}{2}}^{r+\frac{1}{2}} \ln x \, dx - \int_{n}^{n+\frac{1}{2}} \ln x \, dx < \sum_{r=2}^{n} \ln r - \frac{1}{2} \ln n$$

leading to

$$\int_{\frac{3}{2}}^{n} \ln x \, dx < \ln(n!) - \frac{1}{2} \ln n.$$

Sum the second inequality of step 1 from $r = 2$ to $r = n$. We find

$$\frac{1}{2} \ln \big((n - 1)!\big) + \frac{1}{2} \ln(n!) < \int_{1}^{n} \ln x \, dx$$

or

$$\ln(n!) - \frac{1}{2} \ln n < \int_{1}^{n} \ln x \, dx.$$

□

Proof of step 3 Compute the integral in step 2. We find

$$n \ln n - n - \frac{3}{2} \ln \left(\frac{3}{2}\right) + \frac{3}{2} < \ln(n!) - \frac{1}{2} \ln n < n \ln n - n + 1$$

which implies the sought-after inequality.

□

Proof of step 4 It is easily seen that

$$u_n - u_{n-1} = 1 - \left(n - \frac{1}{2}\right)\big(\ln n - \ln(n - 1)\big).$$

Compute the integral in the second inequality in step 1. We obtain (with n in place of r):

$$\frac{1}{2}\big(\ln(n-1)+\ln n\big) < n\ln n - n - (n-1)\ln(n-1) + n - 1$$

or

$$1 < \Big(n-\frac{1}{2}\Big)\big(\ln n - \ln(n-1)\big)$$

so that $u_n < u_{n-1}$. □

Proof of step 5 That the limit exists follows since u_n is decreasing, and it satisfies the estimates of step 3. The lower estimate for A is about 2.43. □

Proof of step 6 Let

$$a_n = \frac{n!}{n^{n+\frac{1}{2}}e^{-n}}.$$

Then we have

$$\frac{a_{2n}}{a_n^2} = \frac{(2n)!}{(2n)^{2n+\frac{1}{2}}e^{-2n}}\,\frac{n^{2n+1}e^{-2n}}{(n!)^2} = \frac{(2n)!}{(n!)^2} = \frac{(2n)!\sqrt{n}}{(n!)^2\,2^{2n}\sqrt{2}}.$$

Wallis's product for π (see the exercises below) is

$$\sqrt{\pi} = \lim_{n\to\infty}\frac{(n!)^2\,2^{2n}}{(2n)!\sqrt{n}}.$$

We conclude that

$$\sqrt{\pi} = \frac{1}{\sqrt{2}}\lim_{n\to\infty}\frac{a_n^2}{a_{2n}} = \frac{A}{\sqrt{2}}$$

so that $A = \sqrt{2\pi}$. □

11.10.2 Exercises

1. Prove Wallis's product for π. You can do this in the following steps. First set
   $I_n = \int_0^{\pi/2} \sin^n x\, dx$.

   (a) Prove that $I_{2m-1} > I_{2m} > I_{2m+1}$.
   (b) Show that
   $$\lim_{m\to\infty}\frac{I_{2m+1}}{I_{2m-1}} = 1.$$

   Hint. Use Wallis's integrals, (see Sect. 8.2.6).

(c) Show that

$$\lim_{m \to \infty} \frac{I_{2m}}{I_{2m-1}} = 1.$$

(d) Deduce Wallis's product by applying Wallis's integrals.

2. With the aid of Stirling's approximation, determine whether the series

$$\sum_{n=0}^{\infty} (-1)^n \frac{(2n)!}{2^{2n} (n!)^2}$$

is absolutely convergent, conditionally convergent or divergent.

11.10.3   *Pointers to Further Study*

$\rightarrow$ Gamma function
$\rightarrow$ Computational mathematics

Chapter 12
Improper Integrals

Hardy in his thirties held the view that the late years of a
mathematician's life were spent most profitably in writing books

J. E. Littlewood

12.1 Unbounded Domains and Unbounded Integrands

The definition of the Riemann–Darboux integral assumes that the function is bounded and its domain is a bounded and closed interval. It is desirable to have an integration theory that can be applied directly if the function is unbounded, or the domain is unbounded, or both. This is accomplished with the Lebesgue integral and deserves properly a book of its own.

It is however possible to enlarge the scope of the Riemann–Darboux integral by introducing improper integrals. These are defined as limits of proper (ordinary) integrals. There is really nothing improper about improper integrals. The name simply reflects the fact they are not defined by the normal method laid down for the Riemann–Darboux integral, but by one that builds on it through an additional limiting procedure.

As examples consider the two integrals:

$$\int_0^1 \frac{1}{\sqrt{x}}\,dx, \qquad \int_0^\infty e^{-x}\,dx.$$

These are *improper integrals* and are defined as limits of normal integrals. Riemann–Darboux integration is not immediately applicable; in the first integral the integrand is unbounded so we cannot form upper or lower sums; in the second the interval is unbounded so we cannot partition it into finitely many bounded intervals. These typify the two primary cases, the only ones we shall consider here.

R. Magnus, *Fundamental Mathematical Analysis*, Springer Undergraduate
Mathematics Series, https://doi.org/10.1007/978-3-030-46321-2_12

Case A. For every $\varepsilon > 0$ the function f is bounded and integrable on the interval $[a + \varepsilon, b]$, but unbounded on $[a, b]$. The improper integral from a to b is defined as the limit

$$\int_a^b f = \lim_{\varepsilon \to 0+} \int_{a+\varepsilon}^b f$$

if the limit exists. The endpoint b can be handled in a similar way if f is integrable on $[a, b - \varepsilon]$ for each ε.

Cauchy's principle gives a necessary and sufficient condition for the limit to exist and be finite: for all $\varepsilon > 0$ there exists $\delta > 0$, such that $\left| \int_x^y f \right| < \varepsilon$ for all x and y that satisfy $a < x < y < a + \delta$.

Case B. For every $L > 0$ the function f is bounded and integrable on the interval $[a, L]$. The improper integral from 0 to ∞ is defined as the limit

$$\int_a^\infty f = \lim_{L \to \infty} \int_a^L f$$

if the limit exists. Integrals with lower limit $-\infty$ are handled in a similar way.

Again we have by Cauchy's principle a necessary and sufficient condition for the limit to exist and be finite: for all $\varepsilon > 0$ there exists $K > 0$, such that $\left| \int_x^y f \right| < \varepsilon$ for all x and y that satisfy $K < x < y$.

In each case, if the limit exists and is a finite number, we say that the improper integral is *convergent*. If the limit is infinite, or does not exist, we say that the integral is *divergent*.

An integral such as $\int_{-\infty}^\infty f$ is *improper at both ends*. We say that it is convergent if it is *convergent at each end separately*. This means that each of the integrals

$$\int_0^\infty f \quad \text{and} \quad \int_{-\infty}^0 f$$

is convergent.

Another example of an integral improper at both ends is

$$\int_0^1 x^{-1/2}(1 - x)^{-1/2}\, dx.$$

This is considered convergent if it is convergent at both ends; for example, if both the integrals

$$\int_0^{1/2} x^{-1/2}(1 - x)^{-1/2}\, dx \quad \text{and} \quad \int_{1/2}^1 x^{-1/2}(1 - x)^{-1/2}\, dx$$

are convergent (the choice of where to split the interval clearly does not matter).

12.1.1 Key Examples of Improper Integrals

We compile a list of improper integrals that can be used as yardsticks for studying the convergence or divergence of a large number of cases. We assume that p is a real number.

(1) $\displaystyle\int_1^\infty \frac{1}{x^p}\,dx$ is convergent (at ∞) if and only if $p > 1$.

(2) $\displaystyle\int_0^1 \frac{1}{x^p}\,dx$ is convergent (at 0) if and only if $p < 1$.

(3) $\displaystyle\int_0^\infty \frac{1}{x}\,dx$ is divergent at both ends.

To check these claims we note that if $p \neq 1$ we have

$$\int_1^R \frac{1}{x^p}\,dx = \frac{x^{1-p}}{1-p}\Big|_1^R = \frac{R^{1-p}}{1-p} - \frac{1}{1-p}$$

and the limit as $R \to \infty$ is a finite number if and only if $p > 1$, whereas the limit as $R \to 0$ is a finite number if and only if $p < 1$.

The case $p = 1$ is exceptional; for

$$\int_1^R \frac{1}{x}\,dx = \ln R$$

and neither the limit as $R \to \infty$ nor that as $R \to 0+$ is finite.

Exercise Calculate the integrals in items 1 and 2 in the cases when they are convergent.

We continue the list with

(4) $\displaystyle\int_1^\infty x^a e^{-bx}\,dx.$

The integral is convergent (at ∞) in the following two sets of cases: if $b > 0$ with no condition on a; or, if $b = 0$ and $a < -1$. In all other cases it is divergent.

(5) $\displaystyle\int_e^\infty (\ln t)^a t^{-b-1}\,dt.$

The integral is convergent under precisely the same conditions as the integral in item 4.

For item 5 we can set $x = \ln t$ and reduce it to item 4, which we now consider in detail. We shall apply a comparison test for improper integrals, analogous to the comparison test for series. The details of this are in the next section, but for now we proceed intuitively.

Firstly, if $b > 0$ we let $0 < c < b$. Then $x^a e^{-(b-c)x}$ tends to 0 as $x \to \infty$. There exists $K > 0$, such that $x^a e^{-(b-c)x} < 1$ for all $x > K$. Hence

$$0 < x^a e^{-bx} < e^{-cx}$$

for all $x > K$. The integral

$$\int_1^\infty e^{-cx}\, dx$$

is convergent by calculation. Now we can apply the comparison test for improper integrals (see the next section) and conclude that the integral in item 4 is convergent.

Secondly, if $b = 0$ the integral reduces to item 1. Thirdly, if $b < 0$ we choose c, such that $0 < c < -b$. Then $\lim_{x \to \infty} x^{-a} e^{(c+b)x} = 0$ and there exists $K > 0$, such that

$$x^a e^{-bx} > e^{cx} \quad \text{for all } x > K.$$

The integral

$$\int_1^\infty e^{cx}\, dx$$

diverges by calculation. Again we can use the comparison test, referring the reader to the next section, and conclude that the integral in item 4 is now divergent.

Exercise Calculate the integrals in items 4 and 5 in the cases when $b > 0$ and a is a non-negative integer.

12.1.2   The Comparison Test for Improper Integrals

There is a certain similarity between improper integrals and infinite series. This is exhibited first of all in the use of convergence tests for integrals, just as in the case of series. We even have a comparison test, already used in the previous section. Just as the comparison test for series is only applicable to positive series, the comparison test for integrals is only applicable to positive integrands.

We shall look at two cases; other cases are similar and the reader should provide the details. In all cases we assume that $f(x) \geq 0$ for all x in the domain. The proofs, as in the case of positive series, are simple applications of the fact that an increasing function bounded above must converge to a finite limit.

Case A. The integral

$$\int_a^b f,$$

where $f(x)$ is unbounded as $x \to a+$, but integrable on $[a + \varepsilon, b]$ for each $\varepsilon > 0$.

The comparison test for case A assumes that we have another function g on $[a, b]$, that is bounded, positive and integrable on $[a + \varepsilon, b]$ for each $\varepsilon > 0$. As in the case of positive series there are two modes of comparison: ordering of functions, and limit comparison. The conclusions are as follows:

(1) If $f(x) \le g(x)$ in $]a, b]$ and $\int_a^b g$ is convergent (at a), then $\int_a^b f$ is also convergent.

(2) If $g(x) \le f(x)$ in $]a, b]$ and $\int_a^b g$ is divergent (at a), then $\int_a^b f$ is also divergent.

(3) If $\lim_{x \to a+} f(x)/g(x)$ exists and is neither 0 nor ∞, then either both integrals are convergent or both integrals are divergent.

Case B. The integral

$$\int_a^\infty f$$

where f is integrable on $[0, L]$ for each $L > 0$.

The comparison test for case B assumes that we have a function g on $[a, \infty[$, positive and integrable on $[a, L]$ for each $L > 0$. There are three conclusions, as follows:

(1) If $f(x) \le g(x)$ in $[a, \infty[$ and $\int_a^\infty g$ is convergent (at ∞), then $\int_a^\infty f$ is also convergent.

(2) If $g(x) \le f(x)$ in $[a, \infty[$ and $\int_a^\infty g$ is divergent, then $\int_a^\infty f$ is also divergent.

(3) If $\lim_{x \to \infty} f(x)/g(x)$ exists and is neither 0 nor ∞, then either both integrals are convergent or both integrals are divergent.

Limit comparison, that is, the third conclusion in both cases, is very useful as the limit can often be found after guessing a suitable comparison function g. The key examples given earlier provide a first catalogue of prospective comparison functions.

12.1.3 Exercises

(1) Test the following improper integrals for convergence. In the first two a comparison function is suggested. After that you are on your own.

(a) $\displaystyle\int_0^1 \frac{1}{\sqrt{x}(1+x^2)}\, dx$; comparison function $1/\sqrt{x}$.

(b) $\displaystyle\int_1^\infty \frac{1}{x\sqrt{1+x^2}}\, dx$; comparison function $1/x^2$.

(c) $\displaystyle\int_1^\infty \frac{1}{x}\sqrt{\frac{x+1}{x^2+1}}\, dx$

(d) $\displaystyle\int_0^1 \frac{1}{x^2(1+x^2)}\, dx$

(e) $\displaystyle\int_0^1 \frac{1}{x}\sqrt{\frac{x+1}{x^2+1}}\, dx$

(f) $\displaystyle\int_0^1 \frac{1}{\sqrt{x}\sqrt{1-x}}\, dx$ (improper at both ends)

(g) $\displaystyle\int_1^2 \frac{\sqrt{x}+1}{\sqrt{x}-1}\,dx$

(h) $\displaystyle\int_1^2 \sqrt{\frac{\sqrt{x}+1}{\sqrt{x}-1}}\,dx$

(i) $\displaystyle\int_0^1 x^p(1-x)^q\,dx$ (depends on p and q; improper at both ends for some values).

(j) $\displaystyle\int_0^\infty e^{-x}x^{p-1}\,dx$ (depends on p; improper at 0 for some values).

(k) $\displaystyle\int_{-\infty}^\infty e^{-t^2}\,dt$.

Note. The integral has the value $\sqrt{\pi}$, see Sect. 12.2 Exercise 2. It is immensely important in probability and statistics. See also Sect. 11.7 Exercise 5 and also this section, Exercise 6.

(l) $\displaystyle\int_0^\infty e^{100x-x^2}\,dx$.

2. Determine for which values of a and b the integral

$$\int_0^1 x^a |\ln x|^b\,dx$$

is convergent at 0.

3. Show that if the integral $\int_a^\infty f$ is convergent at ∞ then

$$\lim_{x\to\infty} \int_x^\infty f = 0.$$

4. Give an example of a positive function f, such that the integral $\int_a^\infty f$ is convergent at ∞ but $f(x)$ does not tend to 0 as $x\to\infty$.

5. (a) Let f be a function, differentiable in an interval A that contains -1 and 1. Show that the limit

$$\lim_{\varepsilon\to0+}\left(\int_{-1}^{-\varepsilon}\frac{f(x)}{x}\,dx+\int_\varepsilon^1 \frac{f(x)}{x}\,dx\right)$$

exists and is a finite number.

(b) Let g be a function, twice differentiable in an interval A that contains -1 and 1. Suppose that g has a simple zero at 0 (that is, $g(0)=0$ but $g'(0)\neq0$) and no other zero in the interval $[-1,1]$. Show that the limit

$$\lim_{\varepsilon\to0+}\left(\int_{-1}^{-\varepsilon}\frac{1}{g(x)}\,dx+\int_\varepsilon^1 \frac{1}{g(x)}\,dx\right)$$

exists and is a finite number.

Note. The results of these calculations are called the Cauchy principal values of the improper integrals $\int_{-1}^{1}(f(x)/x)\,dx$ and $\int_{-1}^{1}(1/g(x))\,dx$, respectively.

6. For $n = 0, 1, 2, \ldots$, we define the function

$$J_n(x) = \int_x^\infty t^{-2n} e^{-t^2}\, dt.$$

(a) Show that for all integers $n \geq 1$ we have

$$J_0(x) = \sum_{k=0}^{n-1} \frac{(-1)^k (2k)!}{2^{2k+1} k!} \frac{e^{-x^2}}{x^{2k+1}} + \frac{(-1)^n (2n-1)!}{2^{2n-1}(n-1)!} J_n(x).$$

Hint. Derive a reduction formula for $J_n(x)$ and then use induction.

(b) Show that the series

$$\sum_{k=0}^\infty \frac{(-1)^k (2k)!}{2^{2k+1} k!} \frac{e^{-x^2}}{x^{2k+1}}$$

is divergent for all $x > 0$.

(c) ($\lozenge$) Show that the series is an asymptotic expansion for $J_0(x)$ as $x \to \infty$, with respect to the asymptotic scale

$$h_k(x) = \frac{e^{-x^2}}{x^{2k+1}} \quad (k = 0, 1, 2, \ldots.).$$

Hint. The definition of asymptotic expansion was given in Sect. 11.9. To verify the details L'Hopital's rule can be helpful.

(d) Show that the remainder term in item (a) is numerically less than the last term used in the series.

Note. The function $\mathrm{erfc}(x) := 2J_0(x)/\sqrt{\pi}$ is called the complementary error function. It is related to the normal error function by $\mathrm{erfc}(x) = 1 - \mathrm{erf}(x)$. According to Sect. 11.7 Exercise 5 the Maclaurin series of $\mathrm{erf}(x)$ converges to $\mathrm{erf}(x)$ for all x and so the same is true of $\mathrm{erfc}(x)$. For large x the convergence is too slow to give a practical method to calculate $\mathrm{erfc}(x)$ and then the asymptotic series can be helpful, as its terms decrease in magnitude up to around $k = x$, and for only moderate x the smallest term would seem to be very small. Most significantly for practical calculation, the remainder or error term is numerically bounded by the last term used of the series.

12.2 Differentiation Under the Integral Sign

In this section we take up a topic that is somewhat overdue, and which in the first instance is concerned only with proper integrals. In doing so we shall need to discuss functions of two variables. Actually we only need the notion of partial derivative, in its simplest manifestation. Given a function of two variables $f(x, y)$ we shall denote by

$$D_1 f(x, y)$$

the result of differentiating with respect to the first variable of f, whilst holding the second constant. We shall also need to differentiate twice; this is written as $D_1^2 f(x, y)$, (that is, $D_1 D_1 f(x, y)$). We will never need $D_2 f(x, y)$, but we add that it is the result of differentiating with respect to the second variable whilst holding the first constant. This will be our furthest excursion into the analysis of functions of two variables.

Suppose we can define a function by the formula

$$F(x) = \int_a^b G(x, y) \, dy.$$

What we are concerned with here is the validity (or otherwise) of the formula

$$F'(x) = \int_a^b D_1 G(x, y) \, dy.$$

It is possible to derive some general results on this, using properties of continuous functions of two variables, off bounds here. However, in practice, most problems of this kind are treated in an *ad hoc* manner, typically using Taylor's theorem to estimate a remainder. We shall develop some useful approaches through a sequence of exercises. The result of Exercise 1 is quite adequate for most applications involving proper integrals.

12.2.1 Exercises

1. Suppose that $[a, b]$ is a bounded interval, and A an open interval. Let $G(x, y)$ be a function defined for x in A and $a \le y \le b$. Suppose that G is twice differentiable with respect to its first variable, whilst holding its second fixed, and that there exists a constant M such that $|D_1^2 G(x, y)| \le M$ for all x in A and y in $[a, b]$. Finally suppose that for each x in A the integrals

$$\int_a^b G(x, y) \, dy \quad \text{and} \quad \int_a^b D_1 G(x, y) \, dy$$

exist. Prove that, for each x in A, we have

$$\frac{d}{dx} \int_a^b G(x, y) \, dy = \int_a^b D_1 G(x, y) \, dy.$$

Hint. Use Taylor's theorem to estimate $G(x + h, y) - G(x) - h D_1 G(x, y)$.

2. Define the functions

$$f(x) = \left(\int_0^x e^{-t^2} \, dt \right)^2$$

and

$$g(x) = \int_0^1 \frac{e^{-x^2(t^2+1)}}{t^2+1} \, dt.$$

(a) Show that $g'(x) + f'(x) = 0$ for all x.
(b) Deduce that $g(x) + f(x) = \pi/4$.
(c) Prove that

$$\int_{-\infty}^{\infty} e^{-t^2} \, dt = \sqrt{\pi}.$$

3. Let

$$f(x) = \int_{-\infty}^{\infty} e^{-t^2} \cos(xt) \, dt.$$

Show that $2f'(x) + xf(x) = 0$. Deduce that $f(x) = \sqrt{\pi} e^{-x^2/4}$.
Hint. Justify the differentiation under the integral sign by using Taylor's theorem. The simple differential equation can be treated like the one in the proof of Proposition 11.17.

4. Let f have derivatives of all orders in an interval A, let c be a point of A and suppose that $f(c) = 0$. Show that

$$f(x) = (x - c) \int_0^1 f'(tx + (1-t)c) \, dt.$$

Deduce, using differentiation under the integral sign, that $f(x)/(x - c)$ extends to a function in A having derivatives of all orders.
Note. The proof of the formula should only need the first derivative of f and its continuity. If f has derivatives up to order m then we know, by Sect. 5.8 Exercise 13, that $f(x)/(x - c)$ has derivatives up to order $m - 1$ (including at c). However, it is quite problematic to obtain the last derivative by the method of the present exercise.

12.3 The Maclaurin–Cauchy Theorem

Improper integrals provide a useful convergence test for positive series, possibly the most useful after the ratio test. It underscores the intimate connection between integrals and series.

The integral test. Let $f : [1, \infty[\to \mathbb{R}$ be positive and decreasing. Then the integral $\int_1^{\infty} f$ is convergent if and only if the series $\sum_{n=1}^{\infty} f(n)$ is convergent.

As an example we can consider the series

$$\sum_{n=2}^{\infty} \frac{1}{n^a (\ln n)^b}.$$

It is convergent if and only if, either $a > 1$, or $a = 1$ and $b > 1$. This is because the change of variables $t = \ln x$ gives

$$\int_2^{\infty} \frac{dx}{x^a (\ln x)^b} = \int_{\ln 2}^{\infty} \frac{e^{(1-a)t}}{t^b} \, dt$$

and the integral obtained is convergent under the selfsame conditions.

The integral test results from the simple observation that $\sum_{k=1}^{n} f(k)$ is an upper sum for the integral $\int_1^{n+1} f$ using a partition by integers, whilst $\sum_{k=2}^{n} f(k)$ is a lower sum for the integral $\int_1^n f$. Hence

$$\int_1^{n+1} f \le \sum_{k=1}^{n} f(k) \le f(1) + \int_1^n f.$$

Letting $n \to \infty$ we obtain

$$\int_1^{\infty} f \le \sum_{k=1}^{\infty} f(k) \le f(1) + \int_1^{\infty} f$$

where we are allowing the value ∞ for one or more of the limits. The comparison embodied in these inequalities is often useful.

The integral test is also an immediate consequence of a most striking result, that makes a sharper comparison between the sum and the integral under the same conditions as the integral test.

Proposition 12.1 (Maclaurin–Cauchy theorem) *Let $f : [1, \infty[\to \mathbb{R}$ be positive and decreasing. Then the limit*

$$L = \lim_{n \to \infty} \left(\sum_{k=1}^{n} f(k) - \int_1^n f \right)$$

exists and $0 \le L \le f(1)$.

Proof Let

$$\phi(n) = \sum_{k=1}^{n} f(k) - \int_1^n f.$$

Then

Fig. 12.1 Maclaurin–Cauchy. Picture of the proof

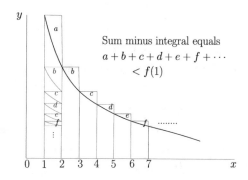

Sum minus integral equals
$a + b + c + d + e + f + \cdots$
$< f(1)$

$$\phi(n) = f(1) + \sum_{k=1}^{n-1} f(k+1) - \sum_{k=1}^{n-1} \int_{k}^{k+1} f$$

$$= f(1) - \sum_{k=1}^{n-1} \int_{k}^{k+1} \big(f(x) - f(k+1) \big) \, dx.$$

Each term in the final sum is positive. We conclude that $\phi(n)$ is decreasing and $\phi(n) \le f(1)$. But we also have

$$\phi(n) = f(n) + \sum_{k=1}^{n-1} f(k) - \sum_{k=1}^{n-1} \int_{k}^{k+1} f$$

$$= f(n) + \sum_{k=1}^{n-1} \int_{k}^{k+1} \big(f(k) - f(x) \big) \, dx.$$

Again each term in the final sum is positive, and therefore $\phi(n) \ge f(n) \ge 0$. We conclude that the limit $L = \lim_{n\to\infty} \phi(n)$ exists, and $0 \le L \le f(1)$. $\square$

The proof is strikingly illustrated in Fig. 12.1.

12.3.1 The Euler–Mascheroni Constant

An important and striking example of the Maclaurin–Cauchy theorem is provided by setting $f(x) = 1/x$. We conclude that

$$\gamma = \lim_{n\to\infty} \left(\sum_{k=1}^{n} \frac{1}{k} - \ln n \right)$$

exists and $0 \le \gamma \le 1$. A more precise value is $\gamma = 0.5772156649....$ As we shall see, this number crops up in unexpected places in analysis and remains somewhat mysterious. It is still not known whether it is rational or irrational.

12.3.2 Exercises

1. Revisit Sect. 3.8 Exercise 10.
2. Study the following series and draw conclusions using the Maclaurin–Cauchy theorem:

(a) $\displaystyle\sum_{n=1}^{\infty} \frac{1}{\sqrt{n}}$

(b) $\displaystyle\sum_{n=1}^{\infty} \frac{1}{n \ln n}$

(c) $\displaystyle\sum_{n=1}^{\infty} \frac{1}{n^2 + 1}.$

3. The series

$$\sum_{n=1}^{\infty} \frac{x}{n^p(1 + nx^2)} \tag{12.1}$$

was studied in Sect. 11.2 Exercise 6. It turned out the series was uniformly convergent with respect to the whole of $\mathbb{R}$ in the case that $p > \frac{1}{2}$. If $0 < p \le \frac{1}{2}$ the series remains pointwise convergent. Prove that in this case the series fails to be uniformly convergent. Use the following steps:

(a) Show that for each $x > 0$ we have

$$\sum_{n=1}^{\infty} \frac{x}{n^p(1 + nx^2)} \ge \int_1^{\infty} \frac{x}{t^p(1 + tx^2)}\, dt.$$

(b) Show that

$$\int_1^{\infty} \frac{x}{t^p(1 + tx^2)}\, dt \ge \frac{x^{2p-1}}{p(1 + x^2)^p}.$$

(c) Let $f(x)$ be the sum of the series (12.1). Show that f cannot be continuous at 0 if $0 < p \le \frac{1}{2}$.

(d) Conclude that if $0 < p \le \frac{1}{2}$ the series, though pointwise convergent, is not uniformly convergent with respect to any interval that contains 0.

3. Show that

$$\gamma = 1 - \int_1^{\infty} \frac{x - [x]}{x^2}\, dx.$$

Hint. Use Sect. 8.1 Exercise 12.

4. Recall the Riemann zeta function (Sects 11.4 Exercise 30 and 11.7 Exercise 15):

$$\zeta(s) = \sum_{n=1}^{\infty} n^{-s}.$$

Show that for $s > 1$ we have

$$\zeta(s) = \frac{1}{2} + \frac{1}{s-1} - sF(s)$$

where

$$F(s) = \int_{1}^{\infty} x^{-s-1}\left(x - [x] - \frac{1}{2}\right) dx.$$

Prove that the function $F(s)$ is differentiable for $s > 0$.

Hint. Use Sect. 8.1 Exercise 12 for the formula. To obtain differentiability at a given $s > 1$ Taylor's theorem can be helpful after guessing what the derivative ought to be.

12.4 Complex-Valued Integrals

Complex-valued functions of a real variable have appeared in a desultory fashion in the present text, never having a section of their own. We recall Sect. 9.2 (under the heading "Logarithm of a complex number") where differentiation of such functions was considered, in particular we differentiated $\text{Log}\,(x + ia)$ and $(x + ia)^m$. In Sect. 11.4 we studied the function e^{ix}, central to the unification of circular functions and the exponential function, and later in the same section we differentiated x^a for a complex power a.

A complex-valued function f of a real variable can be expressed as $f = u + iv$ where u and v are real-valued functions. Concepts such as continuity, boundedness and limit can be defined for them by reference to the real and imaginary parts, just as we defined differentiation in Sect. 9.2. Thus f is said to bounded when u and v are bounded, continuous when u and v are continuous, and we can define $\lim_{x\to c} f(x)$ to be $\lim_{x\to c} u(x) + i \lim_{x\to c} v(x)$ provided the two limits on the right exist and are finite.

A more satisfactory way to extend these concepts to complex-valued functions is to use the modulus of a complex number as a metric assigning a distance between two points in the complex field. Although metrics are definitely not intended to be part of this text, this is what we did when defining the limit of a complex sequence in Sect. 10.1.

Exercise Let $f : A \to \mathbb{C}$ where A is a real number interval. Reformulate the above definitions in terms of the modulus of a complex number as follows:

(a) Let $c \in A$. Show that f is continuous at c if and only if it satisfies the following
    condition: for all $\varepsilon > 0$ there exists $\delta > 0$, such that $|f(x) - f(c)| < \varepsilon$ for all
    x in A that satisfy $|x - c| < \delta$.
(b) Show that f is bounded if and only if the function $|f|$ is bounded.
(c) Let $c \in A$ and let ℓ be a complex number. Show that $\lim_{x \to c} f(x) = \ell$ if and
    only if the following condition is satisfied: for all $\varepsilon > 0$ there exists $\delta > 0$, such
    that $|f(x) - \ell| < \varepsilon$ for all x in A that satisfy $0 < |x - c| < \delta$.

We turn our attention to integrals. Let $f : [a, b] \to \mathbb{C}$ and set $u = \mathrm{Re}\, f$ and $v = \mathrm{Im}\, f$. We define

$$\int f := \int u + i \int v$$

if both integrals on the right-hand side exist. The definition is then extended to
improper integrals; so for example $\int_0^\infty f$ is said to be convergent when the integrals
$\int_0^\infty u$ and $\int_0^\infty v$ are both convergent and then we set

$$\int_0^\infty f = \int_0^\infty u + i \int_0^\infty v.$$

It is easy to prove that the integrals of complex-valued functions satisfy similar
rules to those obeyed by real functions as regards the sum of two functions, and
the product of a function by a complex scalar. In other words integration is a linear
operation over the complex numbers. Moreover the fundamental theorem in its simplest manifestation, Proposition 6.17, extends easily to the case of complex-valued
functions, as does also the rule for integration by parts.

Most importantly we can use Cauchy's principle for a complex integrand, simply
by reinterpreting the absolute value as the modulus. The integral $\int_0^\infty f$ is convergent
if and only if the following condition is satisfied: for all $\varepsilon > 0$ there exists $K > 0$,
such that $\left| \int_x^y f \right| < \varepsilon$ for all x and y that satisfy $K < x < y$.

The extension of Cauchy's principle to the case of a complex integrand is a
straightforward exercise left to the reader

Proposition 12.2 *Let $f : [a, b] \to \mathbb{C}$ be integrable (that is $\mathrm{Re}\, f$ and $\mathrm{Im}\, f$ are both
integrable). Then $|f|$ is integrable and*

$$\left| \int f \right| \leq \int |f|.$$

Proof The proof that $|f|$ is integrable is left to the reader (it may help to recall
Sect. 6.4 Exercise 13).

Let $f = u + iv$ and $\int f = A = a + ib$ (u, v real functions, a, b real numbers).
Now $|A|^2$ is a real number, so that, using the Cauchy–Schwarz inequality (Sect. 2.2
Exercise 17), we find

$$|A|^2 = \bar{A} \int f = \int (\bar{A}u + i\bar{A}v) = \int (au + bv) \quad \text{(it's real!)}$$

$$\leq \int (a^2 + b^2)^{\frac{1}{2}} (u^2 + v^2)^{\frac{1}{2}} = |A| \int |f|.$$

We conclude that $|A| \leq \int |f|$.

The preceding proposition is such a basic tool, and is used so often, that we rarely refer to it in justification.

12.4.1 Absolutely Convergent Integrals

We can learn about the integral $\int_0^\infty f$ by studying the integral $\int_0^\infty |f|$.

Proposition 12.3 *Let* $f : [0, \infty[\to \mathbb{C}$ *be integrable on* $[0, L]$ *for all* $L > 0$. *If the integral* $\int_0^\infty |f|$ *is convergent then so also is the integral* $\int_0^\infty f$.

The proposition applies, with obvious modifications, to other types of improper integrals.

Proof For $0 < x < y$ we have

$$\left| \int_0^y f - \int_0^x f \right| = \left| \int_x^y f \right| \leq \int_x^y |f|.$$

Assume that the integral $\int_0^\infty |f|$ is convergent. Let $\varepsilon > 0$. By Cauchy's principle there exists K, such that

$$\int_x^y |f| < \varepsilon$$

for all x and y that satisfy $K < x < y$. But for the same x and y we then have

$$\left| \int_0^y f - \int_0^x f \right| < \varepsilon$$

by the displayed inequality, and Cauchy's principle tells us that the limit

$$\lim_{x \to \infty} \int_0^x f$$

exists and is finite. $\qquad\square$

The proposition motivates a definition, analogous to the case of infinite series.

Definition If the integral $\int_0^\infty |f|$ is convergent, then the integral $\int_0^\infty f$ is said to be absolutely convergent.

In analogy with series, and because it is convenient to have a term, we can call integrals that are convergent, but not absolutely convergent, conditionally convergent.

Absolutely convergent integrals are generally easier to handle than conditionally convergent ones, because integrals with positive integrands are easier to estimate. In contrast, proving that an integral is conditionally convergent can be tricky. It is striking, and the reader will be shown many examples in the material to follow, that two tools often prove useful in conjunction with studying conditionally convergent integrals. They are the second mean value theorem for integrals (see Sects. 6.8 and 8.1 Exercise 15) and Cauchy's principle.

12.4.2 Exercises

1. Show that a complex integral $\int_0^\infty f$ is absolutely convergent if and only if the two real integrals $\int \mathrm{Re}\, f$ and $\int \mathrm{Im}\, f$ are absolutely convergent.
2. Solve the following indefinite integrals, given that the constant α may be complex:

   (a) $\displaystyle\int e^{\alpha x}\, dx$

   (b) $\displaystyle\int x^m e^{\alpha x}\, dx,$

   where m is a positive integer.

   (c) $\displaystyle\int \frac{1}{(x+\alpha)^m}\, dx,$

   where m is a non-negative integer and $\mathrm{Im}\,\alpha \neq 0$.

   (d) $\displaystyle\int \mathrm{Log}\,(x+i)\, dx.$

4. Solve the integral $\int x^m \cos(\lambda x)\, dx$ (where λ is real) by exploiting the formula $\cos x = \mathrm{Re}\, e^{ix}$.
5. Show that the integral $\int_0^\infty e^{-px}\, dx$ is absolutely convergent for all complex p such that $\mathrm{Re}\, p > 0$, and evaluate it.
6. Show that the integral $\int_{-\infty}^\infty e^{-(x+i)^2}\, dx$ is absolutely convergent.
7. Show that the integral

$$\int_0^\infty \frac{\sin x}{x}\, dx$$

is convergent (at ∞) but that the integral

$$\int_0^\infty \left|\frac{\sin x}{x}\right|\, dx$$

is divergent.

Hint. For the first claim one approach is to integrate by parts. Note that the integral is not improper at 0.

8. (◊) Extend Hölder's inequality to improper integrals (see Sect. 6.4 Exercise 10). Let f and g be functions defined in $[0, \infty[$, integrable on every interval $[0, L]$. Let p and q be positive numbers that satisfy

$$\frac{1}{p} + \frac{1}{q} = 1.$$

Suppose that the integrals $\int_0^\infty |f|^p$ and $\int_0^\infty |g|^q$ are convergent. Then the integral $\int_0^\infty fg$ is absolutely convergent and

$$\int_0^\infty |fg| \leq \left(\int_0^\infty |f|^p \right)^{1/p} \left(\int_0^\infty |g|^q \right)^{1/q}.$$

In the next three exercises we shall explore the improper integral $\int_0^\infty fg$. The results are convergence tests that resemble the variants of Dirichlet's test for series given in Sect. 11.4. The proposed method is to obtain convergence at ∞ by using the second mean value theorem for integrals and Cauchy's principle. Recall that the second mean value theorem asserts that, under the conditions that f is monotonic and g is real valued and integrable on $[a, b]$, there exists ξ in $[a, b]$, such that

$$\int_a^b fg = f(a) \int_a^\xi g + f(b) \int_\xi^b g.$$

In full generality this was proved in Sect. 6.8, and is the reason for marking the exercises with the nugget symbol. Under stronger conditions, for example if f' is continuous and positive (alternatively negative), and g continuous, easier proofs of the second mean value theorem were suggested in the exercises in Sect. 8.1. The reader who has not studied Sect. 6.8 might prefer to adopt these stronger conditions on f and g rather than skip these exercises.

8. (◊) Let f and g be functions with domain $[0, \infty[$, and assume that f is monotonic and that g, which may be complex valued, is integrable on each interval $[0, L]$. Suppose that there exists a constant $K > 0$, such that

$$\left| \int_0^x g \right| < K$$

for all $x > 0$, and suppose that $\lim_{x \to \infty} f(x) = 0$. Show that the integral $\int_0^\infty fg$ is convergent.

9. (◊) Let f and g be functions with domain $[0, \infty[$, and assume that f is monotonic and that g, which may be complex valued, is integrable on each interval $[0, L]$.

Suppose that the integral $\int_0^\infty g$ is convergent and that f is bounded as $x \to \infty$. Show that the integral $\int_0^\infty fg$ is convergent.

The restriction in the second of the above two convergence tests to monotonic and bounded f can be weakened. Obviously it is enough if f can be expressed as a linear combination of functions that satisfy this condition. It turns out that a large class of functions can be expressed as the difference of two bounded monotonic functions, and requiring this of f, along with the stated conditions on g, clearly suffices to guarantee convergence of the integral $\int_0^\infty fg$. This is the content of the following exercise.

In the third item of the exercise we extend the test to allow complex-valued f. Although it could totally supplant the test given in Exercise 9, the latter is still very useful because of its relatively simple and memorable conditions.

10. Prove the following claims, of which the third is a convergence test for the integral $\int_0^\infty fg$ to stand alongside Exercises 8 and 9.

   (a) Let f be real valued with continuous first derivative, and assume that the integral $\int_0^\infty |f'|$ is convergent, and that f is bounded as $x \to \infty$. Then there exist functions g and h, both increasing and bounded as $x \to \infty$, such that
   $$f = g - h.$$
   Hint. Take $g(x) = \int_0^x |f'|$.
   (b) Let f be complex valued with continuous first derivative, and assume that the integral $\int_0^\infty |f'|$ is convergent, and that f is bounded as $x \to \infty$. Then there exist real-valued functions g_1, g_2, h_1 and h_2, all increasing and bounded as $x \to \infty$, such that $f = (g_1 - g_2) + i(h_1 - h_2)$.
   (c) Let f and g be complex-valued functions with domain $[0, \infty[$, assume that g is integrable on each interval $[0, L]$, that f' is continuous and f is bounded as $x \to \infty$. Assume that the integrals $\int_0^\infty g$ and $\int_0^\infty |f'|$ are convergent. Then the integral $\int_0^\infty fg$ is convergent.

   Note. It is possible to go even further in weakening the conditions on the function f. Some clues to this may be found in Sect. 6.5.

11. Here are some examples that illustrate the tests given in the preceding exercises. Show that the following integrals are convergent:

   (a) $\displaystyle\int_0^\infty \frac{\sin x}{\sqrt{x}}\, dx$

   (b) $\displaystyle\int_0^\infty \frac{\sin x \tanh x}{\sqrt{x}}\, dx$

   (c) $\displaystyle\int_0^\infty \frac{\sin x \tanh x \sin(1/x)}{\sqrt{x}}\, dx$

   (d) $\displaystyle\int_0^\infty \frac{\sin x \tanh x \sin(1/x)}{\sqrt{x}\,(\sqrt{x}+i)}\, dx.$

   Note that all four integrals are proper at 0, in spite of the occurrence of $1/\sqrt{x}$ and $1/x$.

12. (a) Show that the function $F(x) = x - [x] - \frac{1}{2}$ is periodic with period 1 and mean 0. For helpful information concerning periodic functions consult Sect. 6.5 Exercise 5.

(b) Show that the integral

$$\int_1^\infty x^{-1}\left(x - [x] - \frac{1}{2}\right) dx$$

is convergent.

13. (◊) Let

$$B = \int_1^\infty x^{-1}\left(x - [x] - \frac{1}{2}\right) dx$$

(see the previous exercise). Use Sect. 8.1 Exercise 12 to show that

$$B = \lim_{n\to\infty}\left(\ln(n!) - \left(n + \frac{1}{2}\right)\ln n + n - 1\right).$$

Deduce that $B = \frac{1}{2}\ln(2\pi) - 1$.

Hint. Use Stirling's approximation to $n!$ (see the eponymous nugget).

Note. This exercise can also be viewed as a proof of Stirling's approximation, somewhat shorter than the one given in Sect. 11.10. If regarded as such then one must determine B without using Stirling's formula; using, for example, Wallis' product for π, as was proposed in Sect. 11.10 in order to determine the number A, related to B by $B = \ln A - 1$.

In the remaining exercises of this section we shall study a remarkable result: the Euler–Maclaurin summation formula. The reason for marking these exercises with the nugget symbol is the important and surprising role played by the sequence of Bernoulli numbers $(B_n)_{n=0}^\infty$. These are defined so that the numbers $B_n/n!$ are the coefficients in the Maclaurin series of the function $x/(e^x - 1)$, which is extended to have the value 1 at $x = 0$. The Bernoulli numbers were studied in Sect. 11.8; the reader is advised to turn back some pages and read about them, if they have not already done so, before proceeding.

14. (◊) Let f be infinitely-often differentiable in an interval A. Let a and b be integers in A such that $a < b$. In a sequence of exercises the reader is invited to prove the Euler–Maclaurin summation formula

$$\sum_{k=a}^b f(k) = \int_a^b f + \frac{f(a) + f(b)}{2} + \sum_{k=2}^m \frac{B_k}{k!}\left(f^{(k-1)}(b) - f^{(k-1)}(a)\right) + R_m,$$

where the constants B_k are universal (that is, independent of f, a, b and m), and R_m is a remainder term. The constants will be identified with the Bernoulli numbers, and the remainder elucidated, in the course of the proof.

We begin by recalling the formula

$$\sum_{k=a}^{b} f(k) = \int_a^b f + \frac{f(a) + f(b)}{2} + \int_a^b \left(x - [x] - \frac{1}{2}\right) f'(x)\, dx$$

which was given in Sect. 8.1 Exercise 12. This will be the case $m = 1$ of the summation formula.

(a) Let $F_1(x) = x - [x] - \frac{1}{2}$. Show that there exists a unique sequence of functions $(F_n)_{n=1}^{\infty}$, in which each function is periodic with period 1 and mean 0, and F_{n+1} is a primitive of F_n for $n = 1, 2, \dots$.
   Hint. Consult Sect. 6.5 Exercise 5.

(b) Show that the summation formula holds with $B_k = (-1)^k k! F_k(0)$, for $k \geq 2$, and

$$R_m = (-1)^{m-1} \int_a^b F_m f^{(m)}.$$

   Hint. Consult Sect. 8.1 Exercise 16.

(c) Show that B_k (for $k \geq 2$) is the kth Bernoulli number.
   Hint. Consider the case $f(x) = x^m$ and compare with the nugget on Bernoulli numbers.

Note that actually the Bernoulli numbers B_k satisfy $B_k = k! F_k(0)$, for $k \geq 2$, in addition to $B_k = (-1)^k k! F_k(0)$, because in fact, as we saw in Sect. 11.8, the Bernoulli numbers B_k with odd k are 0 (except for B_1). This observation can prevent much anguish caused by the appearance of unwanted minus signs.

It also turns out that the function $F_n(x)$ in the Euler–Maclaurin summation formula is a polynomial function of degree n of the periodic function $x - [x]$. Define the functions $P_n(t)$ in terms of the coefficients in the Maclaurin series

$$\frac{x e^{xt}}{e^x - 1} = \sum_{n=0}^{\infty} \frac{P_n(t)}{n!} x^n$$

as shown here. The series converges in an interval $]-r, r[$, the same as the one in which the formula

$$\frac{x}{e^x - 1} = \sum_{n=0}^{\infty} \frac{B_n}{n!} x^n$$

is valid. This is studied in the next exercise.

15. (a) Show that

$$P_n(t) = \sum_{k=0}^{n} \binom{n}{k} B_k t^{n-k}.$$

In particular $P_n(0) = B_n$.

(b) Show that $P'_n(t) = nP_{n-1}(t)$ for $n = 1, 2, \ldots$

(c) Show that $P_n(1) = P_n(0)$ for $n = 2, 3, \ldots$.

(d) Show that the function F_n in the summation formula is given by

$$F_n(x) = \frac{P_n(x - [x])}{n!}.$$

16. Suppose that for each $j \geq 1$ we have $\lim_{x \to \infty} f^{(j)}(x) = 0$ and the function $f^{(j)}$ is monotonic.

(a) Prove, in the notation of the Euler–Maclaurin summation formula, that for each m we have

$$\int_a^\infty \left(x - [x] - \frac{1}{2} \right) f'(x)\, dx = - \sum_{k=2}^m \frac{B_k}{k!} f^{(k-1)}(a) + (-1)^{m-1} \int_a^\infty F_m f^{(m)}.$$

(b) Show that the summation formula can be written as

$$\sum_{k=a}^b f(k) = \int_a^b f + \frac{f(a) + f(b)}{2} + \sum_{k=2}^m \frac{B_k}{k!} f^{(k-1)}(b)$$

$$+ \int_a^\infty \left(x - [x] - \frac{1}{2} \right) f'(x)\, dx - (-1)^{m-1} \int_b^\infty F_m f^{(m)}.$$

This can be useful because the last term tends to 0 as $b \to \infty$.

17. ($\lozenge$) Prove the following generalisation of Stirling's approximation (see Sect. 11.10). For each m we have, as $n \to \infty$:

$$\ln(n!) = n \ln \left(\frac{n}{e} \right) + \frac{1}{2} \ln n + \frac{1}{2} \ln(2\pi) + \sum_{j=2}^m \frac{(-1)^j B_j}{j(j-1)n^{j-1}} + O(n^{-m}).$$

The formula given here is an asymptotic expansion for $\ln(n!)$. The asymptotic scale that appears here is pretty, comprising the sequence of functions

$$n \ln n, \quad n, \quad \ln n, \quad 1, \quad \frac{1}{n}, \quad \frac{1}{n^2}, \quad \frac{1}{n^3}, \quad \ldots$$

exhibiting progressively slower "growth" as $n \to \infty$. For an explanation of the big-O notation and asymptotic expansions see the nugget "Asymptotic orders of magnitude".

12.5   (◊) Integral Transforms

In this section we shall look at the Fourier transform and very briefly at the Laplace transform. These are examples of integral transforms, and are immensely important in applications, both within mathematics, and to science and technology. The reason for including them in this text is that they probably constitute the types of improper integral that the reader is most likely to encounter. Their most important property, that they can be inverted, will not be touched upon; that belongs to the area of further study. We limit the discussion mainly to convergence of the integral, and properties of the transformed function, chiefly continuity, differentiability and decay at infinity. The conclusions will be developed as exercises.

An integral transform is used to transform a given function f into a second function F using the prescription

$$F(y) = \int_a^b K(y, x) f(x) \, dx$$

where the function K, of two variables, is called the kernel of the transform. The integral may be improper, as is the case for both the Fourier transform ($b = \infty = -a$) and the Laplace transform ($a = 0$, $b = \infty$), both considered here.

12.5.1   Fourier Transform

The function

$$F(y) = \int_{-\infty}^{\infty} f(x) e^{-ixy} \, dx$$

is called the Fourier transform of the function f. It can be defined when f is a complex-valued function and the integral is convergent at both ends.

A simple sufficient condition for the existence of the Fourier transform, and the starting point for all studies of the Fourier transform, is that there exists M, such that

$$\int_{-R}^{R} |f| \le M$$

for all $R > 0$. A function f that satisfies this is sometimes called *absolutely integrable*. It simply means that the integral $\int_{-\infty}^{\infty} f$ is absolutely convergent at both ends. By Proposition 12.3 the integral defining the Fourier transform is absolutely convergent for every real y.

The Fourier transform is just one of many ways, though perhaps the most important, to transform one function into another using an integral transform. It is typically applied to find solutions of differential equations defined over the whole line.

12.5.2 Exercises

1. Calculate the Fourier transforms of the following functions:

    (a) $f(x) = 1$ for $a < x < b$, $f(x) = 0$ otherwise.

    (b) $f(x) = e^{-ax}$ for $x > 0$, $f(x) = 0$ for $x < 0$, where $a > 0$ and is a constant.

    (c) $f(x) = e^{-a|x|}$, where $a > 0$ and is a constant.

2. (a) Let $f(x)$ be a real odd function and let $F(y)$ be its Fourier transform. Show that

$$F(y) = \frac{2}{i} \int_0^\infty f(x) \sin(xy)\, dx.$$

    (b) Let $f(x)$ be a real even function and let $F(y)$ be its Fourier transform. Show that

$$F(y) = 2 \int_0^\infty f(x) \cos(xy)\, dx.$$

Note. The integrals $\int_0^\infty f(x) \sin(xy)\, dx$ and $\int_0^\infty f(x) \cos(xy)\, dx$, which can be defined for a function f given on the interval $[0, \infty[$, are called the Fourier sine transform and the Fourier cosine transform of f.

3. Prove that the Fourier transform $F(y)$ of an absolutely integrable function f is continuous and tends to 0 at ∞ and $-\infty$. This can be done in the following steps. We let

$$F_n(y) = \int_{-n}^n e^{-ixy} f(x)\, dx$$

    for each natural number n.

    (a) Show that $F_n(y)$ is a continuous function of y for each n.

    Hint. Use the equality $|e^{it} - 1| = 2|\sin(t/2)|$ and the inequality $|\sin t| \le |t|$.

    (b) Show that $\lim_{n\to\infty} F_n(y) = F(y)$ uniformly with respect to y in $\mathbb{R}$. Deduce that F is continuous.

    (c) Show that $\lim_{y\to\pm\infty} F_n(y) = 0$ for each n.

    Hint. Prove this first in the case that f is a step function. If f is merely integrable it may be approximated in the mean by a step function.

    Note. In fact

$$\lim_{y\to\pm\infty} \int_a^b e^{ixy} f(x)\, dx = 0$$

    for any function f integrable on $[a, b]$, a result known as the Riemann–Lebesgue lemma, although this name is sometimes attached to the conclusion that $\lim_{y\to\pm\infty} F(y) = 0$.

    (d) Deduce that $\lim_{y\to\pm\infty} F(y) = 0$.

12.5.3   Laplace Transform

The Laplace transform is an important tool in applied mathematics, particularly for solving linear differential equations in which time is the independent variable, and solutions that satisfy initial conditions are sought forward in time. We will not go into these applications here; they properly require another setting than the present text.

Suppose that f is a real-valued function with domain $[0, \infty[$, integrable on the interval $[0, L]$ for every $L > 0$. We define the *Laplace transform* of f by the integral

$$F(p) = \int_0^\infty e^{-px} f(x)\, dx$$

the understanding being that $F(p)$ is defined for all p such that the integral is convergent. It is usual to study the Laplace transform for complex p, and only then can its properties be fully appreciated. However, in order to remain within the confines of fundamental analysis we shall restrict our attention to real p.

12.5.4   Exercises (cont'd)

4. Calculate the Laplace transforms of the following functions:

(a) x^n, where n is a non-negative integer.
(b) e^{kx}, where k is a real constant.
(c) $\cos kx$, where k is a real constant.
(d) $\sin kx$, where k is a real constant.
(e) $f(x) = 1$, if $a < x < b$, and otherwise $f(x) = 0$.
   Note. The case of the last function with $b = \infty$ is called the Heaviside unit step at $x = a$.
   It plays an important role in technology.

5. In all items of this exercise we assume that f is integrable on $[0, L]$ for each $L > 0$, and that the integral $\int_0^\infty f$ is convergent.

(a) Show that the integrals $\int_0^\infty e^{-px} f(x)\, dx$ and $\int_0^\infty e^{-px} x f(x)\, dx$ are both convergent for all $p > 0$.

We let

$$F(p) = \int_0^\infty e^{-px} f(x)\, dx, \quad (p \geq 0)$$

$$G(p) = -\int_0^\infty e^{-px} x f(x)\, dx, \quad (p > 0)$$

and for each positive integer n we let

$$F_n(p) = \int_0^n e^{-px} f(x)\, dx$$

$$G_n(p) = -\int_0^n e^{-px} x f(x)\, dx.$$

(b) Show that $G_n(p) = F'_n(p)$.

(c) Show that $\lim_{n\to\infty} G_n(p) = G(p)$ uniformly with respect to the interval $\delta \le p < \infty$ for any given $\delta > 0$, whereas $\lim_{n\to\infty} F_n(p) = F(p)$ uniformly with respect to $0 \le p < \infty$.

   Hint. Use Cauchy's principle and the second mean value theorem.

(d) Show that $F(p)$ is differentiable for $p > 0$ and $F'(p) = G(p)$.

(e) Show that $F(p)$ has derivatives of all orders for $p > 0$ and they are given by

$$F^{(k)}(p) = \int_0^\infty (-x)^k e^{-px} f(x)\, dx.$$

(f) Show that $\lim_{p\to\infty} F(p) = 0$.

(g) Deduce also that

$$\lim_{p\to 0+} \int_0^\infty e^{-px} f(x)\, dx = \int_0^\infty f.$$

Note. The last conclusion here is an Abelian theorem, comparable to Abel's theorem on power series. Imagine that a value is assigned to possibly divergent integrals $\int_0^\infty f$, a kind of summability method (see Sect. 11.5), by computing the limit

$$\lim_{p\to 0+} \int_0^\infty e^{-px} f(x)\, dx,$$

if it exists. The exercise shows that the correct value is assigned to already convergent integrals.

6. Prove that

$$\int_0^\infty \frac{\sin x}{x}\, dx = \frac{\pi}{2}.$$

Hint. Let

$$F(p) = \int_0^\infty e^{-px} \frac{\sin x}{x}\, dx, \quad (p \ge 0),$$

compute $F'(p)$ for $p > 0$, and obtain an explicit formula for $F(p)$.

7. Use the result of the previous exercise, along with some trigonometric identities, to derive the following results:

(a) $\displaystyle\int_0^\infty \frac{\sin x \cos x}{x}\, dx = \frac{\pi}{4}$

(b) $\displaystyle\int_0^\infty \frac{\sin^2 x}{x^2}\, dx = \frac{\pi}{2}$

(c) $\displaystyle\int_0^\infty \frac{\sin^4 x}{x^2}\,dx = \frac{\pi}{4}$

(d) $\displaystyle\int_0^\infty \frac{\sin^4 x}{x^4}\,dx = \frac{\pi}{3}.$

12.5.5   Pointers to Further Study

$\rightarrow$ Fourier analysis
$\rightarrow$ Laplace transforms
$\rightarrow$ Differential equations
$\rightarrow$ Complex analysis

12.6   ($\Diamond$) The Gamma Function

In this nugget we study the Gamma function, arguably the most important of the special functions, and the first to be studied historically. The Gamma function has a way of appearing, as you might say, unexpectedly, in formulas; in particular it is a constituent of some important special functions, notably Bessel functions. There is no better way to begin studying special functions than by learning about the Gamma function.

We shall develop some of its properties through a series of exercises, using only the tools made available in this text. The further study of the Gamma function requires methods that go beyond the fundamental analysis of this text, such as multiple integrals and complex analysis. It is possible to prove some of the properties, normally obtained with ease by more advanced methods, using only fundamental analysis. This requires much ingenuity and effort, and one may ask what is achieved by demonstrating that more powerful methods can be avoided.

The Gamma function $\Gamma(x)$ extends the function $f(n) = (n-1)!$ from the positive integers to the real numbers (with the exclusion of the negative integers and 0), while preserving the characteristic property of the factorial function, $\Gamma(x) = (x-1)\Gamma(x-1)$. It has been defined in many different ways, but by far the simplest is to use the so-called Eulerian integral of the second kind. This defines $\Gamma(x)$ for $x > 0$ by the integral

$$\Gamma(x) = \int_0^\infty t^{x-1}e^{-t}\,dt.$$

The integral, whilst obviously improper because of its upper limit, is also improper at 0 if $x < 1$.

12.6.1 Exercises

1. Show that the integral defining the Gamma function is convergent at both ends if $x > 0$.
2. Show that for $x > 1$ we have

$$\Gamma(x) = (x - 1)\Gamma(x - 1)$$

and deduce that

$$\Gamma(n) = (n - 1)!$$

if n is a positive integer.

Note. The property proved here can be used to extend the Gamma function to negative values of x, except at negative integers, by setting

$$\Gamma(x) = \frac{\Gamma(x + m)}{(x + m - 1)(x + m - 2)...x}$$

using any integer m, such that $x + m > 0$. It does not matter what integer is used; the same value is obtained by using $m + 1$ as by using m.

3. The Gamma function can be used to "tidy up" the coefficients in the binomial series. Let a be a real number. Show that

$$\frac{a(a - 1)...(a - k + 1)}{k!} = \frac{\Gamma(a + 1)}{\Gamma(k + 1)\Gamma(a - k + 1)}.$$

4. Show that $\Gamma(\frac{1}{2}) = \sqrt{\pi}$. Deduce that for all natural numbers n we have

$$\Gamma\left(n + \frac{1}{2}\right) = \frac{(2n)!\sqrt{\pi}}{4^n n!}.$$

5. Show that $\Gamma(x)$ has derivatives of all orders and its nth derivative (for $x > 0$) is given by the formula

$$\Gamma^{(n)}(x) = \int_0^\infty t^{x-1}(\ln t)^n e^{-t}\, dt.$$

Hint. You will have to justify the repeated differentiation under the integral sign. One possibility is to use Taylor's theorem to estimate the quantity

$$t^{x+h-1} - t^{x-1} - ht^{x-1} \ln t,$$

but you will have to cope with the fact that the function t^{x-1}, regarded as a function of x, is decreasing if $0 < t < 1$ and increasing if $t > 1$.

6. (a) Show that for $x > 0$ the Gamma function can be written as the sum of two
series:

$$\Gamma(x) = \sum_{n=0}^{\infty} \frac{(-1)^n}{n!} \frac{1}{n+x} + \sum_{n=0}^{\infty} c_n x^n$$

where

$$c_n = \frac{1}{n!} \int_1^{\infty} t^{-1} (\ln t)^n e^{-t} \, dt.$$

Hint. Split the integral into two: from 0 to 1, and from 1 to ∞. Expand
the integrands in power series and argue that the order of integration and
summation can be interchanged. This is tricky for the improper integral and
is a good example of an argument that is much easier to carry out using the
Lebesgue integral.

(b) Show that the power series $\sum_{n=0}^{\infty} c_n x^n$ has infinite radius of convergence.

(c) Show that the series

$$\sum_{n=0}^{\infty} \frac{(-1)^n}{n!} \frac{1}{n+z}$$

converges for all complex z, except for $z = 0, -1, -2, \ldots$, and that its sum
function, restricted to the real line, has derivatives of all orders on the real
line $\mathbb{R}$ minus the set $\{0, -1, -2, \ldots\}$.

We revisit the example treated in the nugget on asymptotic orders. It has a tenuous
connection to the Gamma function.

7. ($\Diamond$) Recall the function

$$E(x) = \int_0^1 \frac{1}{t} e^{-x/t} \, dt, \quad (x > 0)$$

that was studied in Sect. 11.9.

(a) Show that

$$E(x) = \int_x^{\infty} \frac{e^{-u}}{u} \, du, \quad (x > 0).$$

So $-E(x)$ is an antiderivative for the function e^{-x}/x; in fact the one that
tends to 0 at ∞. It solves the troublesome integral $\int e^{-x}/x \, dx$.

(b) Show that for all $x > 0$ we have the series expansion

$$E(x) = C - \ln x - \sum_{n=1}^{\infty} \frac{(-1)^n x^n}{n \, n!}$$

where C is a certain constant.

(c) Show that

$$C = \int_0^\infty (\ln t) e^{-t}\, dt = \Gamma'(1).$$

Note. $\Gamma'(1)$ is known to be $-\gamma$, where γ is the Euler–Mascheroni constant. See Exercise 8.

(d) The series in item (b) is convergent for all x. For $x = 10$, estimate the size of the 10th term, and conclude that it is far better to use the asymptotic expansion to 10 terms (see Sect. 11.9) than the convergent series to 10 terms.

8. We conclude with some final spectacular results concerning the Gamma function:

   (a) Let $(1/p) + (1/q) = 1$, $p > 1$, $q > 1$. Show that

   $$\Gamma\left(\frac{x}{p} + \frac{y}{q}\right) \le \Gamma(x)^{1/p}\Gamma(y)^{1/q}.$$

   Hint. Use Hölder's inequality, Sect. 12.4 Exercise 7.

   (b) Show that $\ln \Gamma(x)$ is convex on the interval $]0, \infty[$.

   (c) Show that for all $0 < x < 1$ and every natural number n we have

   $$x \ln(n) \le \ln \Gamma(n + x + 1) - \ln(n!) \le x \ln(n + 1).$$

   Hint. Use (b) and compare the chords of $y = \ln \Gamma(x)$ on the intervals $[n, n + 1]$, $[n + 1, n + x + 1]$ and $[n + 1, n + 2]$.

   (d) Deduce from (c) that for $0 < x < 1$ we have

   $$0 \le \ln \Gamma(x) - \ln\left(\frac{n^x n!}{x(x + 1)...(x + n)}\right) \le x \ln\left(1 + \frac{1}{n}\right).$$

   (e) Deduce that

   $$\Gamma(x) = \lim_{n \to \infty} \frac{n^x n!}{x(x + 1)...(x + n)},$$

   not just for $0 < x < 1$, but for all $x > 0$.

   Note. This limit can be rewritten to give Euler's original definition of the Gamma function as an infinite product.

   (f) The result of taking the logarithm on both sides of the limit formula for $\Gamma(x)$ obtained in the previous item, followed by differentiating both sides and then interchanging the derivative and the limit, suggests that the following might be true:

   $$\frac{\Gamma'(x)}{\Gamma(x)} = \lim_{n \to \infty}\left(\ln n - \sum_{j=0}^n \frac{1}{x + j}\right).$$

   Give a rigorous proof by showing that for any $K > 0$ the limit here is attained uniformly with respect to $0 < x < K$.

(g) Deduce that $-\Gamma'(1)$ is the Euler–Mascheroni constant γ.

(h) Prove the theorem of Bohr and Mollerup:

  Let f be a function with domain $]0, \infty[$ that has the following three properties:

  (i) $\ln f(x)$ is convex.

  (ii) $f(x) = (x - 1)f(x - 1)$, $(x > 1)$.

  (iii) $f(1) = 1$.

  Then $f(x) = \Gamma(x)$.

  Hint. Repeat the arguments of items (c), (d) and (e), using f instead of Γ.

12.6.2   *Pointers to Further Study*

→ Special functions

→ Complex analysis

Appendix
Afterword and Acknowledgements

Many readers may have noticed that most, if not all, of the material appearing in this text has been treated elsewhere, probably in dozens of publications. It is inevitable that bits here and there are lifted out of some previous textbooks, whether I like it or no. Therefore I wish to list here the four textbooks that, over more than half a century, have been the ones I have principally turned to when I have needed to check up on some detail of fundamental analysis. No recommendation is implied in this list and it is certain that each has its virtues and faults, neither of which do I wish to elaborate on here. However, it is also certain that some acknowledgement is due. The dates are those of first publication.

(a) G. H. Hardy. A Course of Pure Mathematics. Cambridge University Press, 1908.
(b) E. G. Phillips. A Course of Analysis. Cambridge University Press, 1930.
(c) J. C. Burkill. A First Course in Mathematical Analysis. Cambridge University Press, 1962.
(d) M. Spivak. Calculus. Publish or Perish, 1967.

© The Editor(s) (if applicable) and The Author(s), under exclusive   427
license to Springer Nature Switzerland AG 2020
R. Magnus, *Fundamental Mathematical Analysis*, Springer Undergraduate
Mathematics Series, https://doi.org/10.1007/978-3-030-46321-2

Index

© The Editor(s) (if applicable) and The Author(s), under exclusive
license to Springer Nature Switzerland AG 2020
R. Magnus, *Fundamental Mathematical Analysis*, Springer Undergraduate
Mathematics Series, https://doi.org/10.1007/978-3-030-46321-2

Printed in the United States
By Bookmasters